T0299071

Mathematical Biology and Biological Physics

Mathematical Biology and Biological Physics

Editor

Rubem P Mondaini

Federal University of Rio de Janeiro, Brazil

World Scientific

NEW JERSEY • LONDON • SINGAPORE • BEIJING • SHANGHAI • HONG KONG • TAIPEI • CHENNAI • TOKYO

Published by

World Scientific Publishing Co. Pte. Ltd.
5 Toh Tuck Link, Singapore 596224
USA office: 27 Warren Street, Suite 401-402, Hackensack, NJ 07601
UK office: 57 Shelton Street, Covent Garden, London WC2H 9HE

British Library Cataloguing-in-Publication Data
A catalogue record for this book is available from the British Library.

MATHEMATICAL BIOLOGY AND BIOLOGICAL PHYSICS

ISBN 978-981-3227-87-3

Printed in Singapore

PREFACE

The present book is a collection of papers which have been accepted for publication after a peer review evaluation by the Editorial Board of the BIOMAT Consortium (http://www.biomat.org) and international referees adHoc. These papers have been presented on technical sessions of the BIOMAT 2016 International Symposium, the 16th Symposium of the BIOMAT Series which was held at the Chern Institute of Mathematics — Nankai University, Tianjin, People's Republic of China, from October 30th to November 04th, 2016. On behalf of the BIOMAT Consortium, we thank the Director of the Chern Institute, Prof. Chengming Bai, and the Chair of the BIOMAT 2016 Local Organizing Committee, Prof. Jishou Ruan of Nankai University. Ms. Hongqin Li, the Secretary of the Chern Institute, has used her professional expertise to provide very useful announcements to participants as well as changes on the Scientific Programme during the conference. We are also indebted to all those Chinese research students working on the registration office. Thanks so much for their unvaluable help.

Financial support has been provided by the Chern Institute in terms of fifty fellowships to cover living expenses with full pension accommodation of Keynote Speakers and selected participants. Special thanks are due to our colleagues of the Chern Institute and Nankai University for their hospitality in Tianjin.

The BIOMAT Consortium has succeed once more on its fundamental mission of enhancing the interdisciplinary scientific activities of Mathematical and Biological Sciences on Developing Countries with the organization of the BIOMAT 2016 Intenational Symposium. Participants from Western Europe, Asia, North and South America had the usual opportunity of exchanging scientific feedback of their research fields with their colleagues from P.R. China and excellent joint work is expected to appear in the near future.

The Editor of the book and President of the BIOMAT Consortium is very glad for the collaboration and critical support of his wife Carmem Lucia on the editorial work, from the reception of submitted papers for the peer review procedure of BIOMAT Consortium Editorial Board to the ultimate publication of the Scientific Programme. He also thanks his research student S.C. de Albuquerque Neto from Federal University of Rio de Janeiro for his computational skills and technical expertise with LaTeX versions.

Rubem P. Mondaini
President of the BIOMAT Consortium
Chairman of the BIOMAT 2016 Scientific Advisory Committee

Tianjin, People's Republic of China, November 2016

EDITORIAL BOARD OF THE BIOMAT CONSORTIUM

Professor C.A. Floudas — In Memoriam

All the organizational work of the BIOMAT 2016 International Conference has been dedicated to honour the memory of our dear colleague and friend Christodoulos Achilleus Floudas, who passed away on 14th august 2016. "Chris Floudas", a Stephen C. Macaleer '63 Professor in Engineering and Applied Science, Professor of Chemical and Biological Engineering of Princeton University, Emeritus, and Director of the Texas A&M Energy Institute, Erle Nye '59 Chair Professor for Engineering Excellence. He will be remembered forever as an international scientific leader on the application of methods of Global Optimization to interdisciplinary fields of mathematical and biological sciences. He lectured as a Keynote Speaker on six BIOMAT International Symposia (2003, 2004, 2005, 2006, 2007, 2012). The BIOMAT Consortium is deeply saddened by the passing of this superb human being, a great scientist and scholar. On behalf of the Consortium, its Editorial Board and its Board of Directors, we agree completely with the opinions of his scientific colleagues which testify to his noble character: "He would stand for what he believed no matter the consequences in science, at work and in life. He was honest and upfront and consistent — what you say is what you got" "He was the kind of friend you can count on to stand by you, come what may. In Greek, we call such a person, 'a sword' ".

CONTENTS

COMPREHENSIVE INTRODUCTION TO AGGLOMERATION AND GROWTH MODELS

R. KERNER
Laboratoire de Physique Théorique de la Matière Condensée
Tour 23-13, 5-e étage, Boîte Courrier 121
Université Pierre et Marie Curie, 4 Place Jussieu
75005 Paris, France
E-mail: richard.kerner@upmc.fr

This tutorial lecture is a short introduction to probabilistic models describing agglomeration and growth processes in physics and biology. We show how two apparently different mathematical methods describing tese processes, the stochastic matrices and the Volterra-type equations, lead to similar results and conclusions. Both methods are successfully used while investigating various processes in condensed matter physics, such as growth of crystals and quasi-crystals, and of glass forming, as well as molecular growth, phyllotaxis and homeostasis in biology.

1. Introduction

This article is based on the tutorial lecture delvered at the opening session of the BIOMAT 2016 International held during the fist week of November 2016 at the Chern Mathematical Institute of the Nankai University in Tianjin, China. In this tutorial we shall compare two methods of statistical analysis of dynamical changes and evolutionary trends in both inorganic and biological systems.

When one is dealing with a great number of similar events or objects, we turn naturally to statistical analysis and probability distributions. It is often impossible, (and useless) to keep track of all characteristics of single items in the swarm of data. The precise knowledge of the exact numbers of items found in a given state at a given time is often out of reach experimentally. On the contrary, the average values given by statistical analysis of data, are often the only useful experimental information available. The analysis of dynamical changes and evolutionary trends in chemical or biological systems are given then in terms of probabilities of finding a given state or configuration.

Stochastic matrices transform a probability distribution into another one, keeping the normalization to 1. They display at least one asymptotic régime which can be considered as stable final configuration. The *Lotka-Volterra equations* deal with the numbers of several interacting "populations", but can be also reduced to probabilities. The Lotka-Volterra equations describe the time evolution of various competing with each other, but also interdependent species. They are usually formulated with numbers of specimens of each species, but can be also reformulated using relative probabilities of picking up a particular specimen out of the total population.

In what follows, we shall present several models of dynamics of complex systems, described alternatively by means of stochastic matrices or by non-linear equations of Lotka-Volterra type.

Our aim is to show that in most of the cases, two approaches are either complementary, or equivalent, and may be more or less efficient according to the situation.

2. Stochastic matrices

2.1. *Definitions and properties*

The simplest example of a stochastic matrix operating on a probability distribution is provided by a dychotomic variable, with only two possible issues, say x and y.

Let the probabilities of these issues be given by p_x and p_y; obviously,

$$p_x + p_y = 1$$

because the probabilities must be normalized to 1.

Suppose now that due to some external influence, or by a natural evolutionary trend, these probabilities undergo some change. In a linear approximation the new probabilities are given by the following expression:

$$\begin{pmatrix} p'_x \\ p'_y \end{pmatrix} = \begin{pmatrix} a & b \\ c & d \end{pmatrix} \begin{pmatrix} p_x \\ p_y \end{pmatrix} = \begin{pmatrix} ap_x + bp_y \\ cp_x + dp_y \end{pmatrix}.$$

The new probabilities have to be normalized to 1; this leads to the following algebraic condition:

$$a\, p_x + b\, p_y + c\, p_x + d\, p_y = 1;$$

But this can be also written as:

$$(a + c)\, p_x + (b + d)\, p_y = 1;$$

This must hold for any initial values p_x and p_y, in particular when $p_x = 1$, $p_y = 0$, or $p_x = 0, p_y = 1$; therefore we must have

$$(a + c) = 1 \text{ and } (b + d) = 1$$

In other words, both columns of the 2×2 stochastic matrix must be normalized to 1. It goes without saying that all the entries of stochastic matrix must be non-negative. This result can be readily generalized for the case of $N \times N$ matrices: in order to be stochastic, their columns must be normalized to unity.

Another important feature of stochastic matrices is the presence of at least one *unitary eigenvalue.* Again, this can be checked on the simplest 2×2 stochastic matrix. The characteristic equation can be written as:

$$\det \begin{pmatrix} a - \lambda & b \\ 1 - a & 1 - b - \lambda \end{pmatrix} = 0.$$

which gives

$$\lambda^2 + \lambda\,(b - a - 1) + (a - b) = 0.$$

It is easy to see that $\lambda = 1$ is indeed a solution. All other eigenvalues are of absolute value less than 1.

From the last property it follows that the probabilistic distribution corresponding to the unit eigenvector represents a stable configuration, which is not altered by the action of the stochastic matrix. In our example, the unit eigenvector of the 2×2 stochastic matrix satisfies the equation

$$\begin{pmatrix} a & b \\ 1 - a & 1 - b \end{pmatrix} \begin{pmatrix} p_x \\ p_y \end{pmatrix} = \begin{pmatrix} p_x \\ p_y \end{pmatrix}$$

whose solution is

$$p_x = \frac{b}{1 + b - a}, \quad p_y = 1 - p_x = \frac{1 - a}{1 + b - a}$$

Now we can generalize the simple example presented above, and give a more rigorous description of stochastic matrices.

An $N \times N$ matrix M is a stochastic matrix if its entries are real numbers satisfying the two conditions

$$0 \leq M_{\alpha\beta} \leq 1\,, \quad \forall\, \alpha, \beta = 1, 2, \ldots N \quad \text{and} \quad \sum_{\alpha=1}^{N} M_{\alpha\beta} = 1. \tag{1}$$

If not only the columns, but also the rows of such matric are normalized to 1, we shall call it a *strongly stochastic* matrix (1).

As the immediate consequence, we have

$$\sum_{\alpha=1}^{N} M^2{}_{\alpha\beta} = \sum_{\alpha=1}^{N}\sum_{\gamma=1}^{N} M_{\alpha\gamma}M_{\gamma\beta} = 1. \tag{2}$$

It follows that M^2, as well as all the successive powers M^m, are also stochastic matrices.

2.2. *Linear growth*

One of the best ways to introduce the stochastic matrix approach and its essential properties is the analysis of linear growth, either by agglomeration or by internal multiplication, of simple chains composed of two types of units (molecules, cells, ...), which we shall represent by long and short segments.

The analysis of an elementary agglomeration process can be formulated using linear mappings represented by matrices acting on the set of probabilities. Long one-dimensional chains can be obtained by agglomeration of segments from surrounding medium containing a great number of such building blocks.

Consider a linear growth process in which a very long chain is gradually assembled by sticking at its end two types of segments, long (L) and short (S), one by one, yielding the following growth process as seen from the right side of the chain:

$$SLL \to SLLS \to SLLSL \to SLLSLS \to SLLSLSL \to ...$$

For simplicity, we add new segments to the right side, but a more general situation can be envisaged with growth by agglomeration on both sides; there will be no difference from the statistical point of view.

In the simplest case, let us assume that the probability of agglomeration depends only on the contact interaction between the last segment of the chain and the new one that is about to stick to it, and remain fixed there as the new extreme element, to which another building block, S or L as it may be, shall stick in turn.

Let us denote the relative concentrations of short and long segments respectively, by c and $1-c$, i.e. the total number of fragments being N,

$$c = \frac{N_S}{N} = \frac{N_S}{N_S + N_L} \quad \text{and} \quad 1 - c = \frac{N_L}{N} = \frac{N_L}{N_S + N_L}.$$

A purely statistical probability of an encounter between the end of a chain and a new building block is proportional to the corresponding concentration in the medium; but it should also contain the corresponding Boltzmann-Gibbs factor, which in turn will modify the outcome due to chemical potential barrier.

Linear growth of longer segments can be described with stochastic matrix method. Supposing that long chains of L and S elements are floating in the medium in which the concentration of S blocks is c and that of the L blocks is $(1-c)$. At the ends one may find four configurations of doublets: SS, SL, LS and LL.

New blocks S or L agglomerating at the ends transform the last doublet into a new one. If our aim is to imitate the production of Fibonacci-like chains, we must exclude the possibility of an $S-S$ agglomeration

This leaves the following transformation at the ends:

$$LS + L \to SL, \quad SL + S \to LS,$$

$$SL + L \to LL, \quad LL + S \to LS, \quad LL + L \to LL.$$

the combination $LS + S$ being prohibited.

The probabilities of each transition should be proportional to the concentration of the corresponding blocks (c for the S-block, and $(1-c)$ for the L-block), and Boltzmann factors taking into account the chemical potential for each of newly created pairs:

$$e^{-\eta} \text{ for the LS or SL pair, and } e^{-\varepsilon} \text{ for the LL pair,}$$

Let us denote the probabilities of finding one of the allowed pairs at the end of a chain by

$$p_{LS}, \; p_{SL} \text{ and } p_{LL}$$

The unnormed probability factors can be arranged in a 3×3 matrix:

$$\begin{pmatrix} p'_{SL} \\ p'_{LS} \\ p'_{LL} \end{pmatrix} = \begin{pmatrix} 0 & (1-c)\,e^{-\eta} & 0 \\ c\,e^{-\eta} & 0 & c\,e^{-\eta} \\ (1-c)\,e^{-\varepsilon} & 0 & (1-c)\,e^{-\varepsilon} \end{pmatrix} \begin{pmatrix} p_{SL} \\ p_{LS} \\ p_{LL} \end{pmatrix}$$

After normalization of columns we get the following matrix:

$$\begin{pmatrix} 0 & 1 & 0 \\ \frac{ce^{-\eta}}{ce^{-\eta}+(1-c)e^{-\varepsilon}} & 0 & \frac{ce^{-\eta}}{ce^{-\eta}+(1-c)e^{-\varepsilon}} \\ \frac{(1-c)e^{-\varepsilon}}{ce^{-\eta}+(1-c)e^{-\varepsilon}} & 0 & \frac{(1-c)e^{-\varepsilon}}{ce^{-\eta}+(1-c)e^{-\varepsilon}} \end{pmatrix}$$

Introducing the notation

$$\xi = e^{\varepsilon - \eta}$$

the same matrix can be written as:

$$\begin{pmatrix} 0 & 1 & 0 \\ \frac{c\xi}{c\xi+(1-c)} & 0 & \frac{c\xi}{c\xi+(1-c)} \\ \frac{(1-c)}{c\xi+(1-c)} & 0 & \frac{(1-c)}{c\xi+(1-c)} \end{pmatrix}$$

The characteristic equation of the above matirx is:

$$\lambda^3 - \lambda^2 \frac{c\mu}{c\mu + (1-c)} - \lambda \frac{(1-c)}{c\mu + (1-c)} = 0.$$

It has an obvious singular solution $\lambda = 0$ because we have suppressed one of the agglomeration possibilities (no SS association); and there is an obvious eigenvalue $\lambda = 1$. Its eigenvector represents the asymptotic probability distribution:

$$\begin{pmatrix} p^{\infty}_{SL} \\ p^{\infty}_{LS} \\ p^{\infty}_{LL} \end{pmatrix} = \begin{pmatrix} \frac{c\mu}{2c\mu+(1-c)} \\ \frac{c\mu}{2c\mu+(1-c)} \\ \frac{(1-c)}{2c\mu+(1-c)} \end{pmatrix}$$

2.3. *The Fibonacci sequence*

The Fibonacci sequence can be represented by a chain composed by two types of segments, "long" (L) and "short" (S), defined by the following simple *inflation rule* $L \to LS, \quad S \to L$, One can start with any segment (called "a seed"), and apply the rule iterating it at will. For example, starting from one long segment L, one gets, step after step:

$$L \to LS \to LSL \to LSLLS \to LSLLSLSL \to ... \tag{3}$$

It is well known that as the growth of a Fibonacci chain starting from any initial segment leads to the statistics of L and S segments tending to the limit value

$$\left[\frac{N_L}{N_S}\right]_{\infty} = \tau = \frac{1+\sqrt{5}}{2},$$

known also under the name of the "golden number". One can easily check this assertion by inspecting the numbers N_L and N_S resulting from the subsequent growth steps:

$$(1,0);\ (1,1);\ (2,1);\ (3,2);\ (5,3);\ (8,5);\ (13,8);\ ...$$

These numbers form the *Fibonacci series* (Fig. 1), whose constitutive rule is:

$$a_{n+2} = a_{n+1} + a_n.$$

We can easily follow the approximation to the golden number $\tau \simeq 1.618...$ by taking the fractions of consecutive pairs of Fibonacci numbers:

$$\frac{1}{1} = 1, \quad \frac{2}{1} = 2, \quad \frac{3}{2} = 1.5, \quad \frac{5}{3} = 1.666..., \quad \frac{8}{5} = 1.6, \quad \frac{13}{8} = 1.625, \text{etc.} \tag{4}$$

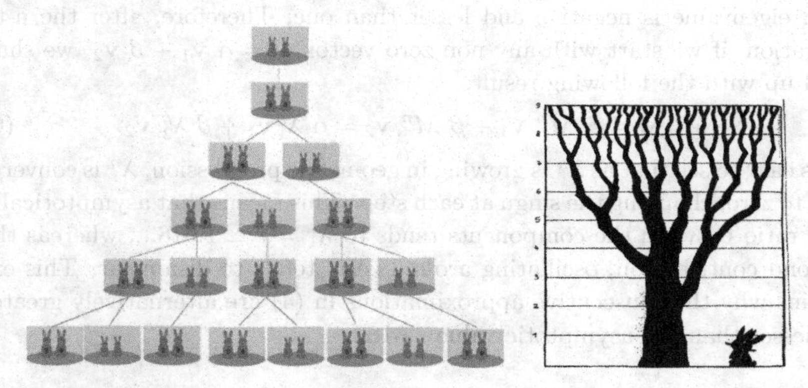

Figure 1. Two examples of growth according to the Fibonacci scheme: proliferation of rabbits (left) and typical growth of a tree (right). The number of rabbits or branches each year follows the scheme $a_{n+2} = a_{n+1} + a_n$. The number of branches grows as 1, 2, 3, 5, 8, 13, 21...

We can represent this process by the action of the so-called *inflation matrix* M on a column representing a couple L and S:

$$\begin{pmatrix} 1 & 1 \\ 1 & 0 \end{pmatrix} \begin{pmatrix} L \\ S \end{pmatrix} = \begin{pmatrix} L+S \\ L \end{pmatrix} \tag{5}$$

Here is an example of successive actions of the inflation matrix:

$$\begin{pmatrix} 1 & 1 \\ 1 & 0 \end{pmatrix} \begin{pmatrix} 1 \\ 0 \end{pmatrix} = \begin{pmatrix} 1 \\ 1 \end{pmatrix}, \quad \begin{pmatrix} 1 & 1 \\ 1 & 0 \end{pmatrix} \begin{pmatrix} 1 \\ 1 \end{pmatrix} = \begin{pmatrix} 2 \\ 1 \end{pmatrix},$$

$$\begin{pmatrix} 1 & 1 \\ 1 & 0 \end{pmatrix} \begin{pmatrix} 2 \\ 1 \end{pmatrix} = \begin{pmatrix} 3 \\ 2 \end{pmatrix}, \quad \begin{pmatrix} 1 & 1 \\ 1 & 0 \end{pmatrix} \begin{pmatrix} 3 \\ 2 \end{pmatrix} = \begin{pmatrix} 5 \\ 3 \end{pmatrix},$$

$$\begin{pmatrix} 1 & 1 \\ 1 & 0 \end{pmatrix} \begin{pmatrix} 5 \\ 3 \end{pmatrix} = \begin{pmatrix} 8 \\ 5 \end{pmatrix}, \quad \begin{pmatrix} 1 & 1 \\ 1 & 0 \end{pmatrix} \begin{pmatrix} 8 \\ 5 \end{pmatrix} = \begin{pmatrix} 13 \\ 8 \end{pmatrix},$$

We see how the same pairs of Fibonacci numbers appear again. The inflation matrix has two eigenvalues, which come from the characteristic equation $\det(M - \lambda I) = \det \begin{pmatrix} 1-\lambda & 1 \\ 1 & -\lambda \end{pmatrix} = 0 \quad \lambda^2 - \lambda - 1 = 0$, whose solutions are $\lambda_1 = \tau \simeq 1.618...$ and $\lambda_2 = 1 - \tau \simeq -0.618...$

Any initial two-valued column "vector" can be projected on two eigenvectors $\mathbf{v}_1$ and $\mathbf{v}_2$ corresponding to the eigenvalues λ_1 and λ_2, so that one has: $M\mathbf{v}_1 = \lambda_1 \mathbf{v}_1, \;\; M\mathbf{v}_2 = \lambda_1 \mathbf{v}_2$.

The first eigenvalue is positive and greater than one, whereas the second eigenvalue is negative and lesser than one; Therefore, after the n-th iteration, if we start with any non-zero vector $\mathbf{w} = \alpha\, \mathbf{v}_1 + \beta\, \mathbf{v}_2$, we shall end up with the following result:

$$M^n\, \mathbf{w} = \alpha\, M^n\, \mathbf{v}_1 + \beta\, M^n\, \mathbf{v}_2 = \alpha\, \lambda_1^n \mathbf{v}_1 + \beta\, \lambda_2^n\, \mathbf{v}_2. \tag{6}$$

It is easy to see that as λ^n is growing in geometric progression, λ^n is converging to zero, changing the singn at each step. This means that asymptotically the ratio between the components tends to $\lambda_1 = \tau \simeq 1.618...$, whereas the second contribution, oscillating around zero, tends to disappear. This explains why the consecutive approximations in (4) are alternatively greater or lesser than the asymptotic value 1.618....

2.4. *Stochastic matrix description*

The inflation matrix acting on *total numbers* N_L and N_S,

$$\begin{pmatrix} 1 & 1 \\ 1 & 0 \end{pmatrix} \begin{pmatrix} N_L \\ N_S \end{pmatrix} = \begin{pmatrix} N_L' \\ N_S' \end{pmatrix} \tag{7}$$

can be replaced by a stochastic matrix acting on the probabilities of picking out an L or an S segment, respectively p_L and p_S:

$$p_L = \frac{N_L}{N_L + N_S}, \quad p_S = \frac{N_s}{N_L + N_S}; \quad \text{obviously, } p_L + p_S = 1. \tag{8}$$

During one step of growth of a Fibonacci chain "from within", when each of the long segments gives rise to the pair LS, and simultaneously all the short segments are transformed into long ones, according to the rule $L-->$ LS, $S-->L$, the corresponding numbers undergo the following change:

$$N_L \to N_L' = N_L + N_S, \quad N_S \to N_S' = N_L. \tag{9}$$

Parallelly, the probabilities of picking up onde or another segment type evolve as follows:

$$p_L \to p_L' = \frac{N_L + N_S}{2N_L + N_S}, \quad p_S \to p_S' = \frac{N_L}{2N_L + N_S}. \tag{10}$$

At this point we can ask the question whether there exists a stationary probability distribution to which such a chain will tend after a very great number of consecutive "blow-ups", i.e. as the infinite limit of this process. The answer is positive, and it is enough to solve the following simple equation to get it:

$$p_L = p_L^{'} \rightarrow \frac{N_L}{N_L + N_S} = \frac{N_L + N_S}{2N_L + N_S}. \tag{11}$$

Dividing by numerators and denominators by N_L and introducing the variable $\mu = N_S/N_L$, we can write

$$\frac{1}{1+\mu} = \frac{1+\mu}{2+\mu} \quad \text{leading to } 2 + \mu = (1+\mu)^2. \tag{12}$$

The resulting quadratic equation $\mu^2 + \mu - 1 = 0$ has two solutions, one of which is negative, and should be rejected. The positive one is

$$\mu = \frac{\sqrt{5}-1}{2} = 0.618..., \tag{13}$$

which is the inverse of the "golden number" $\tau = (\sqrt{5}+1)/2$.

The same process can be described using the concept of stochastic matrix and its properties. Let us compare two columns containing just the old and new probabilities; it is easy to check that they can be related by a linear transformation with appropriate coefficients:

$$\begin{pmatrix} p_L^{'} \\ p_S^{'} \end{pmatrix} = \begin{pmatrix} \frac{N_L+N_S}{2N_L+N_S} & \frac{N_L+N_S}{2N_L+N_S} \\ \frac{N_L}{2N_L+N_S} & \frac{N_L}{2N_L+N_S} \end{pmatrix} \begin{pmatrix} p_L \\ p_S \end{pmatrix}. \tag{14}$$

Using the same variable $\mu = N_S/N_L$ and taking into account that $p_L = 1/(1+\mu)$, $p_S = 1 - p_L = \mu/(1+\mu)$, $p_L^{'} = (1+\mu)/(2+\mu)$, $p_S^{'} = 1/(2+\mu)$, we can rewrite the Eq. (14) as follows:

$$\begin{pmatrix} p_L^{'} \\ p_S^{'} \end{pmatrix} = \begin{pmatrix} \frac{1+\mu}{2+\mu} & \frac{1+\mu}{2+\mu} \\ \frac{1}{2+\mu} & \frac{1}{2+\mu} \end{pmatrix} \begin{pmatrix} p_L \\ p_S \end{pmatrix}. \tag{15}$$

The characteristic equation of the matrix displayed in 15, det $(N - \lambda I) = 9$ yields

$$\lambda^2 - \lambda = 0, \quad \rightarrow \quad \lambda = 0 \text{ or } \lambda = 1.$$

The zero is due to the fact that the stochastic matrix here is singular. The remaining eigenvalue is equal to one, and the corresponding eigenvector is stable under the transformation, therefore it represents the asymptotic state.

To determine the components of this eigenvector, it is enough to solve one equation, e.g. for the first component p_L, because we always have $p_S = 1 - p_L$. Therefore, we can write, for the upper component of Eq. (15),

$$\frac{1+\mu}{2+\mu}\,p_L + \frac{1+\mu}{2+\mu}\,(1-p_L) = \frac{1+\mu}{2+\mu} = p_L. \tag{16}$$

Now, as we had $p_L = 1/(1+\mu)$, again we get the same equation,

$$\frac{1+\mu}{2+\mu} = \frac{1}{1+\mu} \to \mu^2 + \mu - 1 = 0,$$

leading to the same stationary value of $\mu = 0.618$. From this we get the asymptotic and stable probability distribution

$$p_L = \frac{1}{1+\mu} = 0.618.., \quad p_S = 1 - p_L = 0.382...$$

2.5. *Agglomeration of chains*

One can adopt a slightly different point of view, taking into account simultaneous coexistence of different clusters: singlets (i.e. the most elementary building blocks, in our example the S and L segments), the pairs (here LS, SL and LL pairs), triplets SLS, SLL, LSL, LLS and LLL, and so on (Fig. 2). Although all these clusters coexist at the same time, we suppose that they are not in equilibrium. As the time goes by, smaller clusters give way to bigger ones, due to constant agglomeration process taking place.

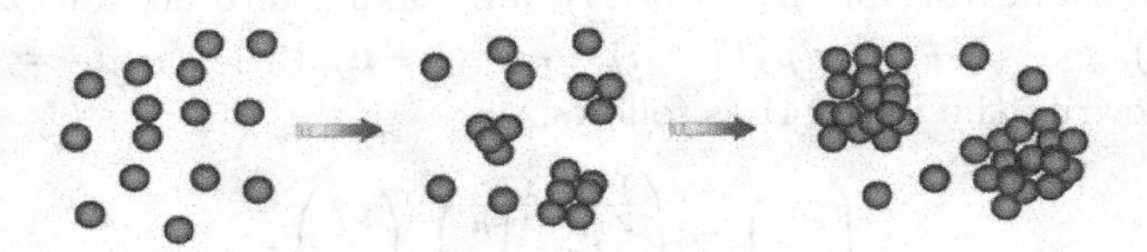

Figure 2. Agglomeration of doublets, triplets and bigger clusters from coexisting smaller items.

Let us denote the relative concentrations of short and long segments respectively, by c and $1-c$. This means that if the total number of segments is $N = N_S + N_L$, then

$$c = \frac{N_S}{N} = \frac{N_S}{N_S + N_L} \quad \text{and} \quad 1 - c = \frac{N_L}{N} = \frac{N_L}{N_S + N_L}.$$

A purely statistical probability of an encounter between two items is given by respective products of the corresponding concentrations:

$$p_{SS} \sim c^2, \quad p_{LS} \sim 2c(1-c), \quad p_{LL} \sim (1-c)^2$$

where the probability p_{LS} takes into account both LS and SL combinations, regardless of their orientation.

It is obvious that with the equality sign in the above formula, the probabilities of binary encounters would be normalized to 1, because $c^2 + 2c(1-c) + (1-c)^2 = [c + (1-c)]^2 = 1^2 = 1$. But when at a given temperature stable bonds between the segments are created, involving certain expense of energy, the probability of clustering should also contain the corresponding Boltzmann-Gibbs factors, which in turn will modify the outcome following the following scheme:

$$p_{SS} \sim c^2\, e^{-\frac{E_{SS}}{kT}}, \quad p_{LS} \sim 2c(1-c)\, e^{-\frac{E_{LS}}{kT}}, \quad p_{LL} \sim (1-c)^2\, e^{-\frac{E_{LL}}{kT}}. \quad (17)$$

Now the sum $p_{SS} + p_{LS} + p_{LL}$ is no longer normalized to 1; in order to find the genuine probability distibution, we must normalize the above expressions by an appropriate factor, to get the correct expressions

$$p_{SS} = \frac{c^2\, e^{-\frac{E_{SS}}{kT}}}{Q}, \; p_{LS} = \frac{2c(1-c)\, e^{-\frac{E_{LS}}{kT}}}{Q}, \; p_{LL} = \frac{(1-c)^2\, e^{-\frac{E_{LL}}{kT}}}{Q} \quad (18)$$

with

$$Q = c^2\, e^{-\frac{E_{SS}}{kT}} + 2c(1-c)\, e^{-\frac{E_{LS}}{kT}} + (1-c)^2\, e^{-\frac{E_{LL}}{kT}}. \quad (19)$$

Of course, once a non-negligible number of doublets is formed, one cannot exclude their collisions with single L or S segments floating around in abundance, and the following formation of *triplets*, i.e. chains like SSS, SSL, SLS, LLS, LSL and LLL. But at the very beginning of the agglomeration process these encounters can be neglected versus singlets' collisions, so we put aside for a while the more complicated processes in order to concentrate on the simple description of the onset of agglomeration.

Let us evaluate the relative concentrations of S and L segments in the doublets thus produced. They are given by the following expressions:

$$p_S^{(1)} = \frac{1}{2}[2p_{SS} + p_{LS}]; \quad p_L^{(1)} = \frac{1}{2}[2p_{LL} + p_{LS}]. \quad (20)$$

Neglecting bigger clusters, we may assume that in the initial stages of agglomeration only singlets and doublets are present. After a given short time Δt, certain number Δn of doublets was formed, and we suppose that

$\Delta n << N$; at the same time, there are $N - 2\Delta n$ singlets left floating around.

Of course, the concentration c of short segments stays almost the same among the remaining singlets as long as $N >> \Delta n$; but it may be very different among the doublets if there is a pronounced preference for agglomeration of certain kind of pairs only.

To compare the variation of concentration we should write the following differential equation:

$$\Delta p_S = \frac{\Delta n \, p_S^{(1)} + (N - 2\Delta n)\, c}{N - \Delta n} - c, \tag{21}$$

which up to the first order in $\Delta n/N$ reduces to

$$\Delta p_S \simeq \left(\frac{\Delta n}{N}\right) [p_S^{(1)} - c]. \tag{22}$$

The quantity $\Delta n/N$ is proportional to the time Δt it took to produce Δn doublets; therefore we can write:

$$\frac{\Delta p_S}{\Delta t} \simeq [p_S^{(1)} - c]. \tag{23}$$

Inserting the explicit expressions, we can write

$$\frac{dp_S}{dt} = \frac{2\, c^2\, e^{-\frac{E_{SS}}{kT}} + 2c(1-c)\, e^{-\frac{E_{LS}}{kT}}}{2[c^2\, e^{-\frac{E_{SS}}{kT}} + 2c(1-c)\, e^{-\frac{E_{LS}}{kT}} + (1-c)^2\, e^{-\frac{E_{LL}}{kT}}]} - c. \tag{24}$$

In order to understand the character of the solutions of the Eq. (24) it is enough to determine the nature of its *singular points*, i.e. particular solutions corresponding to $dp_S/dt = 0$. The denominator Q of fractions defining $p_S^{(1)}$ and $p_L^{(1)}$ is strictly positive, so we can multiply the entire expression by Q and reduce the condition of vanishing of the right-hand side of Eq. (24) to

$$\begin{aligned} &c^2(1-c)\, e^{-\frac{E_{SS}}{kT}} + c(1-c)\, e^{-\frac{E_{LS}}{kT}} \\ &-c^3\, e^{-\frac{E_{SS}}{kT}} - 2c^2(1-c)\, e^{-\frac{E_{LS}}{kT}} - c(1-c)^2\, e^{-\frac{E_{LL}}{kT}} = 0. \end{aligned} \tag{25}$$

which can be put into a simplified form

$$c(1-c)\left[ce^{-\frac{E_{SS}}{kT}} + (1-2c)e^{-\frac{E_{LS}}{kT}} - (1-c)\, e^{-\frac{E_{LL}}{kT}}\right] = 0. \tag{26}$$

There are two obvious solutions $c = 0$ and $c = 1$; besides, the expression in the square brackets can vanish under certain conditions. Let us denote

$$e^{\frac{E_{LS}-E_{SS}}{kT}} = \xi, \qquad e^{\frac{E_{LS}-E_{LL}}{kT}} = \mu; \tag{27}$$

Multiplying the Eq. (26) by $e^{\frac{E_{LS}}{kt}}$, we find that the third singular solution is obtained when x attains the value

$$c = \frac{\xi - 1}{\xi - 1 + \mu - 1}; \quad \text{and obviously} \quad 1 - c = \frac{\mu - 1}{\xi - 1 + \mu - 1} \tag{28}$$

which shows very clearly the invariance of our equations under the simultaneous transformation between c and $1 - c$ and between ξ and μ, in other words, under interchanging the labels L and S, as it should be. It is to be noted that in the particular situation when $\xi = \mu$, which happens at any temperature for $E_{SS} = E_{LL}$, this singular is found exactly in the middle of the segment, $c = 1/2$, which confirms the symmetric character of our equation. When $0 \leq c \leq 1$, we must have either simultaneously $\xi \geq 1$ and $\mu \geq 1$ or simultaneously $\xi \leq 1$ and $\mu \leq 1$. Looking at the definitions of ξ and μ, we see that this means that

$$E_{LS} \geq E_{SS} \quad \text{and} \quad E_{LS} \geq E_{SS} \tag{29}$$

or

$$E_{LS} \leq E_{SS} \quad \text{and} \quad E_{LS} \leq E_{SS}. \tag{30}$$

If neither of these conditions is satisfied, there will be only two singular solutions inside the segment $[0, 1]$ where the probabilities should stay by definition - as a matter of fact, there is always the third solution, but it is either negative or greater than 1, which makes it unacceptable.

It is not difficult to see that for the third solution to be contained inside the segment $[0, 1]$, there are four different situations possible, according to the following classification:

$$\text{(a)} \;\; 1 > \xi > \mu, \quad \text{(b)} \;\; 1 > \mu > \xi, \quad \text{(c)} \;\; 1 < \xi < \mu, \quad \text{(d)} \;\; 1 < \mu < \xi. \tag{31}$$

In the first two cases the points $c = 0$ and $c = 1$ are *repulsive*, i.e. if the initial conditions are close to $p_S = c = 0$, the system will tend to make the p_S grow; the same is true for the opposite case, when p_L is initially close to 0 (which means that p_S is close to 1), the system will tend to make it bigger, too. As a result, both tendencies meet somewhere in between, which means that the corresponding "intermediate" singular solution is an *attractive* point, as shown in Fig. 3.

If the conditions of Eq. (31) are not satisfied, two possibilities remain, when one of the extremal points, $p_S = c = 0$ or $p_S = c = 1$ is repulsive while the other one is attractive, and vice versa. The result of the last possibility is shown in the left diagram of the Fig. 3.

Figure 3. Singular points of Eq. (24): (a) with two solutions only; (b) with a third stable solution at $0 < c < 1$.

3. Agglomeration in two dimensions

3.1. *Planar tilings*

One of the best ways to introduce the stochastic matrix approach and compare it with Lotka-Volterra-like description is the analysis of growth of planar structures, either by agglomeration on the edge, or by free pairing of building blocks.

The analysis of an elementary agglomeration process of equilateral polygons tiling a plane or a sphere can be formulated using linear mappings represented by matrices acting on the set of probabilities, or by the analysis of probabilities of creating just pairs of such polygons at the onset of agglomeration process. Finally, one can use the inflation matrix technique applied to planar structures, which would represent the best imitation of *phyllotaxis* (Fig. 4) in plants, i.e. the creation of two-dimensional quasi regular patterns.

Figure 4. Two examples of phyllotaxis: a sunflower and an artichoke.

The most natural and compact tiling of the plane is the regular hexagonal tiling (Fig. 5), which was noticed already by J. Kepler ([10]):

Euler's rule takes into account the global balance between the average numbers of various faces, edges and vortices of polygons (Fig. 6) forming the lattice (see ([2])).

Assembling (or growing) the polygons on the rim of a bulk such as above

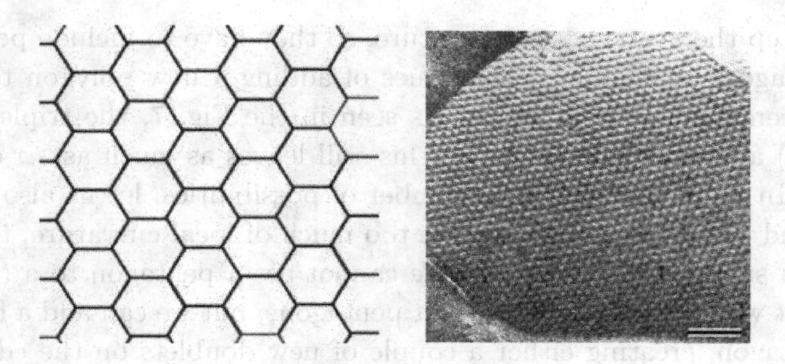

Figure 5. **The hexagonal tiling and its realization: A protein layer.**

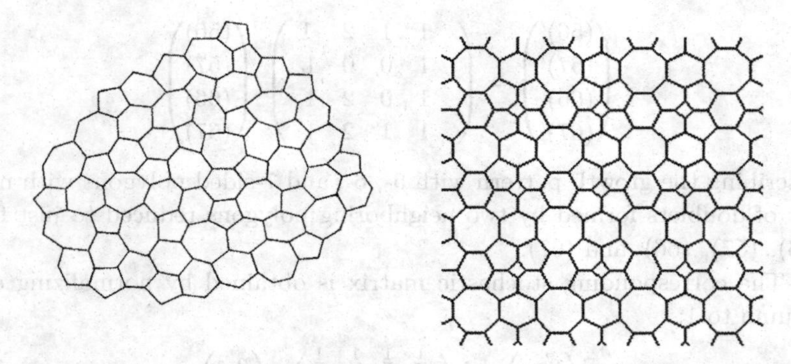

Figure 6. **A random tiling of the plane with 5-, 6- and 7-sided equilateral (but not equiangular) polygons, and a regular tiling with 4, 6 and 8-sided polygons. In order to keep the structure planar, the numbers of 5-sided and 7-sided polygons in the first case must be equal, while the number of 6-sided polygons remains arbitrary. The second example also satisfies Euler's rule.**

can be illustrated by the following scheme:

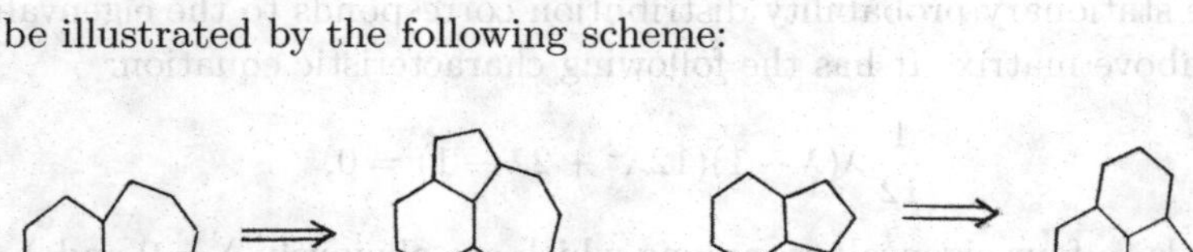

Figure 7. **Two examples of adding a new polygon to an existing cavity between the two assembled ones: (67) + (5) → (56) + (57); (56) + (6) → (56) + (66).**

Let us set up a model of growth of patterns made of 5-, 6-, and 7-sided equilateral (but not equiangular!) polygons ([8]), ([9]). Planar structures should keep the average zero curvature, so they have to include pentagons and heptagons as well ([3]). The issues of adding a new polygon to a cavity between the two other ones. As seen in the Fig. 7, the triplets (555) and (777) are excluded. However, this still leaves as much as *six* different triplets. In order to reduce the number of possibilities, let us also exclude the 55 and 77 pairings which create too much of local curvature, therefore too much stress in the network. We cannot ass a pentagon to a (56) pair because it would create two adjacent pentagons, but we can add a hewagon or a heptagon, creating either a couple of new doublets on the edge, (56) and (66), or (57) and (67), and so on.

This leaves the following sticking schemes available, with statistical factors included:

$$\begin{pmatrix} (56)' \\ (57)' \\ (66)' \\ (67)' \end{pmatrix} = \begin{pmatrix} 1 & 1 & 2 & 1 \\ 1 & 0 & 0 & 1 \\ 1 & 0 & 2 & 1 \\ 1 & 1 & 2 & 1 \end{pmatrix} \begin{pmatrix} (56) \\ (57) \\ (66) \\ (67) \end{pmatrix}$$

describing the growth pattern with 5-, 6-, and 7-sided polygons with number of doublets formed by two neighboring polygons reduced to just four: (56), (57), (66) and (67).

The corresponding stochastic matrix is obtained by normalizing each column to 1:

$$\begin{pmatrix} p'_{56} \\ p'_{57} \\ p'_{66} \\ p'_{67} \end{pmatrix} = \begin{pmatrix} \frac{1}{4} & \frac{1}{2} & \frac{1}{3} & \frac{1}{4} \\ \frac{1}{4} & 0 & 0 & \frac{1}{4} \\ \frac{1}{4} & 0 & \frac{1}{3} & \frac{1}{4} \\ \frac{1}{4} & \frac{1}{2} & \frac{1}{3} & \frac{1}{4} \end{pmatrix} \begin{pmatrix} p_{56} \\ p_{57} \\ p_{66} \\ p_{67} \end{pmatrix}$$

This matrix acts on the *probabilities of doublets.*

The stationary probability distribution corresponds to the eigenvalue 1 of the above matrix. It has the following characteristic equation:

$$\frac{1}{12}\,\lambda(\lambda-1)(12\lambda^2+2\lambda-1)=0, \tag{32}$$

and leads to four eigenvalues, among which are obviously $\lambda = 0$ and $\lambda = 1$. The remaining two eigenvalues are solutions of the quadratic equation $12\lambda^2 + 2\lambda - 1 = 0$, and their absolute values are both smaller than 1.

The stable configuration to which any iteration process will lead is given by the following probability distrubution of doublets and basic polygons:

$$p^{\infty}_{56} = \frac{1}{4},\ p^{\infty}_{57} = \frac{1}{8},\ p^{\infty}_{66} = \frac{3}{8},\ p^{\infty}_{67} = \frac{1}{4},\ \to\ p^{\infty}_{5} = p^{\infty}_{7} = \frac{3}{16},\ p^{\infty}_{6} = \frac{5}{8}.$$

The stochastic matrix description of growth of clusters by agglomeration of basic structural units from the surrounding medium introduces two different concentrations: the probabilities p_i refer to the probability of finding the i-th bloc on the rim of the cluster, and c_i the corresponding concentrations of the same bloc floating around. In many cases, when the *ergodicity* hypothesis is valid, this description can be replaced by simple *pairing* of elementary building blocks Let us consider just the production of pairs from single polygons, like if all the building blocks were floating freely.

If a random three-coordinate network is created by agglomeration at given constant temperature T, the probability distribution of 5, 6 and 7-sided polygons will be affected by the energy barriers related to the local curvature they produce. Let us suppose that the probability of creation of a single polygon of given type is defined; let us consider these probabilities as the input of our model of agglomeration of polygons, and let us denote them by

$$P_5^{(0)}, P_6^{(0)} \text{ and } P_7^{(0)}.$$

By definition, $P_5^{(0)} + P_6^{(0)} + P_7^{(0)} = 1$, so that there are only two independent variables, e.g. P_5 and P_6 whose domain of variation is the triangle (simplex) on the P_5, P_6 plane. The extremal points of the triangle correspond to "pure" configurations with $P_5 = 1, P_6 = P_7 = 0$, or $P_7 = 1, P_5 = P_6 = 0$, or $P_6 = 1, P_5 = P_7 = 0$.

This is the first step towards agglomeration; we shall denote the resulting probability distribution of doublets by $P^{(1)}$. Let us incorporate the energy barriers resulting from the corresponding stresses provoked by local (positive or negative) curvatures into the probabilities of the corresponding doublets.

We shall suppose that the probability of a 5–6 couple, which represents one standard departure from flatness, should contain a Boltzmann factor $e^{-\frac{\Delta E_{56}}{k_B T}} = e^{-\alpha}$; in principle, the same should be true for the probability of the doublet $P_{67}^{(1)}$, but we cannot exclude the possibility of a different energy barrier, giving rise to a Boltzmann factor $e^{-\frac{\Delta E_{67}}{k_B T}} = e^{-\beta}$. We shall consider here a model excluding the pairs (55) and (77), as in the model using the stochastic matrix approach.

The probabilities of doublets should be proportional to the following expressions:

$$P_{56}^{(1)} = \frac{2}{Q} P_5^{(0)} P_6^{(0)} e^{-\alpha}, \quad P_{57}^{(1)} = \frac{2}{Q} P_5^{(0)} P_7^{(0)},$$

$$P_{66}^{(1)} = \frac{1}{Q}\,(P_6^{(0)})^2\,, \quad P_{67}^{(1)} = \frac{2}{Q}\,P_6^{(0)}P_7^{(0)}\,e^{-\beta}\,,$$

where the normalizing factor Q is given by:

$$Q = 2\,P_5^{(0)}P_6^{(0)}\,e^{-\alpha} + 2\,P_5^{(0)}P_7^{(0)} + (P_6^{(0)})^2 + 2\,P_6^{(0)}P_7^{(0)}\,e^{-\beta}.$$

With this at hand, we can calculate the distribution of 5-, 6- and 7-sided polygons in all spontaneously agglomerated doublets. The corresponding set of probabilities will be denoted by $P_k^{(1)}$, and is computed as follows:

$$P_5^{(1)} = \frac{1}{2}\,(P_{56}^{(1)} + P_{57}^{(1)})\,; \qquad P_6^{(1)} = \frac{1}{2}\,(P_{56}^{(1)} + 2\,P_{66}^{(1)} + P_{67}^{(1)})\,;$$

and, obviously,

$$P_7^{(1)} = \frac{1}{2}\,(P_{57}^{(1)} + P_{67}^{(1)}) = 1 - P_5^{(1)} - P_6^{(1)}.$$

In a real agglomeration process, at least at its initial stage, one can observe a mixture of single polygons and freshly created doublets; later on also triplets, quadruplets of agglomerated polygons are created, too, but at the beginning , only single polygons ("singlets") and edge-sharing pairs ("doublets") dominate.

Now a differential system can be set forth, with time parameter replaced by the "step" variable s. Suppose that at certain initial stage of agglomeration one has, in a unit volume, $N - m$ singlets and m doublets. The average probability of finding a k-sided polygon in such sample will be

$$P_k(s) = (1 - s)\,P_k^{(0)} + s\,P_k^{(1)}$$

where $s = m/N$ is a natural parameter describing the progress of agglomeration process at its initial stage (reduced to doublet creation only). The derivative of $P_k(s)$ is therefore

$$\frac{dP_k}{ds} = P_k^{(1)} - P_k^{(0)}$$

and is independent of s in this approximation. Of course, only two out of three above differential equations are independent.

The above differential system, reduced to two ordinary differential equations for independent variables P_5 and P_6 can be written explicitly as follows (we shall replace $P_k^{(0)}$ by P_k for the sake of simplicity):

$$\frac{dP_5}{ds} = \frac{1}{2Q}\left[P_{56}^{(1)} + P_{57}^{(1)}\right] - P_5\,,$$

$$\frac{dP_6}{ds} = \frac{1}{2Q}\left[2\,P_{66}^{(1)} + P_{67}^{(1)}\right] - P_6.$$

The third equation is linearly dependent, because $P_7 = 1 - P_5 - P_6$, $dP_7/ds = -dP_5/ds - dP_6/ds$. The differential equations can be written in a more explicit manner, with clear dependence on the variables P_5 and P_6:

$$\frac{dP_5}{ds} = \frac{1}{2Q}\left[2P_5P_6e^{-\alpha} + 2P_5(1 - P_5 - P_6)\right] - P_5Q,$$

$$\frac{dP_6}{ds} = \frac{1}{2Q}\left[2P_6^2 + 2P_6(1 - P_5 - P_6)e^{-\beta} + 2P_5P_6e^{-\alpha}\right] - P_6Q,$$

with

$$Q = 2P_5P_6e^{-\alpha} + 2P_5P_7 + P_6^2 + 2P_6P_7e^{-\alpha}$$

where $P_7 = 1 - P_5 - P_6$.

We have admitted different energy barriers for creation of a (56) pair and that of a (67) pair, because the stress caused by a positive curvature introduced by a pentagon might be different than the stress of the negative curvature caused by a heptagon. Let us perform the analysis of phase trajectories of this differential system, displayed in Fig. 8. Here the simplified case is considered, with two energy barriers equal, i.e. with $\alpha = \beta$.

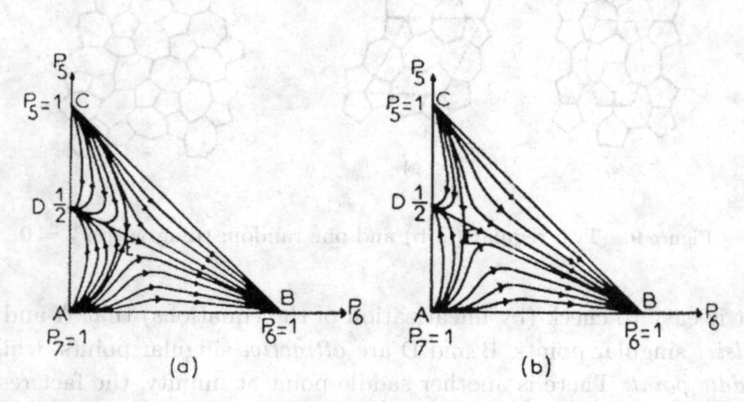

Figure 8. Two phase potraits of probability trajectories describing the agglomeration process of 5-, 6- and 7-sided polygons.

The *singular points*, correspond to constant solutions at which the two derivatives vanish simultaneously. There are five such solutions, three of which are found at the vertices of the simplex of probabilities:

$$A: P_7 = 1, \quad P_5 = 0, \ (P_6 = 0); \quad B: P_6 = 1 \quad (P_5 = P_7 = 0);$$

$$C: P_5 = 1, \quad P_7 = 0, \quad (P_6 = 0);$$

while two remaining ones are:

$$D: \quad P_5 = P_7 = \frac{1}{2} \quad (P_6 = 0),$$

and the fifth one (a saddle point) inside the triangle:

$$E: \quad P_5 = \frac{1}{3 - e^{-\alpha}}, \quad P_6 = \frac{1 - e^{-\alpha}}{3 - e^{-\alpha}}, \quad P_7 = 1 - P_5 - P_6 = \frac{1}{3 - e^{-\alpha}}.$$

The pure 6-polygon tiling correspond to the attractive point B; pure 5's and 7's are obviously impossible (the points A and C are *repulsive*, and the point D on the edge, with $P_5 = P_7$, $P_6 = 0$ is attractive, showing that pure $5 - 7$ tilings are possible. Indeed, regular or random structures can be obtained with only two types of equilateral deformable polygons: (5) and (7), as shown in the figure below (Fig. 9). Such structures correspond to the point (D) in the diagrams displayed in the (8).

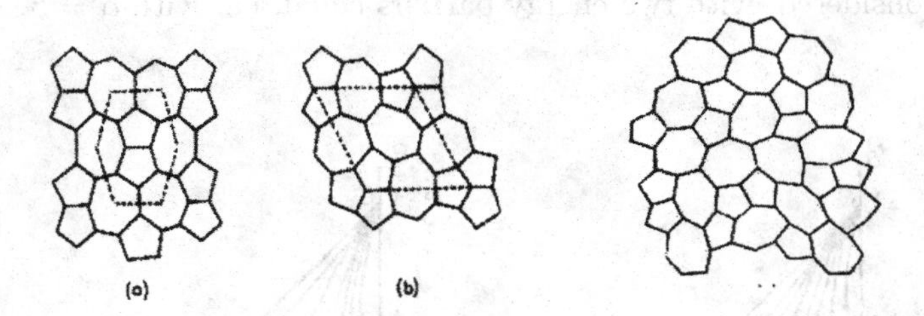

Figure 9. Two regular (a, b) and one random tiling with $P_6 = 0$.

It is easy to check (by linearization of the equations) that A and C are *repulsive* singular points, B and D are *attractive* singular points, while E is a *saddle point.* There is another saddle point at infinity, the fact resulting also from Euler's formula for a sphere, but for the probabilities it has no physical meaning, of course.

When there is an excess of pentagons introduced deliberately, the resulting tiling will display *positive curvature* and eventually close into a structure topologically equivalent to a sphere ([3]).

3.2. *Curved structures*

The simplest example of how a curved surface is being produced via equilateral polygon agglomeration: the case when only 5- and 6-sided polygons assembling.

The structure cannot remain planar because pentagons create local positive curvature. Let us exclude the (55) doublets as creating too much local stress; then only the pairs (56) and (66) can be considered. Adding a (5)- or a (6)-sided polygon to an existing cavity (Fig. 10) can produce only the following results:

$$(56) + (6) \to (56) + (66),$$

$$(66) + (5) \to 2 \times (56),$$

$$(66) + (6) \to 2 \times (66)$$

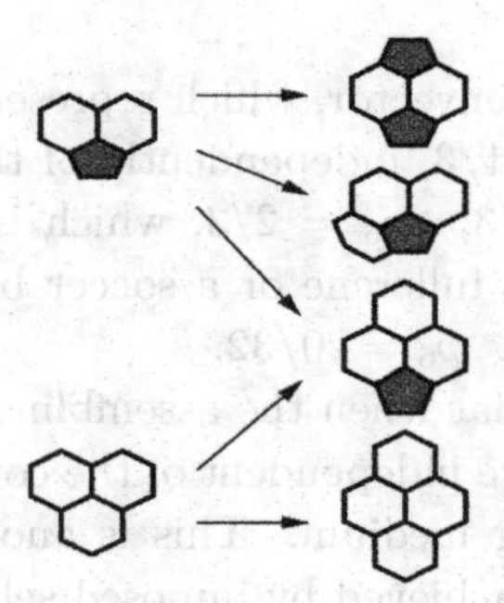

Figure 10. Forming of a new polygon in a (6,5)-type and in a (6,6)-type cavities.

Let us consider the case of structure units reduced to (5)- and (6)-sided polygons Suppose that the concentration of the (5)-sided polygons in the surrounding medium is α, and that of the (6)-sided polygons is $(1 - \alpha)$. Then the probabilities of creating new pairs in cavities will be as follows: From a (56) cavity we can get only the combination (56) + (66), because a pentagon cannot stick to it, as the (55) doublets are prohibited; From a (66) cavity we get either two (56) doublets with probability α, or two (66) doublets with probability $(1 - \alpha)$.

The stochastic matrix of the $\alpha\,(5) + (1 - \alpha)\,(6)$ growth The issues of (5) and (6) polygon agglomeration can be represented by of the following matrix action on the column representing the two pairs, (56) and (66):

$$\begin{pmatrix} (56) \\ (66) \end{pmatrix}' = \begin{pmatrix} 1-\alpha & 2\alpha \\ 1-\alpha & 2(1-\alpha) \end{pmatrix} \begin{pmatrix} (56) \\ (66) \end{pmatrix}$$

Translating these issues into probabilities, we must norm the columns of the matrix so that it becomes a *stochastic matrix*, transforming a probability distribution into another one. The result is:

$$\begin{pmatrix} p'_{56} \\ p'_{66} \end{pmatrix} = \begin{pmatrix} 1/2 & \alpha \\ 1/2 & (1-\alpha) \end{pmatrix} \begin{pmatrix} p_{56} \\ p_{66} \end{pmatrix}$$

It is easy to calculate the eigenvector of the above matrix. The characteristic equation is

$$\det \begin{pmatrix} 1/2 - \lambda & \alpha \\ 1/2 & (1-\alpha) - \lambda \end{pmatrix} = 0$$

which yields $\lambda^2 + \left(\alpha - \frac{3}{2}\right) + \frac{1}{2} - \alpha = 0$, of which $\lambda = 1$ is the obvious solution.

The corresponding eigenvector, which represents the asymptotic statistics, is $p_{56} = 2/3, \;\; p_{66} = 1/3$, independently of the factor α. The polygon statistics is then $p_5 = 1/3, \;\; p_6 = 2/3$, which is not that bad compared with the proportions in a fullerene or a soccer ball: 12 pentagons and 20 hexagons, i.e. $p_5 = 12/32, \;\; p_6 = 20/32$.

This example shows that when the assembling rules are very strict, the resulting configurations are independent of the concentration of elementary blocks in the surrounding medium. This is another example of how the *self-organization* may be achieved by imposed selection rules.

The structures resulting from assembling exclusively hexagons and pentagons cannot remain planar; they display local positive curvature and eventually form closed structures (Fig. 11) like the fullerene molecules or icosahedral viral capsids ([13]), ([14]), ([15]).

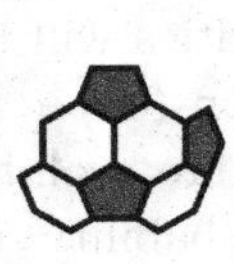

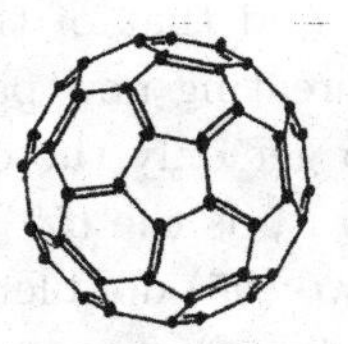

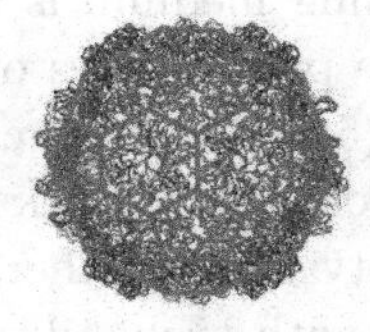

Figure 11. A curved structure with 3 pentagons and 4 hexagons; the fullerene molecule and the *cowpea* virus.

4. Lotka-Volterra equations

4.1. *Classical predator-prey model*

The Lotka-Volterra equations describe the evolution of biological systems with different living organisms, competing for food and space, or even eating each other (predators and prey). The simplest model is given by two species only, the prey x and the predator y. The evolution of their (relative) numbers can be described as follows:

The prey population $X(t)$ increases at a rate $a\,X dt$, proportional to its own number, but is simultaneously killed by predators at a rate $-\beta\, XY dt$; The predator population $Y(t)$ decreases at a rate $-c\,Y dt$, proportional to its own number, but increases at a rate $\gamma XY dt$; which leads to the following differential system:

$$\frac{dX}{dt} = aX - \beta XY, \qquad \frac{dY}{dt} = -cY + \gamma XY.$$

It can be shown that the above system of two equations leads to periodic solutions (Fig. 12). This means that prey and predator populations oscillate with the same period, but with a delayed phase. When the population of predators grows too much, the population of prey decreases; the lack of food provokes the subsequent decrease of predators' number. The diminished predators' population favorizes the rise of number of prey, and so forth...

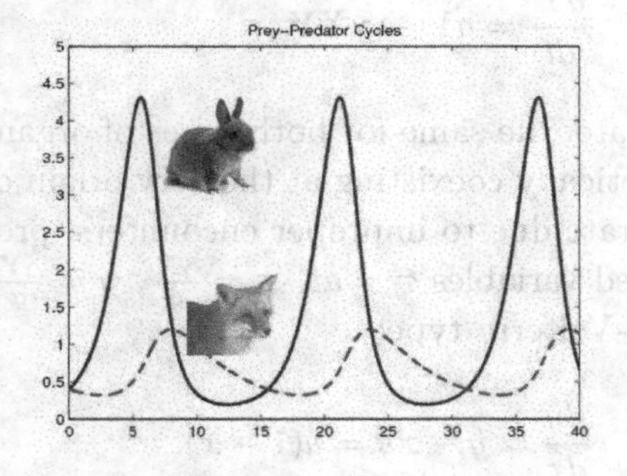

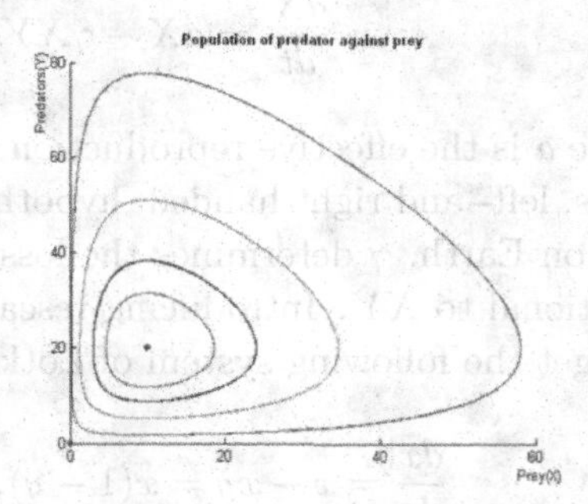

Figure 12. Left: Typical time evolution of total numbers of prey (continuous line) and predator (hatched line); Right: The phase space portrait of the same differential system with closed trajectories.

This classical equations have been used to describe not only biological ecosystems, but also chemical reactions and other collective phenomena in which the number of different items can vary due to binary interactions.

4.2. *The origin of chirality in living organisms*

In the ninetieth century Louis Pasteur discovered a very interesting optical properties of chemical compounds characteristic for living organisms. When crystallized, sugars and certain organic acids extracted from living organisms, polarize light rays in a specific manner. For example, all sugars display the RIGHT chirality, whereas all aminoacids display LEFT chirality. The same sugars and aminoacids can exist as isomers, both left- and right-handed. When synthetized artificially, these isomers are produced at equal rate, 50% of each type.

But actually living organisms can assimilate only the right-handed sugars and the left-handed aminoacids. The "improper" sugars and proteins not only are rejected by living organisms, but can provoke serious dysfunctioning and even death ([6]).

We have no reasons to suppose that at the origin of life on Earth only one type of organic matter was produced; most probably, both types might have appeared simultaneously. However, the coexistence being impossible, one of them inevitably took over. The selection process can be described by aLotka-Volterra type system which we shall present now.

The dynamical coexistence between the two types of living matter can be described by the following simple system of equations of Lotka-Volterra type:

$$\frac{dX}{dt} = aX - \gamma XY, \qquad \frac{dY}{dt} = aY - \gamma XY.$$

Here a is the effective reproduction rate, the same for both types of organisms, left- and right-handed, hypothetically coexisting at the very origin of life on Earth, γ determines the loss rate due to improper encounters, proportional to XY. Introducing rescaled variables $\tau = at$, $x = \frac{\gamma X}{a}$, $y = \frac{\gamma Y}{a}$, we get the following system of Lotka-Volterra type:

$$\frac{dx}{d\tau} = x - xy = x(1-y), \qquad \frac{dy}{d\tau} = y - xy = y(1-x).$$

There are two *singular solutions*, $(0,0)$ and $(1,1)$.

The above differential system, can be solved explicitly via separation of variables. The solutions $x(\tau)$ and $y(\tau)$ are quite cumbersome; however, if we are interested in their asymptotic behavior for large times, To understand the qualitative behavior, it is enough to analyze the trajectories in the (x,y) plane. Dividing the second equation by the first one, we get the following differential equation:

$$\frac{dy}{dx} = \frac{y(1-x)}{x(1-y)}.$$

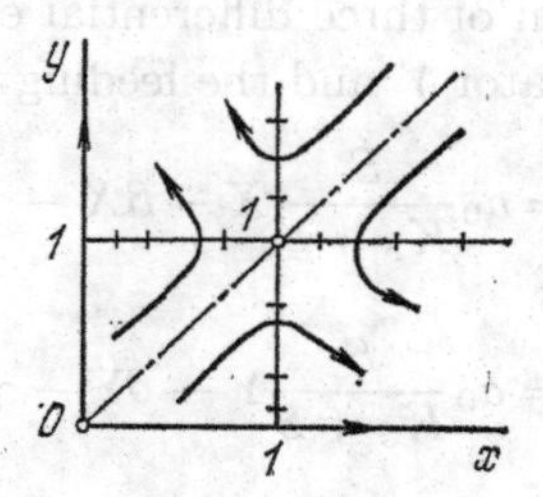

Figure 13. The phase portrait of the two species selection system. The singular solution $[x = 1, y = 1]$ is of the saddle-point type.

The phase portrait shows a singular point $(1, 1)$ of saddle type, and the separatrix $y = x$. Whatever the initial conditions, the trajectories end up either with $x \to \infty$, $y \to 0$, or to the alternative choice, $x \to 0$, $y \to \infty$. The resulting situation is the total domination of one type and disappearing of the second type (Fig. 13).

4.3. *Model with finite substrate*

The simple model exposed above leads to an infinite proliferation of one species, which is not realistic. There is a natural limit to any proliferation, even if it starts with an exponential growth. Later on the population attains a limit when the resources are exhausted ([7]).

Let us improve the model, taking into account the substrate on which two competing species feed. The substrate (neutral organic matter, without any chirality) is used by the left- or right-handed organisms in order to synthetize the appropriate proteins, lipids and sugars. We shall denote the amount of the substrate, which is also a function of time, by $S(t)$.

We shall replace the constant reproduction coefficient a by the expression depending on the amount of substrate S (food, or sun rays for plants) in the following manner:

$$a = a_0 \frac{S}{K_S + S}$$

The substrate itself depends on the number of prey feeding on it; its time evolution can be reasonably described by the following equation; which should be added to the former two:

$$\frac{dS}{dt} = \nu - \sigma\, a_0 \frac{S}{K_S + S}(X + Y).$$

Here is the new system of three differential equations, ruling the time evolution of prey X, predator Y and the feeding substrate S:

$$\frac{dX}{dt} = a_0 \frac{S}{K_S + S} X - \beta X - \gamma XY,$$

$$\frac{dY}{dt} = a_0 \frac{S}{K_S + S} Y - \beta Y - \gamma XY,$$

$$\frac{dS}{dt} = \nu - \sigma\, a_0 \frac{S}{K_S + S}(X + Y).$$

Let us introduce dimensionless variables according to the following substitutions:

$$\tau = \beta t, \quad x = \frac{\gamma X}{\beta}, \quad y = \frac{\gamma Y}{\beta}, \quad z = \frac{\gamma Z}{\beta}, \quad C = \frac{\gamma \nu}{\beta^2}.$$

$$f(z) = a_0 \frac{z}{K_z + z}, \quad \text{with } K_z = \frac{\gamma K_S}{\beta},$$

we arrive at the following system of three equations:

$$\frac{dx}{d\tau} = f(z)\, x - x - xy, \quad \frac{dy}{d\tau} = f(z)\, y - y - xy,$$

$$\frac{dz}{d\tau} = C - \sigma\, f(z)(x + y).$$

The new version takes into account the difference between the growth of the number of items of each species due to the feeding on substrate (the positive terms $\sigma\, f(z)\, x$ and $\sigma\, f(z)\, y$), and the decrease due to natural decay (negative terms $-x$, $-y$), as well as the destructive interaction between the two mutually excluding species, symbolized by negative term $-xy$ present in first two equations.

The amount of substrate proportional to the variable z is supposed to grow naturally with a constant pace C, but is depleted proportionally to the total number of representants of both left- and right-handed species feeding on the substrate with equal voracity (the negative term $-\sigma\, f(z)(x + y)$).

The system of three coupled non-linear equations can be solved only numerically. However, it may be simplified considerably if we suppose that during a long period the substrate is supplied and depleted at a roughly equal pace, remaining on the same level.

This amounts to neglect the time derivative dz/dt:

$$\frac{dz}{d\tau} = C - \sigma\, f(z)(x+y) = 0,$$

enabling us to eliminate the variable z and reduce our system to two equations only:

$$C - \sigma\, f(z)(x+y) = 0 \rightarrow f(z) = \frac{C}{\sigma(x+y)}.$$

The new system takes now the following form:

$$\frac{dx}{d\tau} = x\left[\frac{C}{\sigma(x+y)} - (1+y)\right],$$

$$\frac{dy}{d\tau} = y\left[\frac{C}{\sigma(x+y)} - (1+x)\right]$$

There are four singular points, one unstable (repulsive), at the origin $(x,y) = (0,0)$, two attractive (stable) ones, at $x = \frac{C}{\sigma}, y = 0$ and at $x = 0, y = \frac{C}{\sigma}$, and the saddle point on the separatrix $y = x$.

The resulting phase portrait of the system is displayed in the following Fig. 14, where all the variables remain finite during indefinite time.

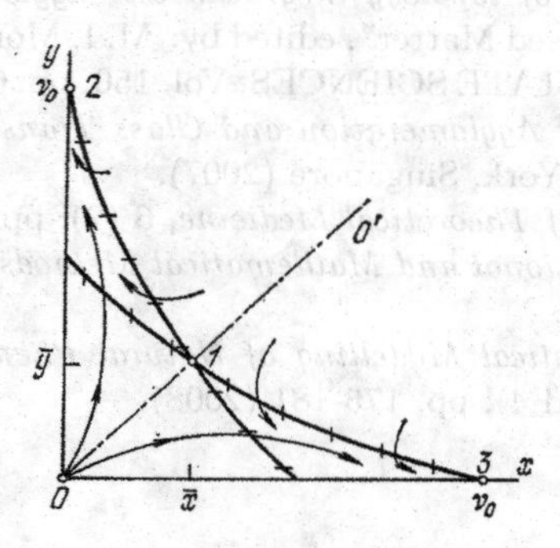

Figure 14. The phase portrait of the two species selection system with limited resources. The stable points are now at finite values of either x or y. There is no possibility of stable coexistence of two opposite chiralities of living matter.

The two methods of analysis of dynamics and statistical evolution of complex systems with interacting parts appear to be complimentary. Both display advantages and shortcomings, so that one or another can be chosen according to the particular nature of the system under consideration. Sophisticated versions of Lotka-Volterra equations are often used to produce models of *homeostasis* in living organisms ([6]), describing stable or unstable equilibrium states in an organism considered as collection of various types of celles, some of them immune, other ones infected, and the influx of external damaging agents (viruses or cancer cells).

More detailed expositions can be found in the following short reference list, and in some papers published in this volume.

References

1. R. Aldrovandi, *Special Matrices in Mathematical Physics*, World Scientific, Singapore (2001).
2. M. C. M. Coxeter, *"Regular polytopes"*, Methuen and C°, London (1948)
3. R. Kerner, *Phil. Mag.*, **47** (2), p. 151 (1983).
4. R. Kerner, *Phys. Rev. B,* **28** (10), pp. 5756-5761 (1983).
5. R. Kerner and R. Aldrovandi, Proceedings of the Conference "BIOMAT-2009", ed. R. P. Mondaini, pp. 87-109, *World Scientific*, 2010.
6. M. Eigen, *Selforganization of matter and the evolution of biological molecules*, Springer-Verlag, *D*ie Natutwissenschaften **58** heft 10 (1971).
7. D. Hartl, A. G. Clark, *Principles of population genetics*, Simauer Ass. Inc. Publishers, Sunderland, Mass (1985).
8. R. Kerner, *C. R. Acad. Sci. Paris*, 304, pp.109-114 (1987).
9. R. Kerner, D. M. dos Santos, *Phys. Rev. B*, **37**, pp. 3881-4000 (1988).
10. J. Kepler, *De nive sexangula*, Teubneri ed., Leipzig (1611).
11. R. Kerner, *The role of topology in growth and agglomeration*, in the book "Topology in Condensed Matter", edited by: M. I. Monastyrsky, SPRINGER SERIES IN SOLID-STATE SCIENCES, Vol. 150, pp. 61-91, published: 2006.
12. R. Kerner, *Models of Agglomeration and Glass Transition*, Imperial College Press, London, New York, Singapore (2007).
13. R. Kerner, *Journal of Theoretical Medicine*, **6** (2), pp. 95-97 (2005).
14. R. Kerner, *Computational and Mathematical Methods in Medicine*, **9** (3 and 4), p. 175-181 (2008).
15. R. Kerner, *Mathematical Modelling of Natural Phenomena*, **6**, No. 6, pp. 136-158 (2011) (3 and 4), pp. 175-181 (2008).

THE PATTERN RECOGNITION OF PROBABILITY DISTRIBUTIONS OF AMINO ACIDS IN PROTEIN FAMILIES

R. P. MONDAINI, S. C. DE ALBUQUERQUE NETO

Federal University of Rio de Janeiro
Centre of Technology, COPPE
Rio de Janeiro - RJ, Brazil
E-mail: Rubem.Mondaini@ufrj.br

A Pattern Recognition of a Probability Distribution of amino acids is obtained for selected families of proteins. The mathematical model is derived from a theory of protein families formation which is derived from application of a Pauli's master equation method.

1. Introduction

The formation and evolution of a protein family is a problem of the same importance as that of protein folding and unfolding. We also think that the last problem will be solved or at least treated on a more general perspective, by concentrating the theoretical research on the joint formation and evolution of the whole set of proteins of each protein family. A probabilistic analysis of a model for protein family formation is then most welcome which is able to unveil the specific nature of this protein family formation process (PFFP) and the consequent folding/unfolding process. A pictorial representation of the PFFP to be seen as a game for the upsurge of life and its homeostasis can be introduced by thinking on **n** consecutive trials of **m** icosahedra, each face of them corresponding to a different amino acid. This sort of ideas has been already introduced on the scientific literature of the Entropy Maximization Principle[1,2]. Our desiderata is then to translate the information contained on biological almanacs (protein databases) in terms of random variables in order to model a dynamics for describing the folding and unfolding of proteins, This means that we think on PFFP instead of the evolution of a single protein as the key to understand the protein dynamics.

In section 2, we introduce a description of the sample space of probability to be used in the calculations of the present contribution[3,4]. We have also made a digression on generalized proposals of joint probabilities which should be used on future generalizations of the present work. In section 3, a Poisson statistical receipt is derived from applications of a Master Equation method in order to derive one adequate probability distribution of our PFFP problem[5,6] in section 4. In section 5, we describe the distribution of amino acids in protein families and the pattern recognition through level curves of the probability distribution. We also determine the domain of the variables for the present probabilistic model. In section 6, we introduce a scheme in terms of graphical cartesian representations of level curves with selected protein families from the Pfam database[7,8,9,10]. We also investigate the possibility of using this representation to justify the classification of protein families into clans. A final section of concluding remarks will comment on the introduction of alternative candidates to usual probability distributions, like the use of joint probabilities and the study of Non-additive entropy measures. All these proposals which aim to improve the pattern recognition method will appear in forthcoming publications.

2. The Organization of the Sample Space of Probability

The fundamental idea of the Protein Family Formation Process as introduced in section 1, is of a dynamical game played by nature. The amino acids which are necessary to participate in the formation of a protein family are obtained from a "universal ribossome deposit" and the process will consist in the distribution on shelves of a bookshelf. In the first stage, $\mathbf{m}$ amino acids are distributed on $\mathbf{m}$ shelves of the first bookshelf. On a second stage, the $\mathbf{m}$ shelves of the second bookshelf should be fulfilled by the amino acids transferred from the first bookshelf and so on. After choosing a database, in order to implement all the future calculations with random variables, we select an array of $\mathbf{m}$ protein domains (rows). There are n_1 amino acids on the first column, n_2 in the second one, $\ldots n_m$ in the mth column. We then select the $(m \times n)$ block where

$$n = \min(n_1, n_2, \ldots, n_m) \tag{1}$$

Another way of introducing a $(m \mathrm{x} n)$ block is to specify the number **n** of columns a priori and to delete all proteins such that $n_r < n$, $r = 1, 2, \ldots, m$ and to also delete the last $(n_r - n)$ amino acids on the remaining proteins. In Fig. 1 below we present an example of a $(m \mathrm{x} n)$ block, which is organized according to this second basic procedure. There is at least one block associated to a protein family.

```
VLLHGPPGCGKTVLANAIANKAQVPFMSISAPSVVSGMSGESEKKIREIFEEARAIAPCL...PDAIDPALRRAGRFDEEIAMAV
ILMIGPTGVGKTEISRRLAKLAGAPFIKIEATKFTEVGYVGRDVESIIRDLVEIGIGLVR...
VLLVGPPGTGKTLLARAVAGEAGVPFFSISGSDFVEMFVGVGASRVRDLFENAKKNAPCI...DVLDPALLRPGRFDRQIMVDR
PVLIGEPGVGKSACVEGLAQAIVRGDVPETLRDKKIYSLDLGSMVAGSRYRGDFEERMKK...LDEYRKYIEKDAALERRFQPIQV
LLLSGPPGAGKTTLAHVAAKHCGYETIEINASDDRSASTLKLKLADALQTRSAFEKQKPK...PLRDVAKIIRMK
PVLIGEPGVGKTAIAEGLAQRIIARDVPESLRD
VLLYGPPGTGKTLLAKAVATECSLNFLSVKGPELINMYIGESEKNVRDIFQKARSARPCV...DLIDPALLRPGRFDKLLYVGV
LCFVGPPGVGKTSLASSIAKALNRKFIRISLGGVKDEADIRGHRRTYIGSMPGRLIDGLK...KVVFVATANRMQPIPPALLDRMEVIELPG
FVFSGPPGTGKTSVARTLATIFHSFGLLPTARVVEASRADLVGEYLGATAIKTNELVDRA...MDRFLASNPGLASRFATRISFPS
LYISGAPGTGKTACLNCVLQEQKALLKGIQTVVINCMNLRSSHAIFPLLGEQLEVPKGNS...NALDLTDRILPRLQAKPHC
ILLFGPPGTGKTLLAKAVATECSMTFLSVKGPELINMYVGQSEENIREVFSRARLAAPCI...LLDQSLLRPGRLDKLVFVGL
MYVSGVPGTGKTATVHEVMRCLQQAADVDQIPSFSFVEINGMKMTDPHQAYVQILQELTG...RHARLVVLTIANTMDLPERVMINRVASRLGLTR
LLINGPKGNGQQYVGAAILNYLEEFNVQNLDLASLVSESSRTIEAAVVQSFMEAKKRQPS...LSDFAFDKNIF
PVLIGEAGVGKTAVVEGLANKIVNAEVPEKLMDKEVIRLDVASLVSGTGIRGQFEERMQQ...TLSEYRKIEKDPALERRLQPVKVN
IIFYGPAGTGKTMSALAMAKSMKKTVLSFDCSKILSKWVGESEQNVRKIFDTYKNIVQTC...LESLDSAFSRRFDYKIEFKK
ILMYGPPGTGKTVMARAVANETGAFFFLINGPEIMSKMAGESESNLRKAFEEAEKNAPSI...DPALRRFGRFDREVDIGV
PVLIGEAGVGKTAIVEGLAQAIVRGDVPDNLRNKRLITLDLALMIAGTKYRGQFEERIKA...IDEYRKHIEKDAALERRFQKVMVAPA
                    ⋮                                           ⋮
```

Figure 1. A block of $(m = 100)\,\mathrm{x}\,(n = 200)$ amino acids from Pfam database.

Let $p_j(a)$ be the probability of occurrence of the amino acid and $a =$ A, C, D, E, F, G, H, I, K, L, M, N, P, Q, R, S, T, V, W, Y in the jth column,

$$p_j(a) = \frac{n_j(a)}{m} \tag{2}$$

where $n_j(a)$ is the number of occurrences of the a-amino acid in the jth column of the $(m \mathrm{x} n)$ block.

We have

$$\sum_a n_j(a) = m\,, \quad \forall j \tag{3}$$

and

$$\sum_a p_j(a) = 1\,, \quad \forall j \tag{4}$$

The probabilities $p_j(a)$ will be considered as the components of a 20-component vector, $\overrightarrow{p_j} = 1, 2, \ldots, n$. These n vectors will be random variables w.r.t. the probability distribution already defined in Eq. (2).

We can also introduce the probability distribution corresponding to the occurrence of amino acids a, b in columns j, k, respectively. We can write,

$$P_{jk}(a,b) = \frac{n_{jk}(a,b)}{m} \tag{5}$$

with $a, b =$ A, C, D, E, F, G, H, I, K, L, M, N, P, Q, R, S, T, V, W, Y.

This is the joint probability distribution and it can be understood in terms of the conditional probability $P_{jk}(a|b)$ as

$$P_{jk}(a,b) = P_{jk}(a|b)p_k(b) \tag{6}$$

From Bayes' law[11] we can write

$$P_{jk}(a,b) = P_{jk}(a|b)p_k(b) = P_{kj}(b|a)pj(a) = P_{kj}(b,a) \tag{7}$$

Analogously to Eqs. (3), (4), we can write

$$\sum_a \sum_b n_{jk}(a,b) = m, \quad \forall j,k, \quad j < k \tag{8}$$

and

$$\sum_a \sum_b P_{jk}(a,b) = \sum_a p_j(a) = 1, \quad \forall j,k, \quad j < k \tag{9}$$

We can take on Eqs. (8), (9):

$$j = 1, 2, \ldots, (n-1), k = (j+1), (j+2), \ldots, n.$$

These random variables will be arranged as $\binom{n}{2} = \frac{n(n-1)}{2}$ square matrices of 20th order.

A straightforward generalization to a multiplet of s amino acids, which occur on s ordered columns, can be done by introducing joint probabilities such as,

$$P_{j_1 j_2 \ldots j_s}(a_1, a_2, \ldots, a_s) = \frac{n_{j_1 j_2 \ldots j_s}(a_1, a_2, \ldots, a_s)}{m} \tag{10}$$

where

$$a_1, a_2, \ldots, a_s = A, C, D, E, F, G, H, I, K, L, M, N, P, Q, R, S, T, V, W, Y \tag{11}$$

and

$$j_1 < j_2 < \ldots < j_s$$

with

$$j_1 = 1, 2, \ldots, (n - s + 1)$$
$$j_2 = (j_1 + 1), (j_1 + 2), \ldots, (n - s + 2)$$
$$\vdots \qquad \vdots \qquad \vdots$$
$$j_s = (j_{s-1} + 1), (j_{s-1} + 2), \ldots, n$$

These random variables $P_{j_1 j_2 \ldots j_s}(a_1, a_2, \ldots, a_s)$ are $\binom{n}{s} = \frac{n!}{s!(n-s)!}$ objects of $(20)^s$ components each. Analogously to Eqs. (8), (9), we can now write,

$$\sum_{a_1} \sum_{a_2} \ldots \sum_{a_s} n_{j_1 j_2 \ldots j_s}(a_1, a_2, \ldots, a_s) = m\,, \quad \forall j_1, j_2, \ldots, j_s \qquad (12)$$
$$j_1 < j_2 < \ldots < j_s$$

and

$$\sum_{a_1} \sum_{a_2} \ldots \sum_{a_s} P_{j_1 j_2 \ldots j_s}(a_1, a_2, \ldots, a_s) = \ldots = \sum_{a_1} \sum_{a_2} P_{j_1 j_2}(a_1, a_2)$$
$$= \sum_{a_1} p_{j_1}(a_1) = 1\,, \quad \forall j_1, j_2, \ldots, j_s\,, \; j_1 < j_2 < \ldots < j_s \qquad (13)$$

Eqs. (10)–(13) will be reserved for future developments. In the present work, we restrict all calculations to simple probabilities given by Eqs. (2)–(4).

3. The Master Equation for Probability Evolution

The temporal evolution of random variables such as the probabilities of occurrence introduced above can be modelled through a master equation approach[5,6].

Let $p\left(n_j(t(a))\right)$ be the probability of occurrences of the amino acids a in the jth column of the $(m \times n)$ block at time $t(a)$. The probability of observing the same amino acid at the jth column after an interval of time Δt is given by:

$$p\left(n_j(t(a) + \Delta t)\right) = \sigma\left(t(a)\right) \Delta t \; p\left(n_{j-1}(t(a))\right) + \left(1 - \sigma\left(t(a)\right) \Delta t\right) \; p\left(n_j(t(a))\right) \quad (14)$$

where $\sigma\left(t(a)\right)$ is the transition probability per unit time between columns $j - 1$ and j.

We now imagine that there is a column $j = 0$ – "the universal ribosome deposit" where all amino acids are present at time $t_0(a)$, as

$$p\left(n_0\left(t_0(a)\right)\right) = 1\,, \quad \forall a \tag{15}$$

This also means that at the initial time $t_0(a)$ no amino acid has been received by the columns $j \neq 0$ – the shelves of amino acid bookshelf (the protein family), as

$$p\left(n_j\left(t_0(a)\right)\right) = 0\,, \quad \forall j \neq 0\,, \quad \forall a \tag{16}$$

After taking the limit $\Delta t \to 0$ on Eq. (14), we get:

$$\frac{\partial p\left(n_j\left(t(a)\right)\right)}{\partial t(a)} = \sigma\left(t(a)\right)\left[p\left(n_{j-1}\left(t(a)\right)\right) - p\left(n_j\left(t(a)\right)\right)\right]\,, \quad j \neq 0 \tag{17}$$

We also have, for $j = 0$,

$$\frac{\partial p\left(n_0\left(t(a)\right)\right)}{\partial t(a)} = -\sigma\left(t(a)\right)\, p\left(n_0\left(t(a)\right)\right) \tag{18}$$

From Eqs. (15) and (18), we have,

$$p\left(n_0\left(t(a)\right)\right) = e^{-v\left(t(a)\right)} \tag{19}$$

where

$$v\left(t(a)\right) = \int_{t_0(a)}^{t(a)} \sigma\left(t'(a)\right)\mathrm{d}t'(a) \tag{20}$$

From Eqs. (16) and (17) we can write for $j = 1$

$$p\left(n_1\left(t(a)\right)\right) = e^{-v(t(a))}\, v\left(t(a)\right) \tag{21}$$

We also have for $j = 2$,

$$p\left(n_2\left(t(a)\right)\right) = e^{-v(t(a))}\left(\int_{t_0(a)}^{t(a)} \frac{\mathrm{d}v\left(t'(a)\right)}{\mathrm{d}t'(a)} v\left(t'(a)\right)\mathrm{d}t'(a)\right) \tag{22}$$

From Eqs. (15), (19), we have

$$v\left(t_0(a)\right) = 0 \tag{23}$$

Eq. (22) will turn into:

$$p\left(n_2\left(t(a)\right)\right) = e^{-v(t(a))}\frac{v^2\left(t(a)\right)}{2} \tag{24}$$

By finite induction on j, we can write the Poisson distribution:

$$p\left(n_j\left(t(a)\right)\right) = e^{-v(t(a))}\frac{v^j\left(t(a)\right)}{j!}\,, \quad \forall j\,, \quad \forall a \tag{25}$$

4. The Distribution of Amino Acids as a Marginal Probability Distribution

We now introduce the marginal probability distributions[11] associated to the Poisson process given by Eq. (25). We have:

$$p_j(t(a)) = \int_{t_0(a)}^{t(a)} p\left(n_j(t'(a))\right) \mathrm{d}t'(a) \tag{26}$$

From Eqs. (25), (26), we can write:

$$p_j(t(a)) = \frac{(-1)^j}{j!} \lim_{\alpha\to 1} \frac{\partial^j}{\partial\alpha^j} \int_{t_0(a)}^{t(a)} e^{-\alpha v(t'(a))} \mathrm{d}t'(a) \tag{27}$$

where α is an auxiliary parameter.

In the present work, we make the assumption that $\sigma(t(a)) \equiv \sigma(a)$, which leads to a linear approximation for $v(t(a))$ through Eq. (20), or,

$$v(t(a)) = \sigma(a)\left(t(a) - t_0(a)\right) \tag{28}$$

We then have from Eqs. (27), (28),

$$p_j(t(a)) = \frac{(-1)^{j-1}}{j!\sigma(a)} \lim_{\alpha\to 1} \frac{\partial^j}{\partial\alpha^j} \left(\frac{e^{-\alpha\sigma(a)(t(a)-t_0(a))} - 1}{\alpha} \right) \tag{29}$$

Let us write now $t_j(a)$ as the time in which the a-amino acid is seen to occur at the jth column of the $(m \times n)$ block.

We write,

$$t_j(a) = t_0(a) + j\Delta(a) \tag{30}$$

where $\Delta(a)$ is the time interval for the transition of the amino acid between consecutive columns,

$$\Delta(a) = t_j(a) - t_{j-1}(a)\,,\; j = 1, 2, \ldots, n \tag{31}$$

The marginal probability distribution function of Eq. (26) should be considered as a two-variable distribution:

$$p_j(t(a)) \equiv p_j(\sigma_j(a), \Delta(a)) = \frac{(-1)^{j-1}}{j!\,\sigma(a)} \lim_{\alpha\to 1} Q_j(\alpha\,; \sigma_j(a)\Delta(a)) \tag{32}$$

where

$$Q_j(\alpha\,;\sigma_j(a)\Delta(a)) = (-1)^j\, j! \left(e^{-\alpha j\,\sigma_j(a)\Delta(a)} \left(\alpha^{-(j+1)} \sum_{m=1}^{j} \frac{\left(j\,\sigma_j(a)\Delta(a)\right)^m \alpha^{-(j-m+1)}}{m!} \right) - 1 \right) \tag{33}$$

we can then write,

$$p_j\big(\sigma_j(a),\Delta(a)\big)=\frac{1}{\sigma_j(a)}\left(1-e^{-j\,\sigma_j(a)\Delta(a)}\sum_{m=1}^{j}\frac{\big(j\,\sigma_j(a)\Delta(a)\big)^m}{m!}\right) \tag{34}$$

Eq. (34) can be written in a more feasible form for future calculations such as

$$p_j\big(\sigma_j(a),\Delta(a)\big)=\frac{1}{\sigma_j(a)}\left(1-\frac{\Gamma\big(j+1,j\,\sigma_j(a)\Delta(a)\big)}{\Gamma(j+1)}\right) \tag{35}$$

where

$$\Gamma(j+1)=\int_0^\infty e^{-z}z^j\mathrm{d}z \tag{36}$$

is the Gamma function[12]. $\Gamma(j+1)=j!$ here, since j is an integer. And

$$\Gamma\big(j+1,j\,\sigma_j(a)\Delta(a)\big)=\int_0^{j\,\sigma_j(a)\Delta(a)} e^{-z}z^j\mathrm{d}z \tag{37}$$

is related to the Incomplete Gamma function[12].

One should note that the real representations of probability distribution functions should be given by the restriction $0\le p_j\big(\sigma(a),\Delta(a)\big)\le 1$ on the surfaces $p_j\big(\sigma(a),\Delta(a)\big)$. In Fig. 2 we present these surfaces for several j-values and the planes $p_j\big(\sigma(a),\Delta(a)\big)=1$ and $p_j\big(\sigma(a),\Delta(a)\big)=0.06$.

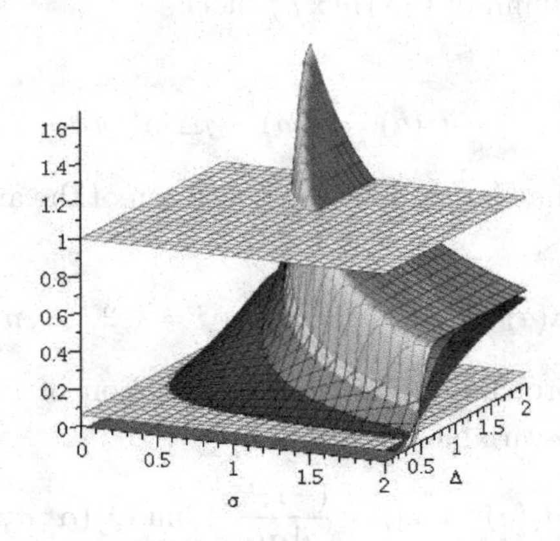

Figure 2. The surfaces $p_j\big(\sigma(a),\Delta(a)\big)$ for $j = 1,6,26,27,35,200$ and the planes $p_j\big(\sigma(a),\Delta(a)\big)=1$, $p_j\big(\sigma(a),\Delta(a)\big)=0.06$.

5. The Domain of the Variables $\sigma(a)$, $\Delta(a)$. The Pattern Recognition Based on Level Curves

In the present section we set up the experimental scenario of the paper as well as the fundamental equations for the statistical treatment of data. We will gave on Table 1 below an example of probability distribution of amino acid occurrence for the $a =$ A amino acid of the Pfam family PF01051 on a (100 x 200) block. The set J of j-values ($j = 1, 2, \ldots, 200$) is partitioned into subsets J_s of values j_s. The j_s-values belonging to a subset J_s do correspond to a unique constant value $M_{J_s}\big(\sigma_{j_s}(A), \Delta(A)\big)$ of the function $p_{j_s}\big(\sigma_{j_s}(A), \Delta(A)\big)$, as

$$p_{j_s}\big(\sigma_{j_s}(A), \Delta(A)\big) = M_{J_s}\big(\sigma_{j_s}(A), \Delta(A)\big) = \frac{n_{J_s}(A)}{100} \tag{38}$$

where $n_{J_s}(A)$ is the number of occurrences of the $a =$ A amino acid on each subset J_s of the j_s-values. These values do correspond to the level curves of the surfaces $p_j\big(\sigma(a), \Delta(a)\big)$.

Table 1. The j_s values on each row of the second column belong to the subset J_s of values for defining a level curve of the surfaces $p_{j_s}\big(\sigma(a), \Delta(a)\big)$. Data obtained from a (100 x 200) block as a representative of the Pfam family PF01051.

$M_{J_s}\big(\sigma_{J_s}(A), \Delta(A)\big)$	J_s
0	17,18,57,179
1/100	2,3,12,38,79,88,97,111,120,157,166,178,180
2/100	5,36,82,92,125,148,172,173,175,176
3/100	7,41,45,46,48,58,81,100,105,119,124,133,136,147,150,151,155,161,162,165,171, 187,197,199
4/100	4,8,23,24,25,60,61,73,74,75,78,80,83,84,86,89,90,91,99,107,112,113,123,126, 127,129,131,135,142,145,149,154,156,168,170,184,188,192,193,198
5/100	9,11,19,34,40,44,56,62,94,96,101,102,104,106,121,134,140,158,159,181,186,189
6/100	1,6,26,27,35,37,39,51,54,95,109,115,118,122,138,146,164,174,177,194,200
7/100	13,20,52,55,63,65,68,70,76,87,103,130,152,167,169,183,185,191
8/100	22,28,47,85,93,110,114,117,139,144,153,160,163,182,190,196
9/100	15,32,42,43,49,59,66,72,98,108,137
10/100	116,143
11/100	64,67,141
12/100	21,30,195
13/100	50,128,132
14/100	33
15/100	31,71
16/100	77
17/100	53
18/100	10,69
21/100	16
26/100	14
30/100	29

In Fig. 3, we represent the histogram corresponding to the probabilities of occurrence of the $a =$ A amino acid which have been obtained from the family PF01051. This should be compared to Fig. 4 in which we advance the representation of the level curves of this amino acid distribution. It seems to exist a clear advantage of the last representation over the histogram and we will intend to represent in the same form the distribution of other amino acids from selected families.

We now undertake a detailed analysis which led to the pictorial representation of Fig. 4 in order to determine the most convenient domain for the variables $\sigma(a)$, $\Delta(a)$.

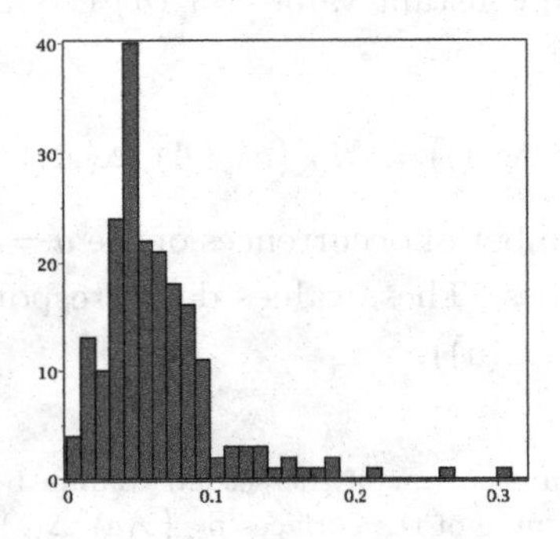

Figure 3. The histogram of the distribution corresponding to Table 1. Amino acid A, Pfam family PF01051.

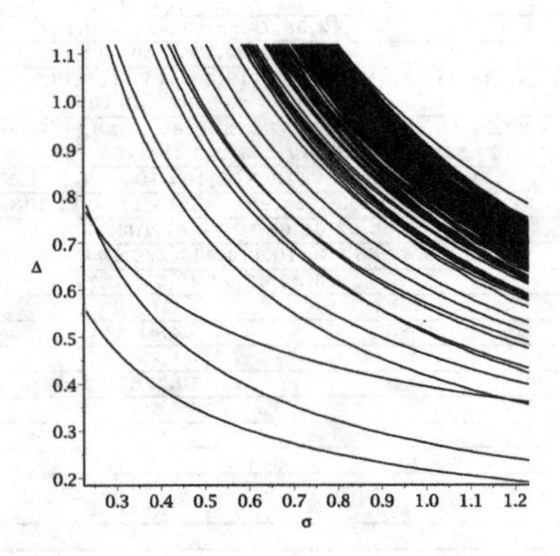

Figure 4. The level curves corresponding to the distribution given in Table 1 and Fig. 3. Amino acid A, Pfam family PF01051.

First of all, we note that the usual analysis of extrema of the surfaces given by Eq. (35) is inconclusive, since from the equations:

$$\frac{\partial p_j\big(\sigma_j(a),\Delta(a)\big)}{\partial\sigma_j(a)}=0\;;\quad\frac{\partial p_j\big(\sigma_j(a),\Delta(a)\big)}{\partial\Delta(a)}=0 \tag{39}$$

we get as the only one solution the point $\big(\sigma(a),\Delta(a)\big)=(0,0)$ and we have:

$$\mathrm{Hess}(0,0)=\det\begin{pmatrix}\left.\frac{\partial^2 p_j\left(\sigma_j(a),\Delta(a)\right)}{\partial^2\sigma_j(a)}\right|_{(0,0)} & \left.\frac{\partial^2 p_j\left(\sigma_j(a),\Delta(a)\right)}{\partial\sigma_j(a)\partial\Delta(a)}\right|_{(0,0)}\\ \left.\frac{\partial^2 p_j\left(\sigma_j(a),\Delta(a)\right)}{\partial\sigma_j(a)\partial\Delta(a)}\right|_{(0,0)} & \left.\frac{\partial^2 p_j\left(\sigma_j(a),\Delta(a)\right)}{\partial^2\Delta(a)}\right|_{(0,0)}\end{pmatrix}=0 \tag{40}$$

We then proceed to an alternative analysis in order to characterize the 2-dimensional domain of variables $\sigma(a)$, $\Delta(a)$ for the modelling process. This is based on the study of the intersection of the $p_j\big(\sigma(a),\Delta(a)\big)$ surfaces by horizontal and vertical planes.

The horizontal planes $h_j(a)$ will determine the level curves which are given by:

$$p_j\big(\sigma_j(a),\Delta(a)\big)=M_j(a)=\frac{1}{\sigma_j(a)}\left(1-\frac{\Gamma\big(j+1,j\,\sigma_j(a)\Delta(a)\big)}{\Gamma(j+1)}\right)\;\forall a\,,\;1\le j\le n \tag{41}$$

where $M_j(a)$ corresponds to a characteristic constant value associated to the subset of j-values of the specific amino acid according to the example given in Table 1.

From Eq. (41), we can write:

$$\mathrm{d}\sigma_j(a)\frac{\partial p_j\big(\sigma_j(a),\Delta(a)\big)}{\partial\sigma_j(a)}+\mathrm{d}\Delta(a)\frac{\partial p_j\big(\sigma_j(a),\Delta(a)\big)}{\partial\Delta(a)}=0 \tag{42}$$

The local minimum points $\big(\sigma_{j\ \min}(a),\Delta_{\min}(a)\big)$ of the level curves are obtained from Eq. (41) and from

$$\frac{\mathrm{d}\Delta(a)}{\mathrm{d}\sigma_j(a)}=0=\sigma_j(a)\Big(\Gamma(j+1)-\big(j\,\Delta(a)\big)^{j+1}\big(\sigma_j(a)\big)^j e^{-j\,\sigma_j(a)\Delta(a)}\Big) \tag{43}$$

We now consider the vertical planes $V_j(a)$, associated to a generic value $\delta_j(a)$. On each vertical plane, there is a vertical curve such that its local maximum is given by $\big(\sigma_{j\ \max}(a),\delta_{j\ \max}(a)\big)$.

These vertical curves are given by

$$p_j\big(\sigma_j(a),\delta_j(a)\big)=\frac{1}{\sigma_j(a)}\left(1-\frac{\Gamma\big(j+1,j\,\sigma_j(a)\delta_j(a)\big)}{\Gamma(j+1)}\right) \tag{44}$$

and their local maxima can be obtained from Eq. (44), and from

$$\frac{\partial p_j\big(\sigma_j(a),\Delta(a)\big)}{\partial\sigma_j(a)} = 0 = -\frac{1}{\sigma_j^2(a)}\left(1-\frac{\Gamma\big(j+1, j\,\sigma_j(a)\Delta_j(a)\big)+\big(j\,\sigma_j(a)\delta_j(a)\big)^{j+1}e^{-j\,\sigma_j(a)\delta_j(a)}}{\Gamma(j+1)}\right) \tag{45}$$

From Eqs. (41), (43)–(45), we can write:

$$M_j(a) = \frac{1}{\sigma_{j\ \min}(a)}\left(1-\frac{\Gamma(j+1, j\,\sigma_{j\ \min}(a)\Delta_{\min}(a))}{\Gamma(j+1)}\right) \tag{46}$$

$$\Gamma(j+1) - \big(j\,\Delta_{\min}(a)\big)^{j+1}\big(\sigma_{j\ \min}(a)\big)^{j}e^{-j\,\sigma_{j\ \min}(a)\Delta_{\min}(a)} = 0 \tag{47}$$

$$p_j(a) = \frac{1}{\sigma_{j\ \max}(a)}\left(1-\frac{\Gamma\big(j+1, j\,\sigma_{j\ \max}(a)\delta_{j\ \max}(a)\big)}{\Gamma(j+1)}\right) \tag{48}$$

$$\Gamma(j+1)p_j(a) - \big(j\,\delta_{j\ \max}(a)\big)^{j+1}\big(\sigma_{j\ \max}(a)\big)^{j}e^{-j\,\sigma_{j\ \max}(a)\Delta_{\max}(a)} = 0 \tag{49}$$

Some remarks should be done before we proceed to determine the points $\big(\sigma_{j\ \min}(a)\Delta_{\min}(a)\big)$, $\big(\sigma_{j\ \max}(a)\delta_{j\ \max}(a)\big)$ from Eqs. (46)–(49). From Eq. (45) we see that the product $\sigma(a)\Delta(a)$ does not depend on the amino acid $\mathbf{a}$, or

$$\sigma_j(a)\delta_j(a) = f(j)\,, \quad \forall(a) \tag{50}$$

Let us suppose the ordering of $\Delta_j(a)$ values for different amino acids but at the same jth column of the $(m\,\mathrm{x}\,n)$ block,

$$\delta_j(a) \geq \delta_j(b) \geq \delta_j(c) \tag{51}$$

We then have from Eq. (50)

$$\sigma_j(a) \leq \sigma_j(b) \leq \sigma_j(c) \tag{52}$$

An example of level curves corresponding to these inequalities can be seen at Fig. 5.

Let us also suppose that there is an amino acid $\mathbf{a}'$ such that for two columns j_1, j_2 we have:

$$\delta_{j_1}(a') \geq \delta_j(a) \geq \delta_{j_2}(a')\,, \quad j_1 \leq j \leq j_2 \tag{53}$$

From Eq. (50), we can also write:

$$\sigma_{j_1}(a') \leq \sigma_j(a) \leq \sigma_{j_2}(a')\,, \quad j_1 \leq j \leq j_2 \tag{54}$$

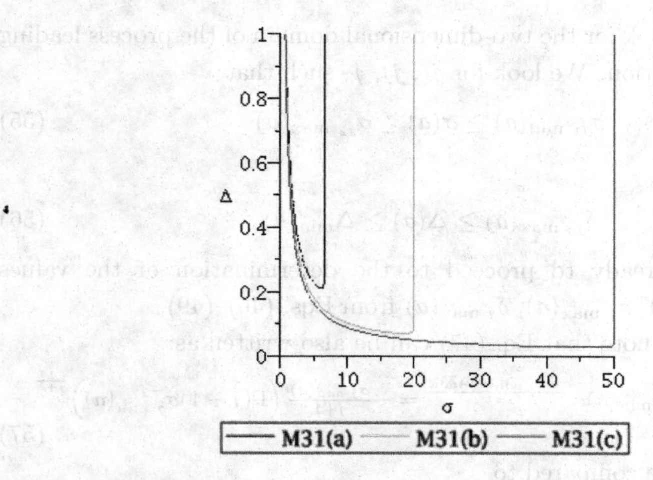

Figure 5. Three level curves corresponding to the $j = 31$ value and the amino acids $a = A$, $b = C$, $c = D$ of the 100 x 200 block of the PF01051 family.

We can take $j_2 = j_{\max} = 200$ and $j_1 = j_{\min} = 1$ in the case of a 100 x 200 block. In Fig. 6 an example of the level curves for $j = 55$ and the amino acids $a' = K$, $a = L$ from PF01051 family.

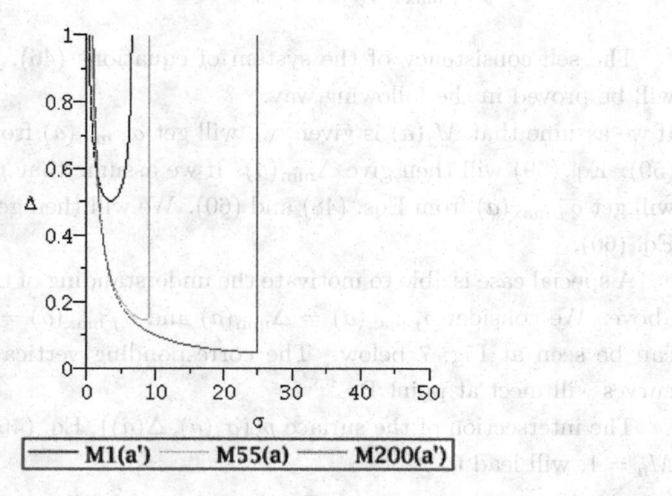

Figure 6. Three level curves corresponding to the $j_1 = 1$, $j_2 = 200$, $j = 55$, $a' = K$, $a = L$.

We can then look for the two-dimensional domain of the process leading to pattern recognition. We look for j_1, j_2, j_3 such that

$$\sigma_{j_1\ \mathrm{min}}(a) \geq \sigma(a) \geq \sigma_{j_2\ \mathrm{max}}(a) \tag{55}$$

and

$$\delta_{j_3\ \mathrm{max}}(a) \geq \Delta(a) \geq \Delta_{\mathrm{min}}(a) \tag{56}$$

We are now ready to proceed to the determination of the values $\sigma_{j\ \mathrm{min}}(a)$, $\Delta_{\mathrm{min}}(a)$, $\sigma_{j\ \mathrm{max}}(a)$, $\delta_{j\ \mathrm{max}}(a)$ from Eqs. (46)–(49).

First of all we note that Eq. (47) can be also written as:

$$-\tfrac{j}{j+1}\sigma_{j\ \mathrm{min}}(a)\Delta_{\mathrm{min}}(a)e^{\frac{-j\,\sigma_{j\ \mathrm{min}}(a)\Delta_{\mathrm{min}}}{j+1}} = -\tfrac{\sigma_{j\ \mathrm{min}}(a)}{j+1}\left(\Gamma(j+1)\sigma_{j\ \mathrm{min}}^{-j}(a)\right)^{\frac{1}{j+1}} \tag{57}$$

and this should be compared to

$$W(z)e^{W(z)} = z \tag{58}$$

where $W(z)$ is the Lambert W Non-injective function[13]. We then have:

$$\Delta_{\mathrm{min}}(a) = -\frac{(j+1)}{j\,\sigma_{j\ \mathrm{min}}(a)}W\left(-\frac{\sigma_{j\ \mathrm{min}}(a)}{j+1}(\Gamma(j+1)\sigma_{j\ \mathrm{min}}^{-j}(a))^{\frac{1}{j+1}}\right) \tag{59}$$

Analogously, we can write from Eq. (49):

$$\delta_{j\ \mathrm{max}}(a) = -\frac{(j+1)}{j\,\sigma_{j\ \mathrm{max}}(a)}W\left(-\frac{\sigma_{j\ \mathrm{max}}(a)}{j+1}(\Gamma(j+1)p_j(a)\sigma_{j\ \mathrm{max}}^{-j}(a))^{\frac{1}{j+1}}\right) \tag{60}$$

The self-consistency of the system of equations (46), (59), (48), (60), will be proved in the following way:
If we assume that $M_j(a)$ is given, we will get $\sigma_{j\ \mathrm{min}}(a)$ from Eqs. (46) and (59). Eq. (59) will then give $\Delta_{\mathrm{min}}(a)$. If we assume that $p_j(a)$ is given we will get $\sigma_{j\ \mathrm{max}}(a)$ from Eqs. (48) and (60). We will then get $\delta_{j\ \mathrm{max}}(a)$ from Eq. (60).

A special case is able to motivate the understanding of the developments above. We consider $\delta_{j\ \mathrm{max}}(a) = \Delta_{\mathrm{min}}(a)$ and $\sigma_{j\ \mathrm{max}}(a) = \sigma_{j\ \mathrm{min}}(a)$. This can be seen at Fig. 7 below. The corresponding vertical and horizontal curves will meet at point P.

The intersection of the surface $p_j(\sigma_j(a), \Delta(a))$, Eq. (46) with the plane $M_n = 1$, will lead to

$$1 - \sigma_{n\ \mathrm{min}}(a) = \frac{\Gamma(n+1, n\,\sigma_{n\ \mathrm{min}}(a)\Delta_{\mathrm{min}}(a))}{\Gamma(j+1)} \tag{61}$$

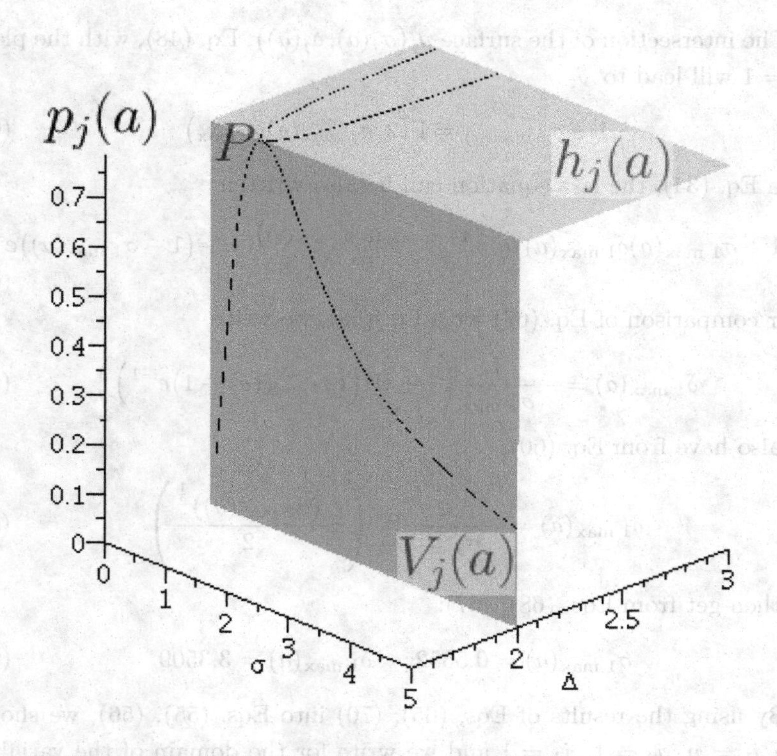

Figure 7. Sketch of the level and vertical curves corresponding to the case $\sigma_{j\ \max}(a) = \sigma_{j\ \min}(a)$ and $\delta_{j\ \max}(a) = \Delta_{\min}(a)$ for a generic amino acid a and a j-value $1 \le j \le n$. $h_j(a)$ and $V_j(a)$ are horizontal and vertical planes, respectively.

We can write from Eqs. (61) and (34), if $n \gg 1$

$$1 \approx \frac{1}{\sigma_{n\ \min}(a)} \tag{62}$$

By applying the Stirling approximation as given by

$$n! \approx e^{-n}(n)^n, \quad n \gg 1 \tag{63}$$

We get from Eqs. (47) and (62):

$$e^{-n}(n)^n - e^{-n\,\Delta_{\min}(a)}\big(n\,\Delta_{\min}(a)\big)^{n+1} \approx 0 \tag{64}$$

For $n = 200$, we can write from Eqs. (62), (64):

$$\sigma_{n\ \min}(a) \approx 1.0000\,; \quad \Delta_{\min}(a) \approx 1.2538 \tag{65}$$

The intersection of the surface $p_j\big(\sigma_j(a), \delta_j(a)\big)$, Eq. (48), with the plane $M_1 = 1$ will lead to

$$1 - \sigma_{j\ \max(a)} = \Gamma\big(2, \sigma_{1\ \max}(a)\delta_{1\ \max}\big) \tag{66}$$

From Eq. (34), the last equation can be also written:

$$-\big(1 + \sigma_{1\ \max}(a)\delta_{1\ \max}(a)\big)e^{-\left(1+\sigma_{1\ \max}(a)\delta_{1\ \max}(a)\right)} = -\big(1 - \sigma_{1\ \max}(a)\big)e^{-1} \tag{67}$$

After comparison of Eq. (67) with Eq. (58), we write,

$$\delta_{1\ \max}(a) = -\frac{1}{\sigma_{1\ \max}}\left(1 + W\Big(\big(\sigma_{1\ \max}(a) - 1\big)e^{-1}\Big)\right) \tag{68}$$

We also have from Eq. (60),

$$\delta_{1\ \max}(a) = -\frac{2}{\sigma_{1\ \max}}W\left(\pm\frac{\big(\sigma_{1\ \max}(a)\big)^{\frac{1}{2}}}{2}\right) \tag{69}$$

We then get from Eqs. (68), (67):

$$\sigma_{1\ \max}(a) = 0.5352\,; \quad \delta_{1\ \max}(a) = 3.3509 \tag{70}$$

By using the results of Eqs. (65), (70) into Eqs. (55), (56), we should have $j_1 = n$, $j_2 = 1$, $j_3 = 1$ and we write for the domain of the variables $\sigma(a)$, $\Delta(a)$

$$1.0000 \geq \sigma(a) \geq 0.5352\,; \quad 3.3509 \geq \Delta(a) \geq 1.2538 \tag{71}$$

6. Final Numerical Calculations and Graphical Representation

In this section we show the proposed graphical representation of the probability occurrences of all amino acids from some selected protein families. These representations do correspond to the level curves of the surface of probability distribution, Eq. (39) and an example has been already presented on Fig. 4 of a cartesian representation for the $a =$ A amino acid of family PF01051. All the necessary essential steps have been derived in the previous sections and we stress once more that the cartesian representations of the level curves is based on the partition of the j-values ($1 \leq j \leq n$) into subsets J_s. Each j-values belonging to a J_s subset is then associated to a constant value $M_{j_s}(\sigma(a), \Delta(a))$ of the probability distribution function, according to Eq. (38).

In order to emphasize the usefulness and efficiency of the cartesian representation of level curves, we shall make a comparison of probability occurrences in terms of histograms, Figs. 8, 9, 10 and level curves, Figs. 11, 12, 13 for the families PF01051, PF03399, which belong to the same Clan CL00123 and PF00060 belonging to the Clan CL00030. We have chosen for representation of four rows and five columns corresponding to Table 2 below. The one-letter code for amino acids are ordered by decreasing hydrophobicities[14], with cells to be read from top left to bottom right, at pH2.

Table 2. The representation scheme of amino acids ordered according to decreasing hydrophobicities, top left to bottom right, at pH2.

$A_{00} = L$	$A_{01} = I$	$A_{02} = F$	$A_{03} = W$	$A_{04} = V$
$A_{10} = M$	$A_{11} = C$	$A_{12} = Y$	$A_{13} = A$	$A_{14} = T$
$A_{20} = E$	$A_{21} = G$	$A_{22} = S$	$A_{23} = Q$	$A_{24} = D$
$A_{30} = R$	$A_{31} = K$	$A_{23} = N$	$A_{33} = H$	$A_{34} = P$

In order to introduce the cartesian representation of the level curves, we restrict the limits of the variables $\sigma(a)$, $\Delta(a)$ of Eq. (72) to each cell of the array given in Table 2. A generic element A_{kl}, $k = 0, 1, 2, 3$, $l = 0, 1, 2, 3, 4$ is associated to the limits:

$$\varphi \frac{\sigma_{1\ \max}(a)}{\sigma_{n\ \min}(a) - \sigma_{1\ \max}(a)} + l \leq \frac{\sigma(a)}{\sigma_{n\ \min}(a) - \sigma_{1\ \max}(a)} \leq \varphi \frac{\sigma_{1\ \max}(a)}{\sigma_{n\ \min}(a) - \sigma_{1\ \max}(a)} + (l + 1) \tag{72}$$

$$\varphi \frac{\delta_{1\ \max}(a)}{\delta_{1\ \max}(a) - \Delta_{\min}(a)} + (3 - k) \leq \frac{\Delta(a)}{\delta_{1\ \max}(a) - \Delta_{\min}(a)} \leq \varphi \frac{\delta_{1\ \max}(a)}{\delta_{1\ \max}(a) - \Delta_{\min}(a)} + (4 - k) \tag{73}$$

where $\sigma_{1\ \max}(a)$, $\sigma_{n\ \min}(a)$, $\delta_{1\ \max}(a)$, $\Delta_{\min}(a)$ are given into Eqs. (65) and (71) and φ is an arbitrary non-dimensional parameter to be chosen as $\varphi = 0.2$ for obtaining a convenient cartesian representation.

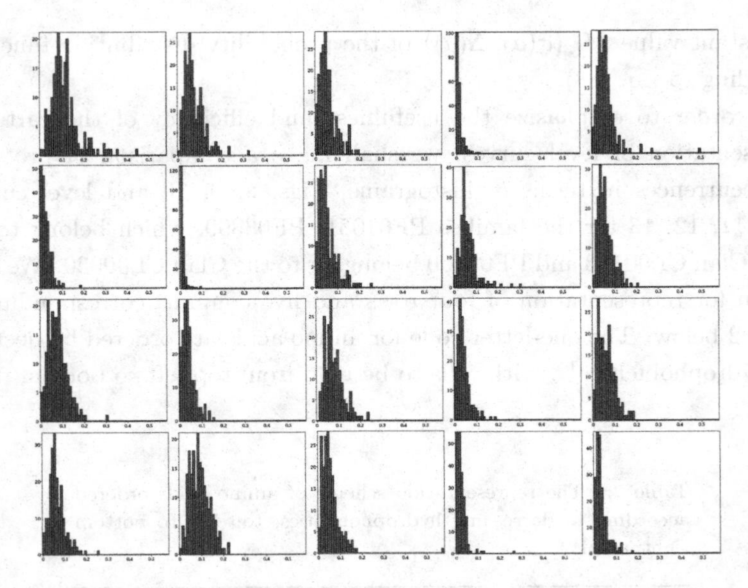

Figure 8. Histograms of the probability occurrences of amino acids for the Pfam protein family PF01051 from Clan CL00123.

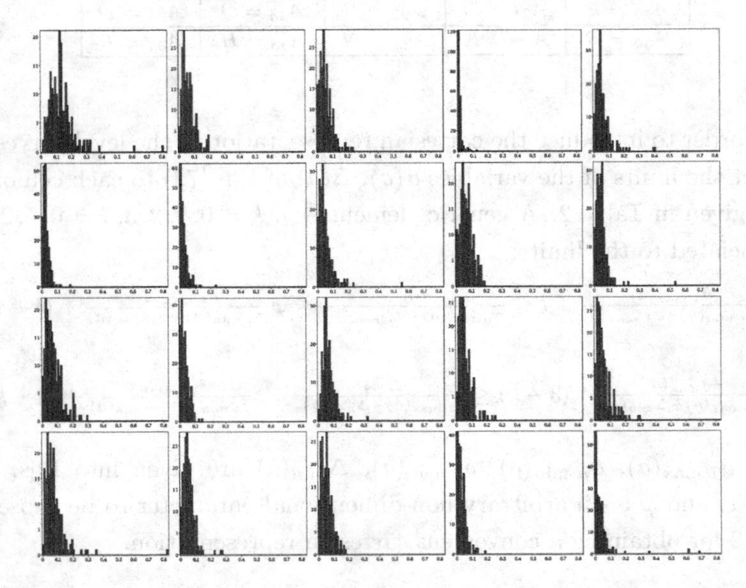

Figure 9. Histograms of the probability occurrences of amino acids for the Pfam protein family PF01051 from Clan CL00123.

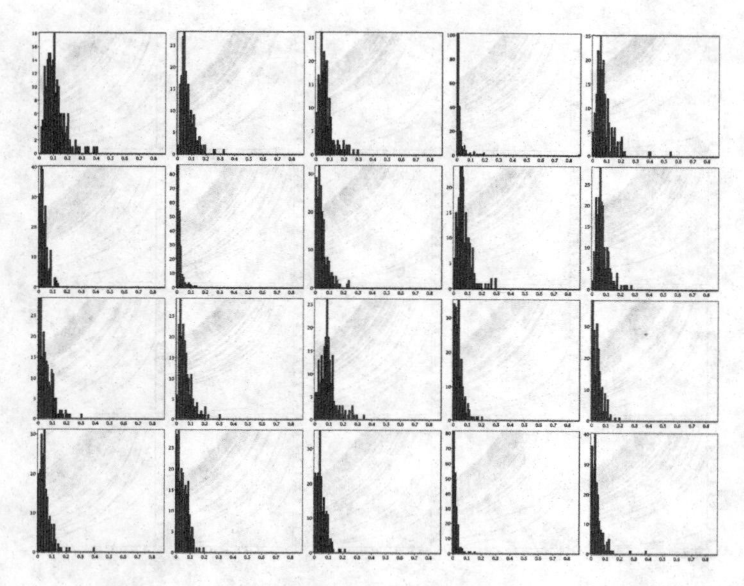

Figure 10. Histograms of the probability occurrences of amino acids for the Pfam protein family PF00060 from Clan CL00030.

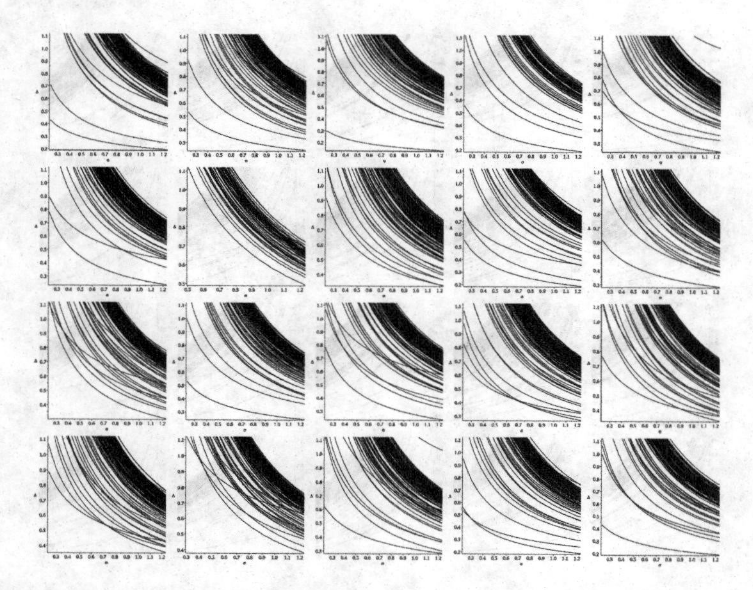

Figure 11. The level curves of the probability occurrences of amino acids for the Pfam protein family PF01051 from Clan CL00123.

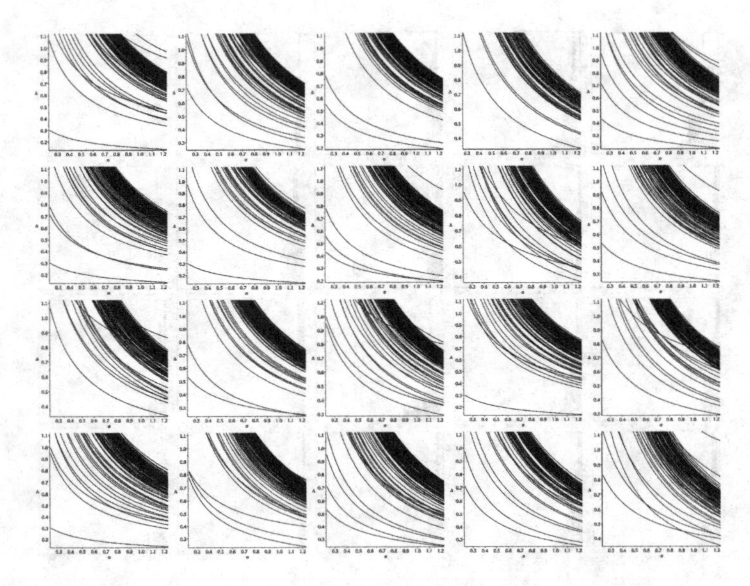

Figure 12. The level curves of the probability occurrences of amino acids for the Pfam protein family PF03399 from Clan CL00123.

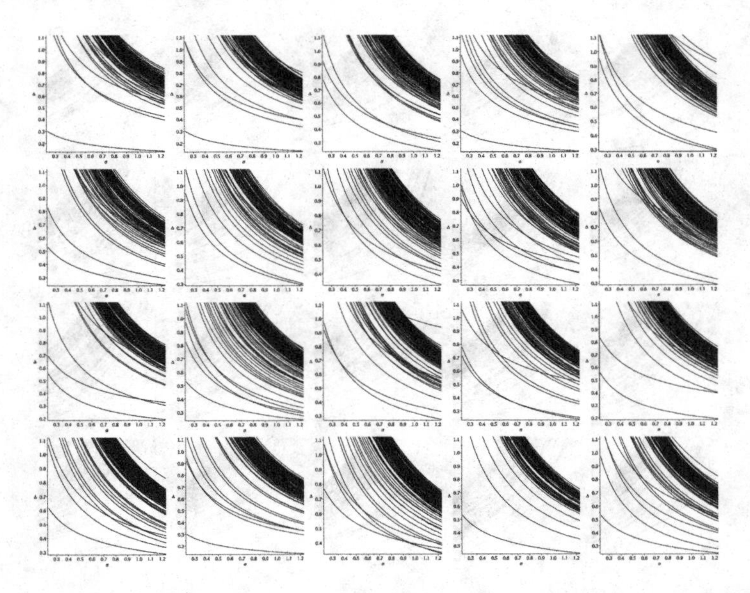

Figure 13. The level curves of the probability occurrences of amino acids for the Pfam protein family PF00060 from Clan CL00030.

The level curves corresponding to the values of $M_{j_s}(\sigma(a), \Delta(a)) = \frac{n_{J_s}(a)}{100}$ (the height of the planes) which intersect the surfaces $p_j(\sigma(a), \Delta(a))$ should be written for each A_{kl} cell as:

$$M_{j_s kl}\big(\sigma(a), \Delta(a)\big) = \frac{1}{\sigma(a) - l\big(\sigma_{n_{\min}}(a) - \sigma_{1_{\max}}(a)\big)} \Bigg(1 - e^{j\left(\sigma(a) - l\left(\sigma_{n_{\min}}(a) - \sigma_{1_{\max}}(a)\right)\right)\left(\Delta(a) - (3-k)(\delta_{1_{\max}}(a) - \Delta_{\min}(a))\right)} \cdot \sum_{m=0}^{j} \frac{\Big(j\big(\sigma(a) - l\big(\sigma_{n_{\min}}(a) - \sigma_{1_{\max}}(a)\big)\big)\big(\Delta(a) - (3-k)\big(\delta_{1_{\max}}(a) - \Delta_{\min}(a)\big)\big)\Big)^m}{m!}\Bigg) \quad (74)$$

where we have used Eq. (34).

7. Concluding Remarks

In the present work we have introduced the idea of protein family formation process (PFFP) and we now stress that the evolution of an "orphan" protein is not a special case of this process[15]. We then consider that proteins do not evolute independently. Their evolution is a collective evolution of all their "relatives" grouped into families. In order to derive a model of probability occurrence of amino acids, we have started from a master equation and we have made a very simple assumption such as that of Eq. (25). We expected to obtain a good method for identifying the Clan association of protein families[4] in terms of level curves of the probability distributions of amino acids. The examples given into section 6, Figs. 11, 12, 13, in despite of their advantage over any conclusion derived from the usual analysis of histograms of Figs. 8, 9, 10, do not seem to lead to an unequivocal conclusion about Clan formation. We then propose the following research lines for future development:

(1) The consideration of other families and clans into the analysis reported in section 6.
(2) The introduction of other assumptions for deriving the probability distributions. The Saddle Point Approximation seems to be a most convenient one.
(3) The consideration of models of distributions based on the joint probabilities of section 2.

(4) The introduction of generalized Entropy Measures as the selected functions of random variables instead of studying the elementary probability distributions of section 2.

Some reports on recent results related to these proposals are now in preparation and will be published elsewhere.

References

1. E. T. Jaynes – Probability Theory - The Logic of Science - Cambridge University Press, 2003.
2. J. Harte – From Spatial Pattern in the Distribution and Abundance of Species to a Unified Theory of Ecology. The Role of Maximum Entropy Methods - Applied Optimization, vol.102, Mathematical Modelling of Biosystems (2008) 243-272, Springer-Verlag, R. P. Mondaini, P. M. Pardalos (eds.).
3. R. P. Mondaini – Entropy Measures based Method for the Classification of Protein Domains into Families and Clans - BIOMAT 2013 (2014) 209-218.
4. R. P. Mondaini, S.C. de Albuquerque Neto – Entropy Measures and the Statistical Analysis of Protein Family Classification - BIOMAT 2015 (2016) 193-210.
5. N. G. Van Kampen – Stochastic Process in Physics and Chemistry, Elsevier B. V. 2007, 3rd edition.
6. W. Bialek – Biophysics - Searching for Principles, Princeton University Press, 2012.
7. R. D. Finn *et al.* – Pfam: Clans, web tools and services - Nucleic Acids Research, 34 (2006) D247-D251.
8. M. Punta *et al.* – The Pfam Protein Families database - Nucleic Acids Research, 40 (2012) D290-D301.
9. R. D. Finn *et al.* – The Pfam Protein Families database - Nucleic Acids Research, 42 (2014) D222-D230.
10. R. D. Finn *et al.* – The Pfam Protein Families database - Nucleic Acids Research, 44 (2016) D279-D285.
11. M. H. DeGroot, M. J. Schervish - Probability and Statistics, 4th edition, Addison-Wesley, 2012.
12. M. Abramowitz, J. A. Stegun – Handbook of Mathematical Functions, Dover Publ., N. York, 9th printing, 1972.
13. D. Veberič – Having Fun with Lambert W function - arχiv: 1003.1628 v1, 8 March 2010.
14. T. J. Sereda *et al.* – Reversed-phase chromatography of synthetic amphipathic α-helical peptides as a model for ligand/receptor interactions. Effect of changing hydrophobic environment on the relative hydrophilicity/hydrophobicity of amino acid side-chains - J. Chromatogr. A 676 (1994) 139-153.
15. R. P. Mondaini, S. C. de Albuquerque Neto – The Pattern Recognition of Probability Distributions of Amino Acids in Protein Families. The Saddle Point approximation (2016) – in preparation.

MODELLING THE ELECTRICAL ACTIVITY OF THE HEART*

R. A. BARRIO[1], I. DOMINGUEZ-ROMAN[1], M. A. QUIROZ-JUAREZ[2], O. JIMENEZ-RAMIREZ[2], R. VAZQUEZ-MEDINA[3], J. L. ARAGON[4]

[1]*Instituto de Física Universidad Nacional Autónoma de México Ciudad de Mexico 01000, D.F., Mexico*

[2]*Escuela Superior de Ingeniería Mecánica y Eléctrica Culhuacán, IPN Santa Ana 1000, 04430, Ciudad de México*

[3]*Centro Mexicano para la Producción más Limpia, IPN, Acueducto S/N, 07340 Ciudad de Mexico*

[4]*Departamento de Nanotecnología, Centro de F'isica Aplicada y Tecnología Avanzada, Universidad Nacional Autónoma de México, Apartado Postal 1-1010 Querétaro 76000, México*

E-mail: barrio@fisica.unam.mx

In cardiac electrical activity, different types of waves meander through the heart. We present a model of the electrical activity of the heart that proposes that the homogeneous wave fronts propagating through the heart are in fact non-linear wave fronts. We use a general set of reaction-diffusion equations known as the BVAM model[1] that presents a wealth of non-linear bifurcations and chaos. We study numerically the dynamics of wave fronts in the model to describe the mechanisms leading to heart failure. We adapt the model to be able to compare the findings with real electrocardiogram signals obtained for various heart anomalies.

1. Introduction

The human heart beats about 2 to 3 billion times in a normal lifespan. The contractions of the heart are triggered by activation potentials which originate from a region in the right atrium, called the sinoatrial node and propagate throughout the entire muscular tissue as spiral waves that allow the organ to act as a pump. For an efficient pumping of blood, a number of about 10^{10} myocytes should contract in a well coordinated and ordered dynamical pattern. Evidently, this dynamical pattern should be extremely energy efficient, robust and adaptable to different situations. This resilience

*This work is supported by Conacyt, México, through project 179616.

is typical of non-linear oscillators, therefore one could conclude that if one wants to model these processes, one needs to have some way to discern which model captures most of the characteristics of the heart motion, in particular, the alterations due to cardiac malfunctions.

As a result of this motions, electrical currents are generated on the surface of the body, producing variations in the electrical potential of the skin surface (of the order of millivolts), which represent the physical information recorded in an electrocardiogram (ECG), widely used for detecting abnormalities of the heart, usually called arrhythmias[2,3]. There are three functional properties of the heart, automatism, excitability and conductivity, which are related to the electric properties of the cardiac fibres, and one should reproduce these faithfully when modelling the dynamics of the heart. It is now widely accepted that the general features of arrhythmias can only be fully understood in the context of nonlinear dynamics[4]. Existing approaches based on real ECG studies, consider some arrhythmias as examples of chaos[5,6,7,8] resulting from nonlinear deterministic dynamics[9]. Experimental studies[10] indicate that fibrillation is a form of spatio-temporal chaos that arises from a normal rhythm through the so-called torus bifurcation, quasi-periodicity or Ruelle–Takens-Newhouse (RTN) scenario[11].

In this paper we propose a model that fulfils the requirements mentioned above. Furthermore, it reveals a surprising-but acceptable fact that the electrical excitations in the heart propagate in space as solitary waves, or solitons. Indeed, this type of waves propagate in a non-dissipative way, which is relevant if one considers that the heart cannot stop working during a life time span. Our model is able to reproduce the peculiar way in which these fronts collide (typical of solitons), as shown by experimental studies[12].

2. Dynamical Model

The Barrio-Varea-Aragon-Maini (BVAM) model[1] is a generic reaction-diffusion system obtained by assuming mass conservation of two morphogens and by retaining up to cubic non-linearities in a Taylor expansion around an equilibrium point. This model has been used to describe a large variety of patterns observed in biological or chemical systems[13], and in particular we have shown that it exhibits a rich variety of dynamics, as Turing spatial patterns, traveling waves, bistabilty, limit cycles and chaos[14]. The dimensionless form of the model reads,

$$\begin{aligned}\frac{\partial u(\mathbf{x},t)}{\partial t} &= D\nabla^2 u + \eta\left[u + g(1-f)v - Cuv - uv^2\right] \\ \frac{\partial v(\mathbf{x},t)}{\partial t} &= \nabla^2 v + \eta\left[g(f+h)v + hu + Cuv + uv^2\right]\end{aligned} \tag{1}$$

where $u(\mathbf{x},t)$ and $v(\mathbf{x},t)$ describe the concentrations of two interacting morphogens at position $\mathbf{x}$ and time t with constant diffusion coefficient ratio $D_u/D_v = D$. The parameter, η gives the scale of the domain, f, g, and h are linear parameters and C is the ratio of the strengths between quadratic and cubic non-linearities.

Through linear analysis of the system one could find several bifurcations, either Turing, Hopf or Turing-Hopf[15]. There is also a bifurcation that produces spatial patterns of droplets of controllable constant radius[16].

Of particular interest to us is a region in parameter space that presents a saddle-node bifurcation near a Hopf bifurcation, as shown in Fig. 1. In this region, by using C as bifurcation parameter, one finds Turing patterns that are stationary but oscillate in time[14] and eventually evolve following the RTN route to chaos[14].

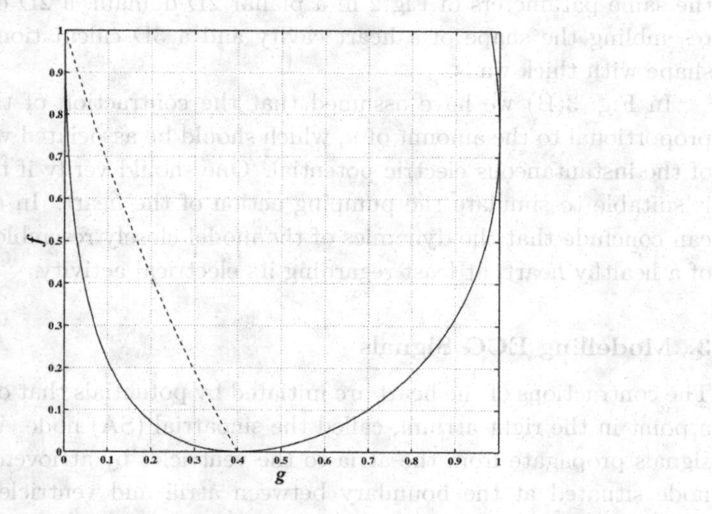

Figure 1. Phase diagram of the BVAM system i the plane (g, f), setting $h = -2.5$, $C = 0$ and $\eta = 0.5$. The continuous line is a Hopf bifurcation and the dotted line is a saddle-node bifurcation. Near this line the system crosses over from a limit cycle on the left to a bistable region on the right.

If one fixes $f \sim 0.7$ one finds a limit cycle on the region to the right of the Hopf bifurcation line (see Fig. 1) and when increasing the bifurcation parameter g, the system becomes bistable when crossing the saddle-node bifurcation line. The wave fronts in this region travel with constant velocity, which in 2 and 3 dimensions depends on the local curvature of the front. This means that the velocity of the front is nil at the points (lines) where the curvature changes sign, so the front rotates around these points, giving the impression of spiral waves.

2.1. *Numerical results*

In order to obtain the patterns in interesting regions of parameter space we have integrated the non-linear system 1 using a simple Euler method, a Conjugate Gradient method and a Finite element method, with periodical boundary conditions in 2 and 3D domains. An example of such calculations in a 2D planar domain is shown in Fig. 2, where one can appreciate that the shape of the front wave is quite complicated, with several Fourier components.

In order to illustrate the appearance of the wave fronts in shapes that resemble the cavities of the heart, in Fig. 3 we show a calculation using the same parameters of Fig 2 in a planar 2D domain, a 2D closed surface resembling the shape of a heart cavity and a 3D calculation in a similar shape with thick walls.

In Fig. 3(B) we have assumed that the contraction of the domain is proportional to the amount of u, which should be associated with the value of the instantaneous electric potential. One should verify if this dynamics is suitable to simulate the pumping action of the heart. In any case, one can conclude that the dynamics of the model closely resembles the motion of a healthy heart, at least regarding its electrical activity.

3. Modelling ECG Signals

The contractions of the heart are initiated by potentials that originate from a point in the right atrium, called the sinoatrial (SA) node. The electrical signals propagate from the atria to the ventricles by atrioventricular (AV) node situated at the boundary between atria and ventricles. Electrical conduction to the ventricles is provided by a specialised conduction system called His-Purkinje complex (HP), which triggers a depolarisation wave in the ventricles, that results in their efficient contraction to pump blood[17].

Notice that the wave shape in Fig. 2 presents sharp peaks resembling

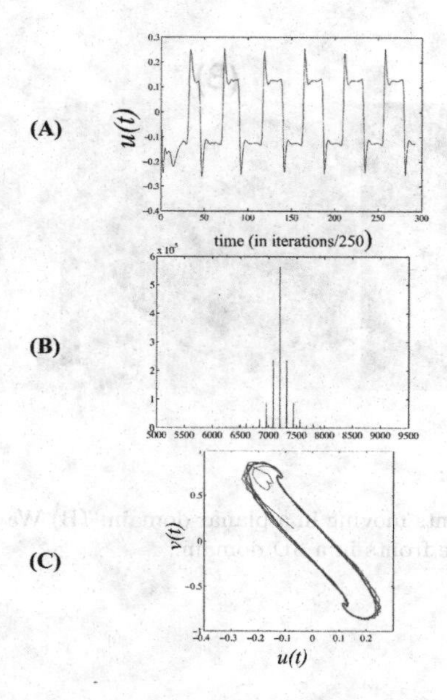

Figure 2. Numerical results using $D = 1$, $h = -2.5$, $C = 0$, $f = 0.5143$, $g = 0.175$, $dt = 0.06$, $dx = 0.2$ and $\eta = 10$. in a rectangular grid of 120×120 with periodic boundary conditions. (A) Time history of u in the central point of the domain. (B) Fourier transform of u. (C) Phase portrait of the calculation. Observe the heteroclinic trajectories that connect the two stable fixed points.

the ones measured in a ECG. In order to simulate the conditions of the waves obtained in an actual ECG we may use our model in a discrete one dimensional domain of only three points, simulating the actual pacemakers of the heart. In these conditions, the one dimensional system allows not only a detailed linear analysis, but also a full non-linear bifurcation analysis. The discretised system contains 6 dynamical variables, but when imposing boundary conditions, two of them are functions of the other 4, thus one ends up with a system of four coupled ordinary differential equations that could be solved numerically. Bifurcation analysis indicates that the suitable control parameter is h. Recently, we proposed a model[18] that reproduces clinically comparable ECG waveforms by the inclusion of an ectopic pacemaker that stimulates the ventricular muscles.

In Fig. 4(A) we show a typical EGC signal from a healthy heart whose main features are indicated by the letters commonly used in medical practice. In Fig. 4(B) we show the numerical calculation of the dynamical behaviour of the three nodes $u_i = 1, 2, 3$ and a linear combination of

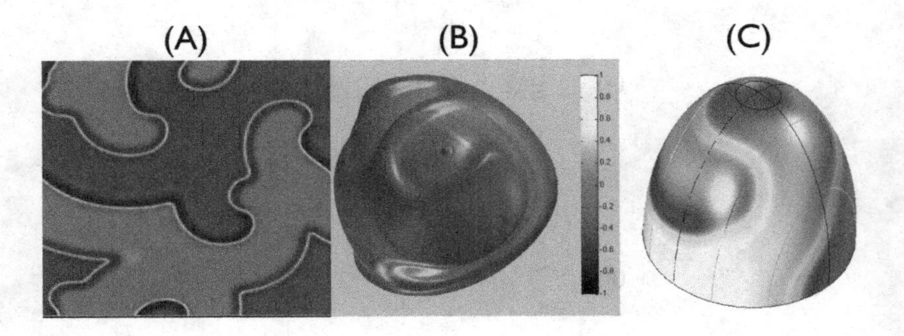

Figure 3. (A) Wave fronts moving in a planar domain. (B) Wave fronts in a parametric closed surface. (C) Wave fronts in a 3D domain.

these $EGS(t) = \alpha_0 + \sum_i \alpha_i u_i$ that reproduce faithfully the shape of an actual ECG. The values used in the figure are $\alpha_0 = 0.004$, $\alpha_1 = -0.076$, $\alpha_2 = -0.024$, $\alpha_3 = 0.008$.

In order to validate the model we followed the bifurcation tree along the route to chaos to simulate various heart failures an diseases. We were able to reproduce various common arrhythmias. In Fig. 5 we show the comparison of the actual ECG signals (taken from Ref.[19]) with the numerical results from our discretised model using a linear combination of the six u_i and v_i variables.

4. Conclusions

We have presented a dynamical model that is able to simulate the electrical activity of the heart under normal and pathological conditions. As a result of this work we conclude that the electrical signals in the heart propagate as solitary wave fronts, which are robust and adaptable to variations of the external conditions. The model presents a bifurcation tree that drives the system from periodic oscillations at increasing frequencies to chaotic pulsations. This feature is useful to validate the model by investigating

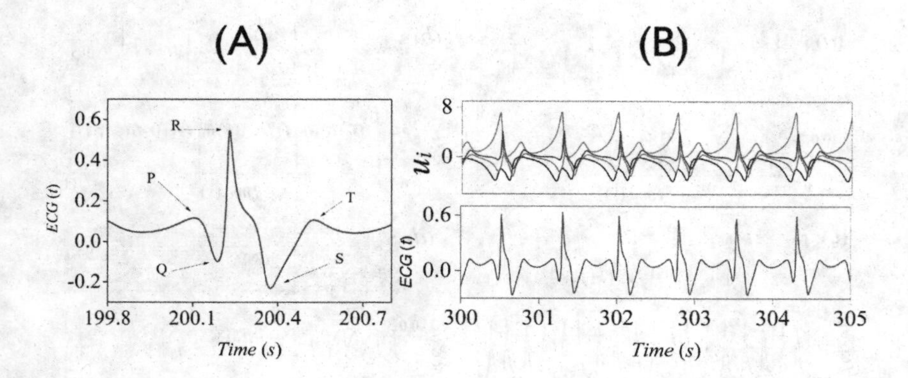

Figure 4. (A) Actual ECG pulse showing the main critical points. (B) Time history of the three u_i variables of our discretised model. In the lower panel we show a suitable linear combination of the three signals that reproduces the shape of an actual ECG.

the route to chaos and comparing it with experimental data from various ways in which the heart fails. To our knowledge this is the first model able to tackle this problem, which is the most important in clinic studies and diagnose.

We simplified the model to a one dimensional system of three coupled pacemakers, whose signals could be combined to reproduce the actual ECG measurements in clinical studies to detect heart failures. The justification of such a procedure to be useful in preventing or curing some of the most common and fatal heart diseases is in progress and will be published soon[20] where a full detailed account of the properties of this linear system will be given. A further advantage of this procedure is that we were able to demonstrate the RTN route to chaos suggested by experimental studies of heart failure in dogs and humans[10].

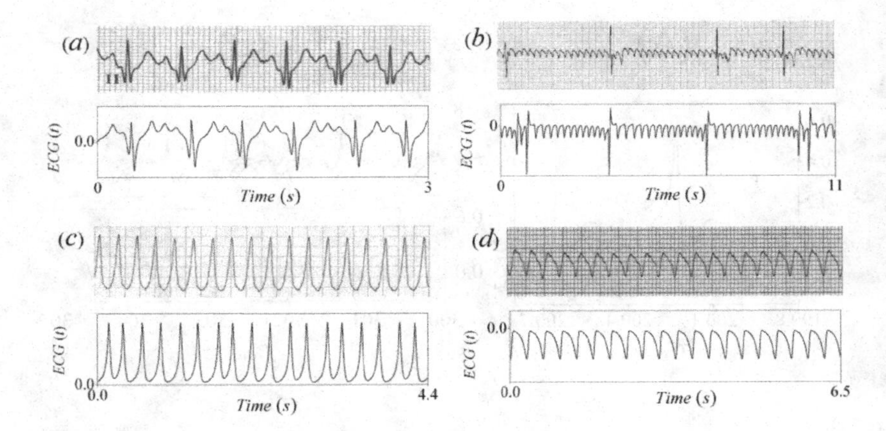

Figure 5. Comparison between real ECG signals (upper panels) and the results of our theoretical model (lower panels). (a) Sinus Tachycardia, (b) Atrial flutter, (c) ventricular Tachycardia and (d) Ventricular flutter.

References

1. R. A. Barrio, C. Varea, J. L. Aragón, and P. K. Maini, A two dimensional numerical study of spatial pattern formation in interacting turing systems, *Bull. Math. Biol.*, 61:483, 1999.
2. E. J. da S. Luz, W. R. Schwartz, G. Cmara-Chvez, and D. Menotti, Ecg-based heartbeat classification for arrhyth- mia detection: A survey, *Computer Methods and Programs in Biomedicine*, 127:144-164, 2016.
3. J. P. Keener and J. Sneyd, *Mathematical Physiology*, vol. 1. Springer, 1998.
4. Z. Qu, G. Hu, A. Garfinkel, and J. N. Weiss, Nonlinear and stochastic dynamics in the heart, *Physics Reports*, 543, no. 2: 61–162, 2014.
5. R. Radhakrishna, D. N. Dutt, and V. K. Yeragani, Nonlinear measures of heart rate time series: influence of posture and controlled breathing, *Autonomic Neuroscience*, 83, 3:148–158, 2000.
6. F. X. Witkowski, L. J. Leon, P. A. Penkoske, W. R. Giles, M. L. Spano, W. L. Ditto, and A. T. Winfree, Spatiotemporal evolution of ventricular fibrillation, *Nature* 392, 6671:78–82, 1998.
7. Y. Zhao, J. Sun, and M. Small, Evidence consistent with deterministic chaos in human cardiac data: surrogate and nonlinear dynamical modeling, *International Journal of Bifurcation and Chaos*, 8, 01:141–160, 2008.
8. L. Glass, M. R. Guevara, A. Shrier, and R. Perez, Bifurcation and chaos in a periodically stimulated cardiac oscillator, *Physica D: Nonlinear Phenomena*, 7, 1:89–101, 1983.
9. D. T. Kaplan and R. J. Cohen, Is fibrillation chaos?, *Circulation Research*, 67, 4:886–892, 1990.

10. A. Garfinkel, P.-S. Chen, D. O. Walter *et al.*, Quasiperiodicity and chaos in cardiac fibrillation, *Journal of Clinical Investigation*, 99, 2:305, 1997.
11. S. Newhouse, D. Ruelle, and F. Takens, Occurrence of strange axioma attractors near quasi periodic flows on t^m, $m > 3$, *Communications in Mathematical Physics*, 64, 1:35–40, 1978.
12. Gil Bub, Alvin Shrier, and Leon Glass, Spiral wave generation in heterogeneous excitable media, *Phys. Rev. Lett.*, 85:058101, 2002.
13. Rafael A. Barrio. Aplicaciones del modelo bvam a sistemas complejos, *Revista Digital Universitaria,*
    URL: http://www.revista.unam.mx/vol.11/num6/art55, 11(6), junio 2010.
14. J. L. Aragón, R. A. Barrio, T. E. Woolley, R. E. Baker, and P. K. Maini, Nonlinear effects on Turing patterns: Time oscillations and chaos, *Physical Review E* 86:026201, 2012.
15. Teemu Lepännen, Computational Studies of Pattern Formation in Turing Systems. Ph.D. Thesis, Helsinki University of Technology, 2004.
16. Tomas E. Wolley, Ruth E. Baker, Philip K. Maini, J. L. Aragón, and R. A. Barrio, Analysis of stationary droplets in a generic turing reaction-diffusion system, *Phys. Rev. E*, 82:051929, 2010.
17. J. P. Keener and J. Sneyd, *Mathematical Physiology*, vol. 1. Springer, 1998.
18. M. Quiroz-Juárez, R. Vázquez-Medina, E. Ryzhii, M. Ryzhii, and J. Aragón, Quasiperiodicity route to chaos in cardiac conduction model, *Communications in Nonlinear Science and Numerical Simulation*, 2016.
19. G. G. Yanowitz, *ECG Libray*, lifeinthefastlane.com/ecg-library.
20. M. Quiroz-Juárez *et al.* (to be published).

HOW THE INTERVAL BETWEEN PRIMARY AND BOOSTER VACCINATION AFFECTS LONG-TERM DISEASE DYNAMICS*

AFFAN SHOUKAT
Agent-Based Modelling Laboratory
York University, Toronto
Ontario M3J 1P3, Canada
E-mail: affans@yorku.ca

AQUINO L. ESPINDOLA
Departamento de Física
Instituto de Ciências Exatas - ICEx
Universidade Federal Fluminense
Volta Redonda, RJ, Brazil, 27.213-145
E-mail: aquinoespindola@id.uff.br

GERGELY RÖST
Bolyai Institute, University of Szeged
6720 Szeged, Hungary
E-mail: rost@math.u-szeged.hu

SEYED M. MOGHADAS
Agent-Based Modelling Laboratory
York University, Toronto
Ontario M3J 1P3, Canada
E-mail: moghadas@yorku.ca

*This work was in part supported by the Natural Sciences and Engineering Research Council of Canada (NSERC), the Mathematics of Information Technology and Complex Systems (MITACS), and the European Research Council StG 259559 and OTKA K109782.

Many vaccine preventable diseases require a booster dose in addition to the primary vaccination in order to provide adequate level of protection. However, the optimal timing of booster dose after primary vaccination is often difficult to determine. Here we propose an agent-based modelling framework to investigate the timing of booster vaccination, based on the efficacy of primary vaccine series and the coverage of booster dose. We show that these factors can significantly affect the long-term epidemiological outcomes of vaccination. We represent our results by simulating the model for the dynamics of *Heamophilus influenzae* serotype b (Hib), using parameter estimates for natural history of the disease, its transmissibility, and the protection efficacy of vaccine. We show that, for estimated vaccine protection efficacy against Hib, if the primary and booster coverages are maintained at high levels, a longer time interval between primary and booster doses leads to better outcomes and potentially elimination of the disease. The length of this time interval is however dependent on the coverage of booster vaccination.

1. Introduction

Vaccination remains the most successful preventive measure against many infectious diseases in modern medicine[1]. Despite successful implementation of vaccination programs, elimination of some disease has proven challenging, with instances of resurgence occurring in different geographic locations[2,3,4]. Several factors have been investigated as potential explicators for persistence and recurrent outbreaks of vaccine-preventable diseases, including incomplete protection efficacy of primary series, inadequate coverage of booster doses, waning immunity over time, and the period of vaccine-induced protection that may be significantly shorter than the average lifetime of the population[5,4]. However, given these factors, the time interval between primary series and booster vaccination may significantly influence both short- and long-term disease outcomes.

In this study, we develop an agent-based modelling (ABM) framework to investigate how the interval between primary and booster vaccination affects disease dynamics. While importance of age at vaccination has been well documented, the optimal timing of booster doses remains unclear in several vaccine-preventable diseases[6,7,4]. For this study, we considered *Haemophilus influenzae* serotype b (Hib), for which primary and booster vaccination has been implemented in routine infant immunization programs in many countries. Yet, the elimination has not been achieved, notwithstanding substantial levels of routine primary vaccine series[2,3,4]. We included key biological and epidemiological determinants of Hib in the model, and parameterize it with the previously published estimates to investigate the effect of timing for booster dose after primary vaccination on

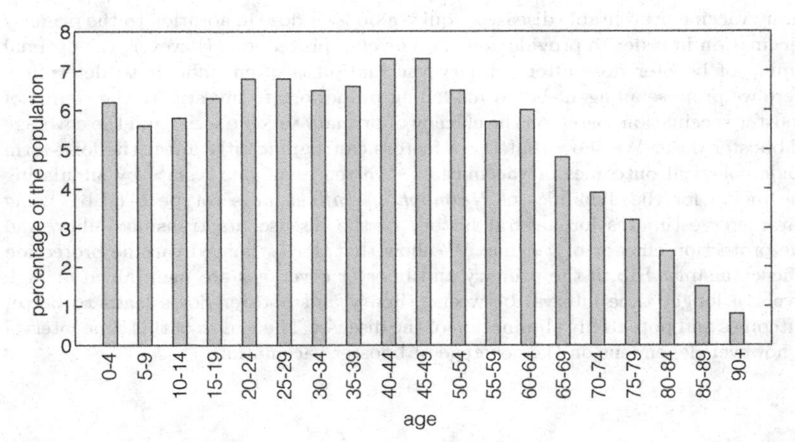

Figure 1. Age distribution of the United Kingdom population according to the 2011 census data: `https://en.wikipedia.org/wiki/Demography_of_the_United_Kingdom`.

the prevalence of disease. We illustrate our results using stochastic Monte-Carlo simulations, and argue that the timing of booster dose may be an essential component of vaccination programs, whose optimal determination depends critically on the protection efficacy of primary vaccine series, and the coverage of booster vaccination.

2. Methodology

2.1. *Agents-environment framework*

The general framework of the model includes two main entities: (i) a simulated environment represented by a two dimensional lattice and (ii) the individuals that are located in the lattice. Each individual is characterized by their time-sensitive information vector, which encapsulates a plethora of information including the current health status, age, and the immune protection level. The size of the lattice was set to 317×317, resulting in an environment with 100489 individuals with age distributions similar to the UK population (Figure 1). The interaction between individuals was modelled through contacts with their immediate neighbours or random contacts on the lattice.

2.2. *Disease model*

In construction of the ABM framework, we consider the disease model with several epidemiological statuses of the individuals, based on which

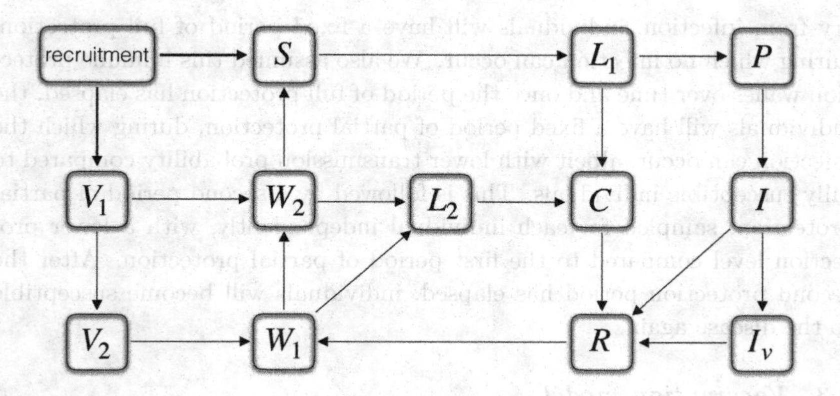

Figure 2. Schematic model diagram for transitions between epidemiological compartments in the disease model of the ABM framework.

the population is divided into several compartments. The disease model includes compartments of susceptible (S), primary latent (L_1), secondary latent (L_2), pre-symptomatic (P), asymptomatic referred to as carriage (C), symptomatic (I), invasive disease (I_v), recovered (R), vaccinated with primary series (V_1), vaccinated with booster dose (V_2), partially protected stage 1 (W_1), and partially protected stage 2 (W_2). Disease transmission occurs as a result of contact between infectious individuals (either in carriage of symptomatic stages), and those with no or incomplete protection (i.e., susceptible or partially protected individuals). The movements of individuals between the epidemiological compartments of the model are schematically represented in Figure 2.

Once exposed to the disease with successful transmission, the individuals will enter one of the latent stages (primary or secondary), depending on whether or not they have any protection level as a result of previous natural infection or vaccination. Individuals in the latent stages are assumed to be non-infectious and therefore cannot transmit the disease. Infected individuals with no pre-existing immune protection will proceed to pre-symptomatic infection and develop either symptomatic disease (with clinical manifestation) or remain asymptomatic (without clinical symptoms) for the entire duration of infectious period. If treatment is not sought, symptomatic individuals may develop invasive disease with potentially fatal outcomes. We assumed that infected individuals with pre-existing immune protection will develop only asymptomatic disease and remain carriage. Following recovery from infection, individuals will have a fixed period of full protection, during which no infection can occur. We also assumed this immune protection wanes over time and once the period of full protection has elapsed, the individuals will have a fixed period of partial protection, during which the infection can occur, albeit with lower transmission probability compared to fully susceptible individuals. This is followed by a second period of partial protection, sampled for each individual independently, with a lower protection level compared to the first period of partial protection. After the second protection period has elapsed, individuals will become susceptible to the disease again.

2.3. *Vaccination model*

For evaluating the effect of vaccination, we implemented primary vaccination for infants within their 6 months of birth[8]. Booster vaccination was offered only to those who had received primary vaccination. Vaccination of infants with primary series will lead to a fixed duration of partial protection, and followed by the second period of partial protection (as described above for natural infection) without booster vaccination. Those who received booster vaccination will be fully protected for a fixed length of time, and follow the same path of partial protection as described for individuals recovered from natural infection.

2.4. *Computational implementation*

The model was calibrated with the reproduction number $R_0 = 1.3$[9] to obtained the baseline transmissibility $\beta = 7 \times 10^{-3}$ in the absence of any interventions (i.e., treatment or vaccination). Simulations were initialized with $C(0) = 15$ for infection. The parameters governing the dynamics of the model were drawn from previously published literature, and are described in Table 1. The model includes both disease induced and natural deaths. All deaths were replaced with newborns, thus maintaining a constant population size. The probability of natural death was calculated from its associated distribution with the median life expectancy of 60 years.

The success of transmission was determined using rejection sampling-based (Bernoulli) trials where the chance of success is defined by a transmission probability distribution. In the absence of pre-existing immune protection, the spread of disease between a susceptible-infectious pair of individuals was calculated using the baseline transmission probability

$$P_0 = 1 - (1 - \eta_p\beta)^{K_p}(1 - \eta_c\beta)^{K_c}(1 - \beta)^{K_s} \quad (1)$$

where K_p is the number of contacts with pre-symptomatic individuals, K_c is the number of contacts with individuals in the stage of carriage, K_s is the number of contacts with symptomatic individuals, and η_p and η_c are respectively the relative reduction of infectiousness in pre-symptomatic and carriage individuals compared with symptomatic individuals. However, the probability P_0 is subject to the immune protection level of the individuals. To account for the effect of immunity in the reduction of disease transmissibility, we calculated the transmission probability by

$$P_{V_1} = (1 - \eta_{v_1})P_0 \tag{2}$$

for primary vaccinated individuals, and by

$$\begin{aligned} P_{W_1} &= (1 - \eta_{w_1})P_0, \\ P_{W_2} &= (1 - \eta_{w_2})P_0, \end{aligned} \tag{3}$$

for individuals in the first and second stages of partial protection, where η_{v_1}, η_{w_1}, and η_{w_2} are respectively the relative reductions of susceptibility to infection (as a results of immune protection) compared with fully susceptible individuals.

2.5. *Parmeterization and timing of booster vaccination*

We parameterize the model with estimates for various epidemiological stage and duration of protection induced by vaccination or natural infection. These parameters and their sources are provided in Table 1. A number of parameters were varied in our simulations to evaluate the effect of the primary vaccine series and the subsequent booster dose. The timing for booster dose was varied between 6 to 48 months after primary vaccination. We assumed three different profiles of booster dose coverage (Figure 3): (i) a fixed coverage the same as the primary vaccine coverage regardless of the time for booster dose (Figure 3, dark grey bars); (ii) a reduced coverage of booster that follows an exponential declining trend (Figure 3, black bars); and (iii) a reduced coverage of booster that follows a declining trend similar to the inverse of the logistic function (Figure 3, light grey bars). The reduction in booster dose coverage may reflect a number of factors, including higher rates of refusal to receive additional vaccine doses following primary series, which may be affected by delay in booster schedule.

Table 1. Description of model parameters and their associated values (ranges) extracted from the published literature.

parameter description	value (range)	source
basic reproduction number	1.3 (1.1 – 1.5)	10,11
coverage of primary vaccination	0.95 (0 – 1)	–
coverage of booster vaccination	variable (0 – 1)	–
latent period	2 (1–3) day	12
pre-symptomatic period	1.5 (1–2) days	12
infectious period of carriage and symptomatic infection	14–70 days	12,13,2
age at completion of primary vaccine series	6 months	8
timing of booster coverage post primary vaccination	variable (0–48) months	–
fixed period of full protection following recovery from infection	2 years	12,2
fixed period of full protection following booster vaccination	6 years	12,2,9,13
fixed period of partial protection stage 1	2 years	12
period of partial protection stage 2	2–8 years	12
level of immune protection following primary vaccination	0.8 (0.6–0.9)	12,2,9,13
level of immune protection during partial protection stage 1	0.6 (0.3–0.9)	12,2
level of immune protection during partial protection stage 2	0.3 (0.1–0.5)	12,2
infectious period following start of treatment in symptomatic infections	2 days	?,12
probability of developing carriage without pre-existing immune protection	0.6–0.9	12,13

3. Results

For each scenario of booster dose coverage (Figure 3), we ran stochastic simulations to obtain the average of 100 independent realizations. The model with vaccination was seeded with the initial conditions given by the components of the endemic equilibrium of the model without vaccination. For the scenarios simulated here, the timing for booster dose after primary vaccination was varied. Figure 4 shows the prevalence of carriage for 28 years following the start of vaccination, when the booster dose was offered with delay of 9 months, 24 months, or 36 months after primary vaccination. For all scenarios, the introduction of vaccination significantly reduced the prevalence of carriage in the absence of vaccination.

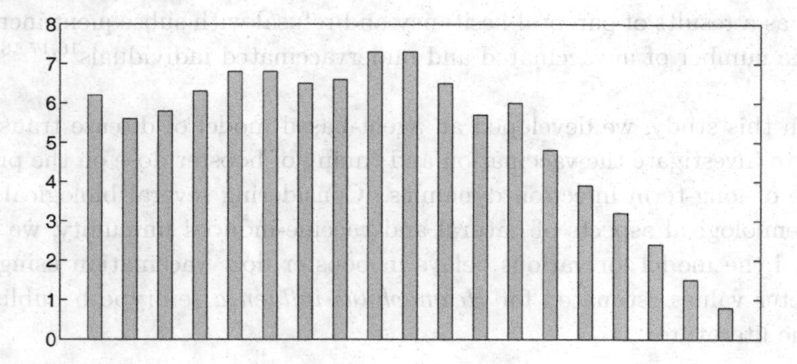

Figure 3. Time-dependent coverage of booster vaccination post primary vaccination.

When the booster dose coverage is maintained at 95% (Figure 4A), we observed minimal differences with increasing delay in booster vaccination, with slight advantage in booster vaccination with longer delay (36 months). However, for a booster coverage that declines exponentially with delay (Figure 4B), we observed that early booster (9 months) outperforms other strategies with longer delays, contrasting the scenario of fixed booster coverage. As delay in timing of booster dose increases, a lower fraction of primary vaccinated individuals receives booster dose which leads to a higher prevalence of infection over time. For the scenario in which the coverage of booster dose reduces over time similar to the inverse of the logistic function (Figure 3), both strategies of early (9 months) and late (36 months) booster vaccination were outperformed by the strategy with an intermediate delay of 24 months. These results suggest that timing of booster vaccination could have a significant impact on the long-term disease outcomes, and the optimal timing depends not only on the protection efficacy of the vaccine, but also on the coverage of the booster dose.

4. Concluding Remarks

There is evidence accumulating that the booster vaccine schedule requires evaluation for several vaccine-preventable diseases to improve the effect of vaccination campaigns[6,7,4]. Previous studies have shown the time interval between the primary vaccination and booster dose could play a vital role in the long-term dynamics of Hib[14], and a large delay in booster vaccination may result in a longer protection period during lifetime[15]. However, a larger delay may lead to a lower fraction of primaries receiving the booster dose as a results of parental hesitancy and refusal with subsequent increase in the number of unvaccinated and undervaccinated individuals[16,17,18].

In this study, we developed an agent-based model of disease transmission to investigate the vaccination and timing of booster dose on the prevalence of long-term infection dynamics. Considering several biological and epidemiological aspects of natural and vaccine-induced immunity, we simulated the model for various delays in booster dose vaccination using parameter values estimated for *Heamophilus influenza* serotype b published in the literature.

Simulation results indicate that the timing of booster dose can significantly influence the prevalence of infection in the population. While we compared the outcomes of different strategies for booster schedule, determination of the optimal timing for booster vaccination remains a challenging task. Although we have shown in a number of scenarios simulated here that the delay in booster vaccination may be beneficial in reducing the disease prevalence, the coverage of booster remains a key determining factor that may be influenced by the delay in booster vaccination[16,17,18]. If the coverage of booster vaccination is expected to drop significantly over time, our simulations illustrate that early booster schedule may be essential to achieve the greatest reduction of the disease prevalence.

The objective of this study was not to determine the optimal timing of a booster dose vaccination, but rather illustrate the complexity of the disease dynamics in the context of vaccination and highlight the challenge in optimizing immunization programs with long-lasting effects in prevention and disease elimination.

The model proposed here could be further evaluated with variability is parameter space to better understand the sensitivity of outcomes with respect to the infection and vaccination specific parameters. This understanding could help to more accurately quantify the benefit of different vaccination programs and booster schedules.

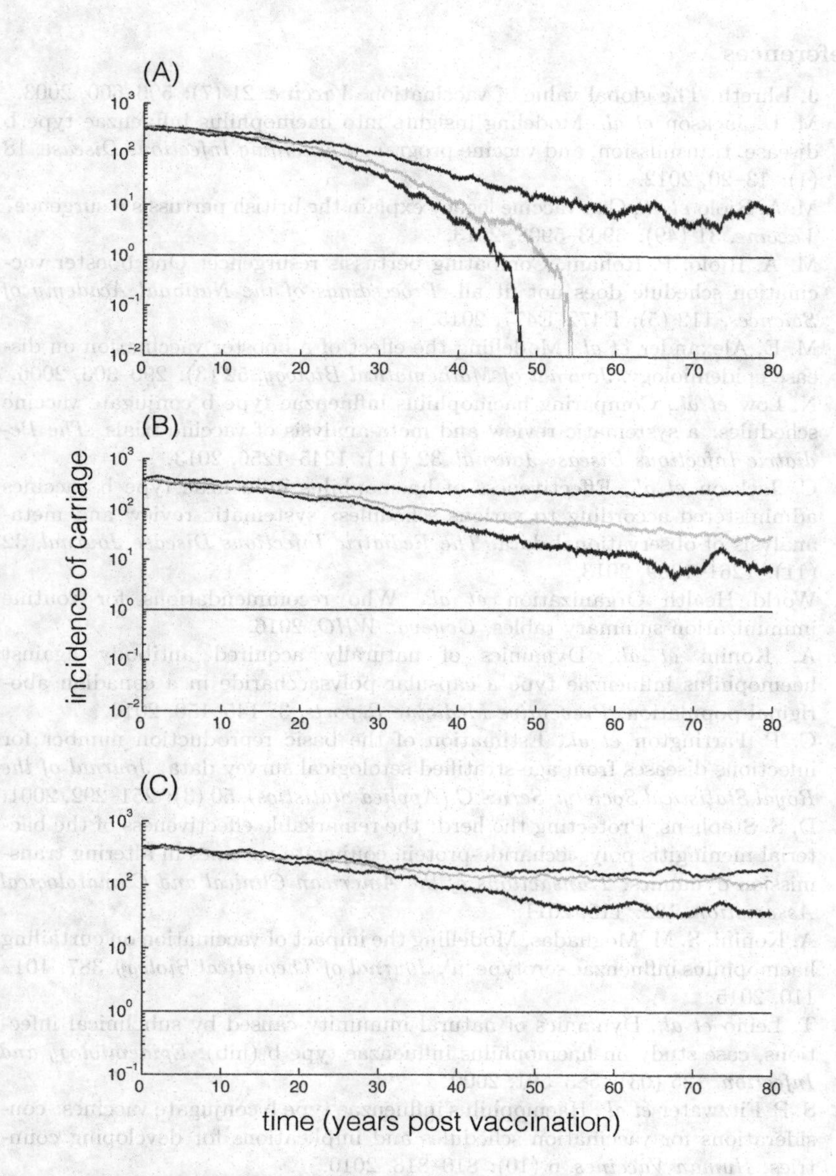

Figure 4. Incidence of carriage with three different booster vaccination coverage scenarios: (A) Fixed (95%) coverage; (B) declining coverage similar to the inverse of logistic function; and (C) exponentially declining coverage (as illustrated in Figure 3). Delay in booster dose is 9 months (bold black curves), 24 months (gray curves), and 36 months (black curve).

References

1. J. Ehreth, The global value of vaccination. *Vaccine*, 21 (7): 596–600, 2003.
2. M. L. Jackson *et al.*, Modeling insights into haemophilus influenzae type b disease, transmission, and vaccine programs. *Emerging Infectious Disease*, 18 (1): 13–20, 2012.
3. M. A. Riolo *et al.*, Can vaccine legacy explain the british pertussis resurgence? *Vaccine*, 31 (49): 5903–5908, 2013.
4. M. A. Riolo, P. Rohani, Combating pertussis resurgence: One booster vaccination schedule does not fit all. *Proceedings of the National Academy of Sciences*, 112 (5): E472–E477, 2015.
5. M. E. Alexander *et al.*, Modelling the effect of a booster vaccination on disease epidemiology. *Journal of Mathematical Biology*, 52 (3): 290–306, 2006.
6. N. Low *et al.*, Comparing haemophilus influenzae type b conjugate vaccine schedules: a systematic review and meta-analysis of vaccine trials. *The Pediatric Infectious Disease Journal*, 32 (11): 1245–1256, 2013.
7. C. Jackson *et al.*, Effectiveness of haemophilus influenzae type b vaccines administered according to various schedules: systematic review and meta-analysis of observational data. *The Pediatric Infectious Disease Journal*, 32 (11): 1261–1269, 2013.
8. World Health Organization *et al.*, Who recommendations for routine immunization-summary tables. *Geneva: WHO*, 2016.
9. A. Konini *et al.*, Dynamics of naturally acquired antibody against haemophilus influenzae type a capsular polysaccharide in a canadian aboriginal population. *Preventive Medicine Reports*, 3: 145–150, 2016.
10. C. P. Farrington *et al.*, Estimation of the basic reproduction number for infectious diseases from age-stratified serological survey data. *Journal of the Royal Statistical Society: Series C (Applied Statistics)*, 50 (3): 251–292, 2001.
11. D. S. Stephens, Protecting the herd: the remarkable effectiveness of the bacterial meningitis polysaccharide-protein conjugate vaccines in altering transmission dynamics. *Transactions of the American Clinical and Climatological Association*, 122: 115, 2011.
12. A. Konini, S. M. Moghadas, Modelling the impact of vaccination on curtailing haemophilus influenzae serotype 'a'. *Journal of Theoretical Biology*, 387: 101–110, 2015.
13. T. Leino *et al.*, Dynamics of natural immunity caused by subclinical infections, case study on haemophilus influenzae type b (hib). *Epidemiology and Infection*, 125 (03): 583–591, 2000.
14. S. P. Fitzwater *et al.*, Haemophilus influenzae type b conjugate vaccines: considerations for vaccination schedules and implications for developing countries. *Human Vaccines*, 6 (10): 810–818, 2010.
15. N. Charania, S. M. Moghadas, Modelling the effects of booster dose vaccination schedules and recommendations for public health immunization programs: the case of haemophilus influenzae serotype b. *BMC Public Health*, in review, 2016.
16. S. B. Omer *et al.*, Vaccine refusal, mandatory immunization, and the risks

of vaccine-preventable diseases. *New England Journal of Medicine*, 360 (19): 1981–1988, 2009.

17. E. Dubé *et al.*, Vaccine hesitancy: an overview. *Human Vaccines & Immunotherapeutics*, 9 (8): 1763–1773, 2013.
18. E. C. Briere *et al.*, Prevention and control of haemophilus influenzae type b disease: recommendations of the advisory committee on immunization practices (acip). *MMWR Recommendations & Reports*, 63 (RR-01): 1–14, 2014.

STABILITY AND HOPF BIFURCATION IN A MULTI-DELAYED ECO-EPIDEMIOLOGICAL MODEL

DEBADATTA ADAK AND NANDADULAL BAIRAGI*

Centre for Mathematical Biology and Ecology
Department of Mathematics, Jadavpur University
Kolkata-700032, India
E-mail: nbairagi@math.jdvu.ac.in

Eco-epidemiology is comparatively a new branch of mathematical biology that considers both the ecological and epidemiological issues simultaneously. Parasites play significant role in trophic interactions and can regulate both predator and prey populations. Mathematical models might be of great use in predicting different system dynamics because models have the potential to predict the system response due to different changes in system parameters. In this paper, we study an eco-epidemiological system where prey population is infected by some micro parasites and predator population consumes disproportionately large number of infected prey due to their lower fitness. The infection process is not instantaneous but mediated by a fixed incubation delay. A constant time delay is incorporated in the disease transmission term to represent this incubation delay. A second delay is considered to incorporate the reproduction time of predator after consuming prey. We study the stability and instability of the endemic equilibrium point of this multi-delayed eco-epidemiological system. Considering delay as a parameter, we investigate the effect of both delays on the stability of the coexisting equilibrium. It is observed that Hopf bifurcation and stability switching occurs when the delay crosses some critical value. Using normal form theory and the center manifold theorem, the explicit formulae which determine the stability and direction of the bifurcating periodic solutions are determined. We show how the amplitude and period of bifurcating periodic solutions, and the system dynamics change with the length of two delay parameters. Analytical results are illustrated with an example.

1. Introduction

Eco-epidemiology has become an important research field in recent past in the field of mathematical ecology. It deals with the issues of predator-prey systems in presence of infection [1-9]. It is shown that disease makes the dynamics complex of a predator-prey system which is otherwise simple.

*This work is supported by DST, India; Ref. No. SB/EMEQ-046/2013.

Different biological properties play important role in the system dynamics. Delay, for example, is an inherent property of biological systems and plays significant role in the qualitative behavior of the system. For a predator-prey model, predator's reproduction delay is frequently used to represent the time required for the reproduction of predator after consuming the prey. On the other hand, disease transmission delay or incubation delay is equally important from epidemic point of view. It is defined as the length of time from the moment of receiving an infection and symptoms appear. In this paper, we analyze an eco-epidemiological system which consider both the incubation delay and the reproduction delay.

We consider the following eco-epidemiological model, which is a combination of a predator-prey model and a SI-type epidemic model, based on [1]:

$$\begin{aligned} \frac{dS}{dt} &= rS\left(1-\frac{S+I}{K}\right)-\lambda SI, \\ \frac{dI}{dt} &= \lambda SI-\frac{mIP}{a+I}-\mu I, \\ \frac{dP}{dt} &= \frac{m\alpha IP}{a+I}-dP. \end{aligned} \tag{1}$$

Here the state variables S, I, P represent, respectively, the densities of susceptible prey, infected prey and predator populations at time t. Susceptible prey grows logistically to the carrying capacity K with intrinsic growth rate r. Only susceptible prey can reproduce but both susceptible and infected preys contribute to the carrying capacity. Infection spreads horizontally from infected to susceptible prey following mass action law with rate constant λ. Predator population consumes infected prey only as they are weakened due to infection and easier to catch. Predation is followed by type II response function with maximum consumption rate m and half-saturation constant a. The parameter α measures the reproductive gain and d measures the death rate of predator. The total death rate of infected prey is measured by μ which contains both the natural and infection related death. For more details readers are referred to [10].

Recently, researchers are inclined to explore the effects of multi-delays in different biological models [11, 12, 13]. However, mathematical analysis becomes more complicated in case of multi-delayed system than the corresponding single-delayed system. Parasite is passed from one infected prey to another susceptible prey, but the infection process is not instantaneous and there always exists a time lag, known as incubation period. Similarly, reproduction of predator after consuming the prey is not instantaneous but mediated by some time lag, known as reproduction delay. To make the model biologically more realistic, one has to consider both the incubation delay and reproduction delay. In this study, we consider both delays in (1) and the model reads

$$\begin{aligned}
\frac{dS}{dt} &= rS(t)\left(1 - \frac{S(t)+I(t)}{K}\right) - \lambda S(t)I(t),\\
\frac{dI}{dt} &= \lambda \int_{-\infty}^{t} S(u_1)I(u_1)F_1(t-u_1)du_1 - \frac{mI(t)P(t)}{a+I(t)} - \mu I(t), \qquad (2)\\
\frac{dP}{dt} &= m\int_{-\infty}^{t} \frac{I(u_2)P(u_2)}{a+I(u_2)}F_2(t-u_2)du_2 - dP(t).
\end{aligned}$$

It is assumed here that the number of actively infected prey populations at time t is arising from the contacts of actual population of susceptible and infected preys at time $(t-u_1)$. Similarly, it is assumed that the reproduction of predator at time t is arising from consuming the infected preys at time $(t-u_2)$. Here u_i is distributed according to a probability distribution function $F_i(u_i)$, known as delay kernel (or memory function), defined by $F_i(u_i) = \frac{\chi_i^{p_i+1}{u_i}^{p_i}}{p_i!}e^{-\chi_i u_i}$, where $\chi_i(>0)$ is a constant and p_i is a non-negative integer, known as order of the delay. If the kernel $F_i(u_i)$ is normalized, i.e., $\int_0^\infty F_i(u_i)du_i = 1$, then mean delay is $\bar{T}_i = \int_0^\infty F_i(u_i)u_i\, du_i = \frac{p_i+1}{\chi_i}$, which becomes the discrete delay τ_i for a δ-function kernel $F_i(u_i) = \delta(u_i - \tau_i)$, defined to be zero except at $u_i = \tau_i$, ensures that $\bar{T}_i = \tau_i$ $(i = 1, 2)$ [14] and (2) becomes

$$\begin{aligned}
\frac{dS}{dt} &= rS\left(1 - \frac{S+I}{K}\right) - \lambda SI,\\
\frac{dI}{dt} &= \lambda S(t-\tau_1)I(t-\tau_1) - \frac{mIP}{a+I} - \mu I, \qquad (3)\\
\frac{dP}{dt} &= \frac{m\alpha I(t-\tau_2)P(t-\tau_2)}{a+I(t-\tau_2)} - dP.
\end{aligned}$$

All parameters are assumed to be positive. Bairagi *et al.* [10] studied the system (3) with $\tau_2 = 0$. Sun *et al.* [15] considered $\tau_1 = 0$ and type I response function mIP with a reproduction delay of predator. Bairagi [16] studied the system (3) with $\tau_1 = 0$. The general observation of these studies is that there exists a critical value of the delay parameter below which the coexistence equilibrium is stable and above which it is unstable.

A Hopf bifurcation takes place at the critical value of the delay parameter. In this paper, we analyze the dynamic behavior of system (3) around its coexistence equilibrium when $\tau_1 \geq 0$ and $\tau_2 \geq 0$.

The paper is arranged in the following sequence. Preliminary results like existence, positivity and boundedness of solutions are presented in the next section. Section 3 deals with existence of equilibrium solutions and permanence of the system. Stability analysis in absence and presence of delay is performed in Section 4. Direction and stability of Hopf bifurcation are presented in Section 5. The next section gives one example to illustrate the model and its analytical results. The paper concludes in Section 7 with a discussion.

2. Preliminary results

Let $X = \mathcal{C}([-\tau, 0], \mathbb{R}^3_+)$, where $\tau = \max\{\tau_1, \tau_2\}$, $\tau > 0$, be the Banach space of continuous real-valued functions on the interval $[-\tau, 0]$ with norm $\|\varrho\| = sup_{-\tau \leq \theta \leq 0} |\varrho(\theta)|$ for $\varrho \in \mathcal{C}([-\tau, 0], \mathbb{R}^3_+)$. By the standard theory of functional differential equations [17], for any $\varrho \in \mathcal{C}([-\tau, 0], \mathbb{R}^3_+)$, there exists a unique solution

$$Y(t, \varrho) = (S(t, \varrho), I(t, \varrho), P(t, \varrho))$$

of (3) which satisfies $Y_0 = \varrho$. The initial conditions are given by

$$S(\theta) = \varrho_1(\theta), I(\theta) = \varrho_2(\theta), P(\theta) = \varrho_3(\theta), \theta \in [-\tau, 0], \tag{4}$$

where $\varrho = (\varrho_1, \varrho_2, \varrho_3) \in \mathbb{R}^3_+$ with $\varrho_1(\theta) \geq 0$, $\varrho_2(\theta) \geq 0$, $\varrho_3(\theta) \geq 0$, $\theta \in [-\tau, 0]$ and $\varrho_1(0) > 0$, $\varrho_2(0) > 0$, $\varrho_3(0) > 0$.

2.1. *Positivity and boundedness of the solutions*

We now show that solutions of system (3) are positive for all $t > 0$. From the first equation of (3), we have

$$S(t) = S_0 e^{\int_0^t \left\{ r\left[1 - \frac{S(\nu_1) + I(\nu_1)}{K}\right] - \lambda I(\nu_1) \right\} d\nu_1}. \tag{5}$$

It implies that $S(t) > 0$ for all $t > 0$ whenever $S(0) > 0$. Again, the second equation of (3) gives

$$\begin{aligned} \dot{I}(t) &\geq \lambda S(t - \tau_1) I(t - \tau_1) - \mu I \\ \Rightarrow I(t) &\geq I(0)\, e^{-\mu t} + \int_0^t \lambda S(\rho - \tau_1) I(\rho - \tau_1) e^{-\mu(t-\rho)} d\rho. \end{aligned} \tag{6}$$

Third equation of (3) yields

$$P(t) = P(0)e^{-dt} + \int_0^t \frac{m\alpha I(\rho-\tau_2)P(\rho-\tau_2)}{a+I(\rho-\tau_2)} e^{-d(t-\rho)} d\rho. \tag{7}$$

Using (4), we have $I(t) \geq 0$ and $P(t) \geq 0$ for all $t \geq 0$. Hence all solutions of (3) are positively invariant.

We now show that solutions of system (3) are ultimately bounded. The first equation of (3) gives

$$\dot{S} = rS\left(1 - \frac{S+I}{K}\right) - \lambda SI \leq rS\left(1 - \frac{S}{K}\right). \tag{8}$$

This gives $S(t) \leq \frac{\hat{A}K}{\hat{A}-e^{-rt}}$, where $\hat{A}$ is a constant and can be determined from the initial condition. Therefore,

$$\lim_{t->\infty} S(t) \leq K. \tag{9}$$

Hence $S(t)$ is bounded and defined on $[0, \infty)$.

Define a function

$$V(t) = \alpha S(t-\tau_1) + \alpha I(t) + P(t+\tau_2). \tag{10}$$

The derivative of V along the solutions of (3) is given by

$$\begin{aligned}\dot{V} &= r\alpha S(t-\tau_1)\left(1 - \frac{S(t-\tau_1)+I(t-\tau_1)}{K}\right) - \alpha\lambda S(t-\tau_1)I(t-\tau_1)\\ &\quad + \alpha\lambda S(t-\tau_1)I(t-\tau_1) - \frac{m\alpha IP}{a+I} - \alpha\mu I + \frac{m\alpha IP}{a+I} - dP(t+\tau_2)\\ &\leq 2r\alpha S(t-\tau_1) - \left[r\alpha S(t-\tau_1) + \mu\alpha I(t) + dP(t+\tau_2)\right],\end{aligned}$$

Assuming $\gamma = \min\{r, \mu, d\}$ and from (9), we obtain

$$\dot{V} + \gamma V \leq 2r\alpha K.$$

Hence, $\lim_{t->\infty} V(t) \leq \frac{2r\alpha K}{\gamma}$. So $V(t)$ is bounded for all $t \geq 0$. Evidently, $I(t)$ and $P(t)$ are also bounded. Hence all solutions of system (3) are bounded for $t \geq 0$ in the region Γ_L defined by

$$\Gamma_L = \{(S, I, P) \in X \times R_+^3 \times X :\ 0 \leq S(t) \leq K,\ 0 \leq I(t) \leq \tfrac{2rK}{\gamma},\ 0 \leq P(t) \leq \tfrac{2r\alpha K}{\gamma}\}.$$

Moreover, the region Γ_L is positively invariant with respect to system (3) and hence the model is well posed. Summarizing we obtain the following proposition.

Proposition 2.1. *All solutions of system (3) are positively invariant and uniformly bounded in* Γ_L*, where*

$$\Gamma_L = \left\{ (S, I, P) \in \mathbb{R}^3_+ \,|\, 0 \leq S(t) \leq K,\ 0 \leq I(t) \leq \frac{2rK}{\gamma},\ 0 \leq P(t) \leq \frac{2r\alpha K}{\gamma} \right\}$$

and $\gamma = \min\{r, \mu, d\}$.

3. Equilibria and permanence of the system

The model system (3) has four equilibria.

(i) The trivial equilibrium $E_0(0,0,0)$ always exists.
(ii) The infection-and predator-free equilibrium point $E_1 = (K, 0, 0)$ also always exists.
(iii) The predator-free equilibrium $E_2 = (\hat{S}, \hat{I}, 0)$, where $\hat{S} = \frac{\mu}{\lambda}$, $\hat{I} = \frac{r(\lambda K - \mu)}{\lambda(r + \lambda K)}$, exists if $\lambda > \frac{\mu}{K}$, i.e., when the basic reproductive number $R_0 > 1$, where $R_0 = \frac{\lambda K}{\mu}$.
(iv) The coexistence or interior equilibrium $E^* = (S^*, I^*, P^*)$, where $S^* = K - \frac{ad(r+\lambda K)}{r(m\alpha - d)}$, $I^* = \frac{ad}{m\alpha - d}$, $P^* = \frac{(a+I^*)(\lambda S^* - \mu)}{m}$, exists if

$$(i)\ \lambda > \frac{\mu}{K}, \quad (ii)\ m > \frac{d}{\alpha} + \frac{ad\lambda(r + \lambda K)}{r\alpha(\lambda K - \mu)}. \tag{11}$$

One can clearly see that existence of E^* assures the existence of E_2, but not the vice versa. Conversely, if E_2 does not exist then E^* also does not exist. We, therefore, always assume that the predator-free equilibrium E_2 exists. In this study, we are interested to observe the dynamics of E^*, where all populations coexists.

3.1. *Permanence of the system*

We have shown that solution of system (3) exists and is bounded in Γ_L, implying that the system is dissipative. We now show that system (3) is permanent if $\lambda > \frac{\mu}{K}$ and $m > \frac{d}{\alpha} + \frac{ad\lambda(r+\lambda K)}{r\alpha(\lambda K - \mu)}$. First we present some definitions due to Hale and Waltman [18].

Definition 3.1. System (3) is said to be uniformly persistent if there exists $\vartheta > 0$ (independent of initial conditions) such that every solution $(S(t), I(t), P(t))$ with initial condition (4) satisfies

$$\lim_{t->\infty} \inf S(t) \geq \vartheta, \quad \lim_{t->\infty} \inf I(t) \geq \vartheta, \quad \lim_{t->\infty} \inf P(t) \geq \vartheta.$$

Definition 3.2. System (3) is said to be permanent if there exists a compact region $\Gamma_0 \in \text{int } \Gamma_L$ such that every solution of system (3) with initial condition (4) will eventually enter and remain in Γ_0.

Theorem 3.1. *The system (3) is permanent whenever E^* exists.*

Let $X = \mathcal{C}([-\tau, 0], \mathbb{R}^3_+)$ be the Banach space of continuous functions mapping the interval $[-\tau, 0]$ into $\mathbb{R}^3_+$, equipped with supremum norm d. Let, $T(t)$, $t \geq 0$ be the family of solution operators corresponding to system (3). Then $T(t) : X \to X$, $t \geq 0$ is a C_0-semigroup on X, i.e., $T(0) = Identity$, $T(t+s) = T(t)T(s)$ for $t, s \geq 0$ and T is continuous. Define

$$X^0 = \Big\{(\nu_1, \nu_2, \nu_3) \in X : \nu_1(\theta) > 0, \; \nu_2(\theta) > 0, \; \nu_3(\theta) > 0, \theta \in [-\tau, 0], \\ \tau = \max\{\tau_1, \tau_2\}\Big\}$$

and

$$X_0 = X \setminus X^0 = \Big\{(\nu_1, \nu_2, \nu_3) \in X : \nu_1(\theta) > 0, \; \nu_2(\theta) > 0, \; \nu_3(\theta) = 0, \\ \theta \in [-\tau, 0], \; \tau = \max\{\tau_1, \tau_2\}\Big\}.$$

Therefore, by Proposition 2.1, the metric space X is the closure of the open set X^0 such that $X = X^0 \bigcup X_0$, where X_0 is the boundary of X^0 and $X^0 \subset X$, $X_0 \subset X$, $X^0 \cap X_0 = \phi$. First we show that X^0 is positively invariant for $T(t)$. From (3), we can write

$$\frac{dP}{dt} \geq - \, dP(t), \; \forall \, t \geq 0. \tag{12}$$

From (4), we have $P(0, \varrho) = \varrho_3(0) > 0$. Then it follows that

$$P(t, \varrho) \geq \varrho_3(0) \; e^{-dt} > 0, \; \forall \, t \geq 0. \tag{13}$$

Thus, X^0 is positively invariant. Moreover, the C_0-semigroup $T(t)$ on X satisfies

$$T(t) : \; X^0 \; \to \; X^0, \; T(t) : \; X_0 \to X_0. \tag{14}$$

Let $T_\rho(t) = T(t)|_{X_0}$ and define G_ρ as the global attractor for $T_\rho(t)$. We also define $\omega-$limit set as

$$\omega(\sigma) := \{\sigma_1 \in X \mid \text{there exists a sequence } t_n \to \infty \text{ as } n \to \infty \text{ with} \\ T(t_n)\sigma \to \sigma_1 \text{ as } n \to \infty\}.$$

Lemma 3.1. *Let $T(t)$ satisfies (14) and we have the followings:*

(i) *There exists a* $t_0 \geq 0$ *such that* $T(t)$ *is compact for* $t > t_0$.
(ii) $T(t)$ *is point dissipative in* X.
(iii) $\overline{G_\rho} = \cup_{z \in G_\rho} \omega(z)$ *is isolated and has an acyclic covering* $\overline{N}$. *Then* $\overline{N} = \cup_{i=1}^{k} N_i$, *where each* N_i *is pairwise disjoint, compact and isolated invariant set for* T_ρ *and it is also an invariant set for* T. *Also,* $\overline{N}$ *being an acyclic covering no subset of* N_i *forms a cycle.*
(iv) $W^s(N_i) \cap X^0 = \phi$, *for* $i = 1, 2, \ldots, n$, *where* W^s *is stable or attracting set of an compact invariant set and it is defined as* $W^s(N_i) = \{x_1 \mid x_1 \in X,\ \omega(x_1) \neq \phi,\ \omega(x_1) \subset N_i\}$.

Then, X_0 *is a uniform repeller with respect to* X^0. *Which implies that there exists a* $\delta > 0$ *such that for any* $z \in X^0$, $\liminf_{t \to \infty} d(T(t)z, X_0) \geq \delta$, *where* d *is the distance of* $T(t)z$ *from* X_0.

Proof. Using this lemma, we show that the boundary planes of $\mathbb{R}^3_+$ repel the positive solutions of (3) uniformly. Moreover, if $X^0 = \text{int } X$, it will be sufficient to show that there exists a $\delta_0 \geq 0$ such that for any solution u_t of system (3) starting from X^0, $\lim_{t \to \infty} \text{int } d(u_t, X^0) \geq \delta_0$. To do so, we verify that the conditions of Lemma 3.1 are satisfied. Note that the bound of solution of (3) does not depend on initial condition (4). Thus, for any bounded set D in X, the positive orbit $\Gamma^+(D) = \cup_{t>0} T(t)D$ through $D \subset X$ is bounded in X. Therefore, $T(t)$ is asymptotically smooth and for any nonempty bounded set $D \subset X$ for which $T(t)D \subset D$, there is a compact set $D_0 \subset D$ such that D_0 attracts D. Let us define, $\mathbb{A} = \cap_{\sigma \in G_\rho} \omega(\sigma)$, where G_ρ as the global attractor for $T_\rho(t)$ restricted to X_0, i.e., $T_\rho(t) = T(t)|_{X_0}$. We will show that $\mathbb{A} = \{E_2\}$. One can see that coexistence equilibrium E^* exists only if E_2 is feasible. Thus following the existence condition of E_2, we assume $\lambda > \frac{\mu}{K}$. Now E_2 is a non trivial solution in X_0, where $S(t) = \hat{S},\ I(t) = \hat{I},\ P(t) = 0$. If $(S(t), I(t), P(t))$ is a solution of (3) initiating from X_0 then $S(t) \to \hat{S},\ I(t) \to \hat{I}, P(t) \to 0$ as $t \to \infty$. It is obvious that E_2 is isolated invariant. Therefore, $\{E_2\}$ is a compact and isolated invariant set in X. Hence, the covering is simply $\{E_2\}$, which is acyclic, implying that there is no orbit that connects E_2 to itself in X_0. Further, from Proposition 2.1, $T(t)$ is point dissipative in X. Thus, first three conditions of Lemma 3.1 are satisfied.

On the contrary, we assume there exists a positive solution $(\tilde{S}, \tilde{I}, \tilde{P})$ of (3) such that $(\tilde{S}, \tilde{I}, \tilde{P}) \to (\hat{S}, \hat{I}, 0)$ as $t \to \infty$. We choose $\zeta > 0$ small enough and $t_0 > 0$ sufficiently large such that

$$\begin{cases} \hat{S} - \zeta < \tilde{S} < \hat{S} + \zeta \text{ and } \hat{I} - \zeta < \tilde{I} < \hat{I} + \zeta \text{ for } t_0 - \tau \le t < t_0, \\ \hat{S} - \zeta > \tilde{S} \text{ and } \hat{I} - \zeta > \tilde{I} \text{ for } t \ge t_0. \end{cases} \tag{15}$$

Then for $t \ge t_0$, we have

$$\begin{aligned} \frac{dI}{dt} &\ge \lambda(\hat{S} - \zeta)(\hat{I} - \zeta) - \mu(\hat{I} - \zeta), \\ \frac{dP}{dt} &= \frac{m\alpha(\hat{I} - \zeta)\tilde{P}}{a + (\hat{I} - \zeta)} - d\tilde{P}. \end{aligned} \tag{16}$$

We now define a matrix V_ζ such that

$$V_\zeta = \begin{pmatrix} \lambda(\hat{S} - \zeta) - \mu & 0 \\ \frac{am\alpha\tilde{P}}{[a+(\hat{I}-\zeta)]^2} & \frac{m\alpha(\hat{I}-\zeta)}{a+(\hat{I}-\zeta)} - d \end{pmatrix}.$$

According to Perron-Frobenious theorem, as V_ζ admits non negative off-diagonal elements, there exists a maximum eigenvalue and its corresponding positive vector. Now the eigenvalues of the characteristic equation corresponding to V_ζ are $v_1 = \lambda(\hat{S} - \zeta) - \mu$ and $v_2 = \frac{m\alpha(\hat{I}-\zeta)}{a+(\hat{I}-\zeta)} - d$. From (16), we obtain $\mu \ge \lambda(\hat{S} - \zeta)$. Upon substitution in v_1, one can clearly see that $v_1 \le 0$ when $t \ge t_0$. However, the other eigenvalue $v_2 = \frac{m\alpha(\hat{I}-\zeta)}{a+(\hat{I}-\zeta)} - d > 0$ if $m > \frac{d}{\alpha} + \frac{ad\lambda(r+\lambda K)}{r\alpha(\lambda K-\mu)}$ when $t \ge t_0$. Therefore, the characteristic equation associated with (16) will have a positive eigenvalue if $\lambda > \frac{\mu}{K}$ and $m > \frac{d}{\alpha} + \frac{ad\lambda(r+\lambda K)}{r\alpha(\lambda K-\mu)}$.

Now we consider the system

$$\begin{aligned} \frac{dI}{dt} &= \lambda(\hat{S} - \zeta)(\hat{I} - \zeta) - \mu(\hat{I} - \zeta), \\ \frac{dP}{dt} &= \frac{m\alpha(\hat{I} - \zeta)\tilde{P}}{a + (\hat{I} - \zeta)} - d\tilde{P}. \end{aligned} \tag{17}$$

Let $u = (u_1, u_2)$ be the eigenvector associated to the positive eigenvalue and $l > 0$ be small enough such that

$$lu_1 < \tilde{I}(t_0 + \varrho), \; lu_2 < \tilde{P}(t_0 + \varrho) \text{ for } \varrho \in [-\tau, 0].$$

Let $(I(t), P(t))$ be a solution of (17) satisfying $I(t) = lu_1$, $P(t) = lu_2$ for $t_0 - \tau \le t \le t_0$. Since the semiflow of (17) is monotone and $V_\zeta u > 0$, we get $I(t)$ and $P(t)$ are strictly increasing and $I(t) \to \infty$, $P(t) \to \infty$ as $t \to \infty$ [19, 20]. We have $\tilde{I} \ge I(t)$, $\tilde{P} \ge P(t)$ for $t > t_0$. Hence we obtain

$\tilde{I} \to \infty, \tilde{P} \to \infty$ as $t \to \infty$, but this contradicts that the solutions of the system (3) are bounded. Thus, from the well-posedness of the solutions of system (3), we get X_0 repels the positive solutions of system (3) uniformly. Hence from well-posedness and Lemma 3.1, one can say that system (3) is permanent provided $\lambda > \frac{\mu}{K}$ and $m > \frac{d}{\alpha} + \frac{ad\lambda(r+\lambda K)}{r\alpha(\lambda K-\mu)}$, i.e. when the coexistence equilibrium exists. □

4. Stability of the coexistence equilibrium

After linearizing about coexistence equilibrium $E^*(S^*, I^*, P^*)$, system (3) can be expressed in the following form:

$$\begin{aligned}\frac{dS}{dt} &= A_1S(t) + A_2I(t),\\ \frac{dI}{dt} &= A_3I(t) + A_4P(t) + B_1S(t-\tau_1) + B_2I(t-\tau_1),\\ \frac{dP}{dt} &= A_5P(t) + C_1I(t-\tau_2) + C_2P(t-\tau_2),\end{aligned} \tag{18}$$

where

$$A_1 = -\frac{rS^*}{K},\ A_2 = -\left(\frac{rS^*}{K} + \lambda S^*\right),\ A_3 = -\left(\mu + \frac{maP^*}{(a+I^*)^2}\right),\ A_4 = -\frac{d}{\alpha},$$
$$A_5 = -d, B_1 = \lambda I^*, B-2 = \lambda S^*,\ C_1 = \frac{ma\alpha P^*}{(a+I^*)^2},\ C_2 = d.$$

The corresponding characteristic equation is given by

$$\begin{aligned}&(\xi^3 + A\xi^2 + B\xi + C) + (D\xi^2 + E\xi + F)\ e^{-\xi\tau_1}\\ &\qquad + (G\xi^2 + H\xi + J)\ e^{-\xi\tau_2} + (L\xi + M)\ e^{-\xi(\tau_1+\tau_2)} = 0,\end{aligned} \tag{19}$$

where

$$\begin{aligned}&A = -(A_1 + A_3 + A_5), B = A_1A_3 + A_3A_5 + A_5A_1, C = -A_1A_3A_5,\\ &D = -B_2, E = B_2A_5 - A_2B_1 + A_1B_2, F = A_5(A_2B_1 - A_1B_2),\\ &G = -C_2, H = A_1C_2 - A_4C_1 + C_2A_3, J = A_1(A_4C_1 - C_2A_3),\\ &L = B_2C_2, M = C_2(A_2B_1 - A_1B_2).\end{aligned}$$

Case 1. $\tau_1 = 0,\ \tau_2 = 0$

Characteristic equation (19) in this case becomes

$$\xi^3 + (A + D + G)\xi^2 + (B + E + H + L)\xi + (C + F + J + M) = 0. \tag{20}$$

One can easily verify that the characteristic equation (20) is identical with that of [10] and [16]. We state the following theorem regarding the stability of E^* in absence of delays [10, 16].

Theorem 4.1. *If*

(i) $R_0 > 1$ *or* $\lambda > \frac{\mu}{K}$ *and*

(ii) $\underline{m} < m < \overline{m}$, *where* $\underline{m} = \max\left\{\frac{d\lambda K}{r\alpha}, \frac{d}{\alpha} + \frac{ad\lambda(r+\lambda K)}{r\alpha(\lambda K-\mu)}\right\}$ *and* $\overline{m} = \min\left\{\frac{\mu}{\alpha}, \frac{d}{\alpha} + \frac{2ad\lambda(r+\lambda K)}{r\alpha(\lambda K-\mu)}\right\}$,

system (3) is locally asymptotically stable around E^* *in absence of both delays and unstable otherwise.*

We now reproduce some definitions given in [21, 22].

Definition 4.1. The equilibrium E^* is called asymptotically stable if there exists a ϖ such that

$$\sup_{-\tau\leq\theta\leq 0} [|\psi_1(\theta) - S^*|, |\psi_2(\theta) - I^*|, |\psi_3(\theta) - P^*|] \leq \varpi$$

implies

$$\lim_{t\to\infty} (S(t), I(t), P(t)) = (S^*, I^*, P^*),$$

where $\tau = \max\{\tau_1, \tau_2\}$, $\tau > 0$ and $(S(t), I(t), P(t))$ is the solution of system (3) which satisfies the initial condition (4).

Definition 4.2. The equilibrium E^* is called absolutely stable if it is asymptotically stable for all delays $\tau_i \geq 0$ and conditionally stable if it is stable for $\tau_i, (i = 1, 2)$ in some finite interval.

To analyze the system in presence of delay, we always assume that the system is stable around E^* in absence of delay and therefore conditions of Theorem 4.1 are always satisfied.

Case 2. $\tau_1 > 0,\ \tau_2 = 0$

When $\tau_1 > 0$ and $\tau_2 = 0$, Eq. (19) takes the form

$$\begin{aligned} \xi^3 + [(A+G) + De^{-\xi\tau_1}]\xi^2 + [(B+H) + (E+L)e^{-\xi\tau_1}]\xi \\ + [(C+J) + (F+M)e^{-\xi\tau_1}] = 0. \end{aligned} \tag{21}$$

To determine whether delay causes instability, we have to find whether the characteristic equation (21) possesses purely imaginary roots of the form $\xi = i\omega_1$, $\omega_1 \in \mathbb{R}_+$. By putting $\xi = i\omega_1$ $(\omega_1 \in \mathbb{R}_+)$ in Eq. (21) and simplifying, we obtain

$$\begin{aligned} \omega_1^6 + [(A+G)^2 - 2(B+H) - D^2]\,\omega_1^4 + [(B+H)^2 \\ - 2(A+G)\,(C+J) - (E+L)^2]\,\omega_1^2 + (C+J)^2 = 0. \end{aligned} \tag{22}$$

Defining $\omega_1^2 = h$, Eq. (22) takes the form

$$\begin{aligned}\mathbb{F}(h) = h^3 + [(A+G)^2 - 2(B+H) - D^2]\, h^2 + [(B+H)^2 \\ - 2(A+G)\,(C+J) - (E+L)^2]\, h + (C+J)^2 = 0.\end{aligned} \tag{23}$$

As $(C+J)^2 > 0$, Eq. (23) will either have no positive root or two positive roots. If (23) has two positive roots say, $h_{+,1} > 0$ and $h_{+,2} > 0$, we will obtain two positive values of ω_1, given by $\omega_{1,1} = \sqrt{h_{+,1}}$ and $\omega_{1,2} = \sqrt{h_{+,2}}$. Eventually, Eq. (21) will have two pairs of purely imaginary roots and E^* will switch its stability twice through two Hopf bifurcations at $\tau_{1,j}^*$, $j = 1, 2$, where

$$\begin{aligned}\tau_{1,j}^* &= \frac{1}{\omega_{1,j}} \times \cos^{-1}\left(\frac{\Lambda_1}{\Lambda_2}\right), \\ \Lambda_1 &= [\omega_{1,j}^3 - \omega_{1,j}(B+H)](E+L)\omega_{1,j} - [(F+M) - D\omega_{1,j}^2] \\ &\quad [(A+G)\omega_{1,j}^2 - (C+J)], \\ \Lambda_2 &= [(F+M) - D\omega_{1,j}^2]^2 + (E+L)^2\omega_{1,j}^2.\end{aligned} \tag{24}$$

The transversality condition at $\xi = i\omega_{1,j}$, $\tau_1 = \tau_{1,j}^*$, $j = 1, 2$ is given by

$$\left[Re\left(\frac{d\xi}{d\tau_1}\right)\right]^{-1} = \frac{\Lambda_3}{\Lambda_2} = \frac{\mathbb{F}'(\omega_{1,j}^2)}{[(F+M) - D\omega_{1,j}^2]^2 + (E+L)\omega_{1,j}^2} \neq 0.$$

where $\Lambda_3 = 3\omega_{1,j}^4 + 2[(A+G)^2 - 2(B+H) - D^2]\omega_{1,j}^2 + [(B+H)^2 - 2(A+G)(C+J) - (E+L)^2]$. The sign of the transversality condition determines the direction of the Hopf bifurcation. If it is positive, Hopf bifurcation occurs in the forward direction (stability to instability) and it occurs in the backward direction (instability to stability) when transversality condition is negative. However, if Eq. (23) has no positive roots then Eq. (21) will have no purely imaginary root and E^* will remain stable for all $\tau_1 > 0$. We state the following theorem concerning the stability of E^* when $\tau_1 > 0$, $\tau_2 = 0$ [10].

Theorem 4.2. *Assume that $\lambda > \frac{\mu}{K}$ and $\underline{m} < m < \overline{m}$ are satisfied.*

(i) If Eq. (23) has no positive root then E^* will be absolutely stable for all $\tau_1 \geq 0$.
(ii) If Eq. (23) has two positive roots then E^* is locally asymptotically stable for $\tau_1 < \tau_{1,1}^*$ and $\tau_1 > \tau_{1,2}^*$. E^* is unstable for $\tau_1 \in (\tau_{1,1}^*, \tau_{1,2}^*)$. A Hopf bifurcation in the forward direction occurs at $\tau_1 = \tau_{1,1}^*$ and another Hopf bifurcation occurs in the backward direction at $\tau_1 = \tau_{1,2}^*$.

Case 3. $\tau_2 > 0,\ \tau_1 = 0$

When $\tau_2 > 0,\ \tau_1 = 0$, Eq. (19) takes the form

$$\begin{aligned} &\xi^3 + (A+D)\xi^2 + (B+E)\xi + (C+F) + [G\xi^2 + (H+L)\xi \\ &\quad + (J+M)]e^{-\xi\tau_2} = 0. \end{aligned} \tag{25}$$

To derive the conditions for delay induced stability switching of E^*, we put $\xi = \pm i\omega_2$, $\omega_2 \in \mathbb{R}_+$ in Eq. (25) and obtain

$$\begin{aligned} &\omega_2^6 + [(A+D)^2 - 2(B+E) - G^2]\,\omega_2^4 + [(B+E)^2 - 2(C+F)\,(A+D) \\ &\quad + 2G(J+M) - (H+L)^2]\,\omega_2^2 + [(C+F)^2 - (J+M)^2] = 0. \end{aligned}$$

Setting $\omega_2^2 = g$, we have

$$\begin{aligned} \mathbb{G}(g) = g^3 &+ [(A+D)^2 - 2(B+E) - G^2]\,g^2 + [(B+E)^2 \\ &- 2(C+F)\,(A+D) + 2G(J+M) - (H+L)^2]\,g \\ &+ [(C+F)^2 - (J+M)^2] = 0. \end{aligned} \tag{26}$$

Eq. (26) may have k positive roots, $k = 1,2,3$ or no positive root. For k positive roots namely $g = g_{+,k},\ k = 1,2,3$, we will get k positive values of ω_2 given by $\omega_{2,k} = \sqrt{g_{+,k}},\ k = 1,2,3$. Therefore, Eq. (25) will have k pairs of purely imaginary roots and E^* will switch its stability through Hopf bifurcations at $\tau_2 = \tau^*_{2,k},\ k = 1,2,3$, where

$$\begin{aligned} \tau^*_{2,k} &= \frac{1}{\omega_{2,k}} \times cos^{-1}\left(\frac{\Lambda_4}{\Lambda_5}\right), \\ \Lambda_4 &= [(H+L) - G(A+D)]\omega_{2,k}^4 + [G(C+F) + (J+M)(A+D) \\ &\quad - (B+E)(H+L)]\omega_{2,k}^2 - (C+F)(J+M), \\ \Lambda_5 &= [G\omega_{2,k}^2 - (J+M)]^2 + (H+L)^2\omega_{2,k}^2. \end{aligned} \tag{27}$$

The transversality condition at $\xi = i\omega_{2,k},\ \tau_2 = \tau^*_{2,k}, k = 1,2,3$ is given by

$$\left[Re\left(\frac{d\xi}{d\tau_2}\right)\right]^{-1} = \frac{\Lambda_6}{\Lambda_5} = \frac{\mathbb{G}'(\omega_{1,k}^2)}{[G\omega_{2,k}^2 - (J+M)]^2 + (H+L)^2\omega_{2,k}^2} \neq 0.$$

where, $\Lambda_6 = 3\omega_{2,k}^4 + 2[(A+D)^2 - 2(B+E) - G^2]\omega_{2,k}^2 + [(B+E)^2 - 2(C+F)\,(A+D) + 2G(J+M) - (H+L)^2]$. If Eq. (26) has no positive root then E^* will not change its stability. Since E^* was stable in absence of delay, it will remain in stable state for all $\tau_2 > 0$. We state the following theorem for the stability of E^* when $\tau_2 > 0,\ \tau_1 = 0$ [16].

Theorem 4.3. *Assume that* $\lambda > \frac{\mu}{K}$ *and* $\underline{m} < m < \overline{m}$ *are satisfied.*

(i) If Eq. (26) has no positive root then E^* will be absolutely stable for all $\tau_2 \geq 0$.

(ii) If Eq. (26) has k positive roots then E^* will switch its stability k times through Hopf bifurcation at each of $\tau_2 = \tau_{2,k}^*$, $k = 1, 2, 3$.

From previous analysis, one should notice that there may exist more than one critical values of τ_1(or τ_2) and more than one stable ranges of τ_1 (or τ_2) within which E^* is stable. To study the case where $\tau_1 > 0$, $\tau_2 > 0$, we first fix τ_1 from its stability range where $\tau_1 > 0$, $\tau_2 = 0$ and then find the stability range of τ_2 treating τ_1 as a constant. In a similar way, we can first fix τ_2 from any of its stability ranges where $\tau_2 > 0$, $\tau_1 = 0$ and then determine the stability range of τ_1 treating τ_2 as a constant.

Case 4. $\tau_2 > 0, \ \tau_1 > 0$

First we fix τ_1 from any of its stable range where E^* is conditionally stable for $\tau_1 > 0$, $\tau_2 = 0$ and treat τ_2 as a free parameter. Putting $\xi = i\omega$ in (19), we have

$$\begin{aligned}&(C - A\omega^2) + (F - D\omega^2)\ cos(\omega\tau_1) + E\omega\ \ sin(\omega\tau_1) + (J - G\omega^2)cos(\omega\tau_2)\\&\quad + H\omega\ sin(\omega\tau_2) + M\ cos[\omega(\tau_1 + \tau_2)] + L\omega\ sin[\omega(\tau_1 + \tau_2)] = 0,\\[1em]&(B\omega - \omega^3) + E\omega\ cos(\omega\tau_1) - (F - D\omega^2)\ sin(\omega\tau_1) + H\omega\ cos(\omega\tau_2)\\&\quad - (J - G\omega^2)\ sin(\omega\tau_2)L\omega\ cos[\omega(\tau_1 + \tau_2)] - M\ sin[\omega(\tau_1 + \tau_2)] = 0.\end{aligned} \tag{28}$$

(28) can be expressed as

$$\begin{aligned}\kappa_1\ cos(\omega\tau_2) + \kappa_2\ sin(\omega\tau_2) &= \kappa_3,\\ \kappa_2\ cos(\omega\tau_2) - \kappa_1\ sin(\omega\tau_2) &= \kappa_4.\end{aligned} \tag{29}$$

where

$$\begin{aligned}\kappa_1 &= (J - G\omega^2) + M\ cos(\omega\tau_1) + L\omega\ sin(\omega\tau_1),\\ \kappa_2 &= H\omega - M\ sin(\omega\tau_1) + L\omega\ cos(\omega\tau_1),\\ \kappa_3 &= (A\omega^2 - C) - (F - D\omega^2)\ cos(\omega\tau_1) - E\omega\ sin(\omega\tau_1),\\ \kappa_4 &= (\omega^3 - B\omega) - E\omega\ cos(\omega\tau_1) + (F - D\omega^2)sin(\omega\tau_1).\end{aligned}$$

Eliminating τ_2 from Eq. (29), one gets

$$\mathbb{J}(\omega) = \omega^6 + F_1\omega^4 + F_2\omega^2 + F_3 + 2F_4\ sin(\omega\tau_1) + 2F_5\ cos(\omega\tau_1) = 0, \tag{30}$$

where

$$F_1 = A^2 - 2B + D^2 - G^2,\ \ F_2 = B^2 + E^2 - H^2 - L^2 - 2AC - 2FD + 2JG, F_3 = (C^2 + F^2) - (J^2 + M^2), F_4 = HM\omega + (\omega^3 - B\omega)(F - D\omega^2) - E\omega(A\omega^2 - C) - L\omega(J - G\omega^2), F_5 = (B\omega - \omega^3)E\omega - M(J - G\omega^2) - HL\omega^2 - (A\omega^2 - C)\ (F - D\omega^2).$$

Clearly, if $F_3 + 2F_5 < 0$, i.e., if $(C + F)^2 < (J + M)^2$ then Eq. (30) will have at least one positive root $\widehat{\omega}$. For this $\widehat{\omega}$, (19) will have a pair of purely imaginary roots of the form $\xi = \pm i\widehat{\omega}$ and E^* will change its stability. A Hopf bifurcation will occur at $\tau_2 = \tau_2^*(\tau_1)$, where

$$\tau_2^*(\tau_1) = \frac{1}{\widehat{\omega}} \arccos\left(\frac{\kappa_1\kappa_3 + \kappa_2\kappa_4}{\kappa_1^2 + \kappa_2^2}\right), \tag{31}$$

where $\kappa_1, \kappa_2, \kappa_3, \kappa_4$ have to be calculated from (29) with $\omega = \widehat{\omega}$. The transversality condition for $\omega = \widehat{\omega},\ \tau_2 = \tau_2^*(\tau_1)$ is verified as

$$Re\left[\left(\frac{d\xi}{d\tau_2}\right)^{-1}\right] = \frac{\epsilon_9(\epsilon_1 + \epsilon_3 + \epsilon_5 + \epsilon_7) + \epsilon_{10}(\epsilon_2 + \epsilon_4 + \epsilon_6 + \epsilon_8)}{\epsilon_9^2 + \epsilon_{10}^2} \neq 0, \tag{32}$$

where

$$\begin{aligned}
\epsilon_1 &= B - 3\omega^2,\ \ \epsilon_2 = 2A\omega,\\
\epsilon_3 &= (D\omega^2\tau_1 - F\tau_1 + E)\ cos(\omega\tau_1) + (2D\omega - E\tau_1\omega)\ sin(\omega\tau_1),\\
\epsilon_4 &= (2D\omega - E\tau_1\omega)\ cos(\omega\tau_1) - (D\omega^2\tau_1 - F\tau_1 + E)\ sin(\omega\tau_1),\\
\epsilon_5 &= (H + G\omega^2\tau_2 - J\tau_2)\ cos(\omega\tau_2) - (2G\omega - H\omega\tau_2)\ sin(\omega\tau_2),\\
\epsilon_6 &= (2G\omega - H\omega\tau_2)\ cos(\omega\tau_2) - (H + G\omega^2\tau_2 - J\tau_2)\ sin(\omega\tau_2),\\
\epsilon_7 &= [L - M(\tau_1 + \tau_2)]\ cos[\omega(\tau_1 + \tau_2)] - L\omega(\tau_1 + \tau_2)\ sin[\omega(\tau_1 + \tau_2)],\\
\epsilon_8 &= [M(\tau_1 + \tau_2) - L]\ sin[\omega(\tau_1 + \tau_2)] - L\omega(\tau_1 + \tau_2)\ cos[\omega(\tau_1 + \tau_2)],\\
\epsilon_9 &= \omega(J - G\omega^2)\ sin(\omega\tau_2) - H\omega^2\ cos(\omega\tau_2) + M\omega\ sin[\omega(\tau_1 + \tau_2)]\\
&\quad - L\omega^2\ cos[\omega(\tau_1 + \tau_2)],\\
\epsilon_{10} &= \omega(J - G\omega^2)\ cos(\omega\tau_2) + H\omega^2\ sin(\omega\tau_2) + M\omega\ cos[\omega(\tau_1 + \tau_2)]\\
&\quad + L\omega^2\ sin[\omega(\tau_1 + \tau_2)].
\end{aligned}$$

Signs of the transversality condition indicate direction of the Hopf bifurcation. We summarize our results in the following theorem.

Theorem 4.4. *Assume that* $\lambda > \frac{\mu}{K}$, $\underline{m} < m < \overline{m}$ *and conditions of Theorem 4.2(ii) are satisfied. Corresponding to a positive root* $\omega = \widehat{\omega}_1$ *of Eq. (30), there exists a critical value* $\tau_2^*(\tau_1)$ *of* τ_2*, where* τ_1 *is taken from the interval where* E^* *is stable for* $\tau_1 > 0,\ \tau_2 = 0$*, such that* E^* *will change its stability at* $\tau_2 = \tau_2^*(\tau_1)$*. A Hopf bifurcation will occur at* $\tau_2 = \tau_2^*(\tau_1)$*.*

5. Direction and stability of Hopf bifurcation

In the previous section, we have analyzed the conditions of delay induced stability switching of the coexistence equilibrium E^*. It is shown that there may exist Hopf bifurcation for some critical values of the delay parameters. We now determine the direction and stability of the bifurcating periodic solutions with respect to τ_2 when τ_1 is in its stable range by using the normal form method and central manifold theorem presented in [23]. Following Theorem 4.2(ii), we assume that E^* is stable for $\tau_1 (\in [0, \tau_{1,1}^*)$ and a Hopf bifurcation occurs at $\tau_1 = \tau_{1,1}^*$ when $\tau_2 = 0$. So we take $\tau_{10}^* \in (0, \tau_{1,1}^*)$. For this value of τ_{10}^*, following Theorem 4.4, E^* undergoes a Hopf bifurcation at $\tau_2 = \tau_2^*(\tau_{10}^*)$. We denote $\tau_2^*(\tau_{10}^*) = \tau_{20}^*$.

Let $\tau_2 = \tau_{20}^* + \psi$, so that $\psi = 0$ is the Hopf bifurcation value of system (3). Define $U_1(t) = S(t) - S^*$, $U_2(t) = I(t) - I^*$, $U_3(t) = P(t) - P^*$. We rescale the time delay $t \longrightarrow \frac{t}{\tau_2}$, so that system (3) can be written as

$$\dot{U}(t) = L_\psi U_t + F(\psi, U_t), \tag{33}$$

where

$$U(t) = (U_1(t), U_2(t), U_3(t))^T \in \mathcal{C}([-\tau, 0], \mathbb{R}_+^3),$$

$$L_\psi \phi = (\tau_{20}^* + \psi)\left(\overline{L}\,\phi(0) + \overline{M}\,\phi\left(-\frac{\tau_{10}^*}{\tau_{20}^*}\right) + \overline{N}\,\phi(-1)\right), \tag{34}$$

$$F(\psi, \phi) = (\tau_{20}^* + \psi)\,(F_1, F_2, F_3)^T,$$

with

$$\phi(\theta) = (\phi_1(\theta), \phi_2(\theta), \phi_3(\theta))^T \in C([-1, 0], \mathbb{R}_+^3),$$

$$\overline{L} = \begin{pmatrix} A_1 & A_2 & 0 \\ 0 & A_3 & A_4 \\ 0 & 0 & A_5 \end{pmatrix}, \; \overline{M} = \begin{pmatrix} 0 & 0 & 0 \\ B_1 & B_2 & 0 \\ 0 & 0 & 0 \end{pmatrix}, \; \overline{N} = \begin{pmatrix} 0 & 0 & 0 \\ 0 & 0 & 0 \\ 0 & C_1 & C_2 \end{pmatrix}$$

and

$$
\begin{aligned}
&F_1 = a_{11}\phi_1^2(0) + a_{12}\phi_1(0)\phi_2(0),\\
&F_2 = b_{21}\phi_2^2(0) + b_{22}\phi_2^2(0)\phi_3(0) + b_{23}\phi_1\left(-\frac{\tau_{10}^*}{\tau_{20}^*}\right)\phi_2\left(-\frac{\tau_{10}^*}{\tau_{20}^*}\right)\\
&\quad + b_{24}\phi_2^3(0) + \cdots,\\
&F_3 = c_{31}\phi_2^2(-1) + c_{32}\phi_2^2(-1)\phi_3(-1) + c_{33}\phi_2^3(-1)\\
&\quad + c_{34}\phi_2^3(-1)\phi_3(-1) + \cdots,\\
&a_{11} = -\frac{r}{K},\quad a_{12} = -\left(\frac{r}{K} + \lambda\right),\\
&b_{21} = \frac{amP^*}{(a+I^*)^3},\quad b_{22} = \frac{am}{(a+I^*)^3},\quad b_{23} = \lambda,\quad b_{24} = -\frac{amP^*}{(a+I^*)^4},\\
&c_{31} = -\frac{am\alpha P^*}{(a+I^*)^3},\quad c_{32} = -\frac{am\alpha}{(a+I^*)^3},\quad c_{33} = \frac{am\alpha P^*}{(a+I^*)^4},\\
&c_{34} = \frac{am\alpha}{(a+I^*)^4}.
\end{aligned}
$$

Hence, according to Riesz representation theorem, there exists a 3×3 matrix function $\eta(\theta,\psi) : [-1,0] \longrightarrow \mathbb{R}_+^3$ whose elements are of bounded variation such that

$$
L_\psi\phi = \int_{-1}^{0} d\eta(\theta,\psi)\ \phi(\theta),\quad \phi \in C([-1,0],\mathbb{R}_+^3). \tag{35}
$$

In fact, we can choose

$$
\eta(\theta,\psi) = \begin{cases} (\tau_{20}^* + \psi)\ (\overline{L} + \overline{M} + \overline{N}), & \theta = 0,\\ (\tau_{20}^* + \psi)\ (\overline{M} + \overline{N}), & \theta \in \left[-\frac{\tau_{10}^*}{\tau_{20}^*}, 0\right)\\ (\tau_{20}^* + \psi)\ \overline{N}, & \theta \in \left(-1, -\frac{\tau_{10}^*}{\tau_{20}^*}\right),\\ 0, & \theta = -1. \end{cases} \tag{36}
$$

Again, for $\phi \in C([-1,0],\mathbb{R}_+^3)$, we define

$$
A(\psi)\phi = \begin{cases} \frac{d\phi(\theta)}{d\theta}, & -1 \le \theta < 0,\\ \int_{-1}^{0} d\eta(\theta,\psi)\ \phi(\theta), & \theta = 0, \end{cases}
$$

and (37)

$$
R(\psi)\phi = \begin{cases} 0, & -1 \le \theta < 0,\\ F(\psi,\phi), & \theta = 0. \end{cases}
$$

Using these definitions, system (33) is transformed to the following operator equations

$$\dot{U} = A(\psi)U_t + R(\psi)U_t \ , \tag{38}$$

where $U_t = U(t+\theta) = (U_1(t+\theta), U_2(t+\theta), U_3(t+\theta))$ for $\theta \in [-1, 0]$. Define $\mathbb{R}^3_* = \{$The three dimensional space of row vectors$\}$. Then, for $\varphi \in C^1([0,1], \mathbb{R}^3_*)$, we define the adjoint operator A^* of A as

$$A^*(\varphi) = \begin{cases} -\frac{d\varphi(s)}{ds}, & 0 < s \leq 1, \\ \int_{-1}^{0} d\eta^T(s,0)\ \varphi(-s), & s = 0, \end{cases} \tag{39}$$

and a bilinear inner product

$$\langle \varphi(s), \phi(\theta) \rangle = \bar{\varphi}(0)\phi(0) - \int_{\theta=-1}^{0} \int_{\nu=0}^{\theta} \bar{\varphi}(\nu - \theta)\ d\eta(\theta)\ \phi(\nu)\ d\nu, \tag{40}$$

where $\eta(\theta) = \eta(\theta, 0)$. Hence, $A(0)$ and $A^*(0)$ are adjoint operators. Note that $\pm i\omega_2^* \tau_{20}^*$ are eigenvalues of both $A(0)$ and $A^*(0)$.

Let $q(\theta) = (1, q_2, q_3)^T\ e^{i\omega_2^* \tau_{20}^* \theta}$ be the eigenvector of $A(0)$ associated with the eigenvalue $i\omega_2^* \tau_{20}^*$ and $q^*(s) = \Delta(1, q_2^*, q_3^*) e^{i\omega_2^* \tau_{20}^* s}$ be the eigenvector of $A^*(0)$ associated with eigenvalue $-i\omega_2^* \tau_{20}^*$. Using the definition of A and A^* and from (34) & (35), we obtain

$$\begin{aligned} q_2 &= \frac{A_2}{i\omega_2^* - A_1}, \quad q_3 = \frac{A_2 C_1 e^{-\omega_2^* \tau_{20}^*}}{(i\omega_2^* - A_5)\ (i\omega_2^* - A_1)}, \\ q_2^* &= -\frac{A_1 + i\omega_2^*}{B_1 e^{i\omega_2^* \tau_{10}^*}}, \quad q_3^* = -\frac{A_4(A_1 + i\omega_2^*)}{B_1 e^{i\omega_2^* \tau_{10}^*}(A_5 + C_2 e^{i\omega_2^* \tau_{20}^*} + i\omega_2^*)}. \end{aligned} \tag{41}$$

Hence, form (40), we get

$$\begin{aligned} \bar{\Delta} = [1 &+ q_2 \bar{q}_2^* + q_3 \bar{q}_3^* + \tau_{10}^* \bar{q}_2^* (B_1 + q_2 B_2) e^{-i\omega_2^* \tau_{10}^*} \\ &+ \tau_{20}^* \bar{q}_3^* (C_1 q_2 + c_2 q_3) e^{-i\omega_2^* \tau_{20}^*}]^{-1}, \end{aligned} \tag{42}$$

so that $\langle q^*, q \rangle = 1$ and $\langle q^*, \bar{q} \rangle = 0$ are satisfied. Following the algorithms given in [23] and computational procedures presented in [24], we deduce the following coefficients to determine the direction and stability of Hopf bifurcation:

$$g_{20} = 2\tau_{20}^*\bar{\Delta}\{a_{11} + q^{(2)}(0) + [q^{(2)}(0)]^2\bar{q}_2^* + \bar{q}_2^* b_{23} q^{(1)}(-\frac{\tau_{10}^*}{\tau_{20}^*})q^{(2)}(-\frac{\tau_{10}^*}{\tau_{20}^*})$$
$$+ \bar{q}_3^* c_{31}[q^{(2)}(-1)]^2,$$
$$g_{11} = \tau_{20}^*\bar{\Delta}\Big\{2a_{11} + \bar{q}^{(2)}(0) + q^{(2)}(0) + 2q^{(2)}(0)\bar{q}^{(2)}(0)\bar{q}_2^*$$
$$+ \bar{q}_2^* b_{23}[q^{(1)}(-\frac{\tau_{10}^*}{\tau_{20}^*})\bar{q}^{(2)}(-\frac{\tau_{10}^*}{\tau_{20}^*}) + \bar{q}^{(1)}(-\frac{\tau_{10}^*}{\tau_{20}^*})q^{(2)}(-\frac{\tau_{10}^*}{\tau_{20}^*})]$$
$$+ 2c_{31}\bar{q}_3^* q^{(2)}(-1)\bar{q}^{(2)}(-1)\Big\},$$
$$g_{02} = 2\tau_{20}^*\bar{\Delta}\{a_{11} + \bar{q}^{(2)}(0) + [\bar{q}^{(2)}(0)]^2\bar{q}_2^* + \bar{q}_2^* b_{23}\bar{q}^{(1)}(-\frac{\tau_{10}^*}{\tau_{20}^*})\bar{q}^{(2)}(-\frac{\tau_{10}^*}{\tau_{20}^*})$$
$$+ \bar{q}_3^* c_{31}[\bar{q}^{(2)}(-1)]^2\},$$
$$g_{21} = \tau_{20}^*\bar{\Delta}\Big\{a_{11}[4W_{11}^{(1)}(0) + 2W_{20}^{(1)}(0)] + W_{20}^{(1)}(0)\bar{q}^{(2)}(0) + 2W_{11}^{(1)}(0)q^{(2)}(0)$$
$$+ 2W_{11}^{(2)}(0) + W_{20}^{(2)}(0) + \bar{q}_2^*[4W_{11}^{(2)}(0)q^{(2)}(0) + 2W_{20}^{(2)}(0)\bar{q}^{(2)}(0)]$$
$$+ \bar{q}_2^* b_{22}[2\bar{q}^{(3)}(0)[q^{(2)}(0)]^2 + 4q^{(3)}(0)q^{(2)}(0)\bar{q}^{(2)}(0)]$$
$$+ \bar{q}_2^* b_{23}\Big[2q^{(1)}(-\frac{\tau_{10}^*}{\tau_{20}^*})W_{11}^{(2)}(-\frac{\tau_{10}^*}{\tau_{20}^*}) + \bar{q}^{(1)}(-\frac{\tau_{10}^*}{\tau_{20}^*})W_{20}^{(2)}(-\frac{\tau_{10}^*}{\tau_{20}^*})$$
$$+ W_{20}^{(1)}(-\frac{\tau_{10}^*}{\tau_{20}^*})\bar{q}^{(2)}(-\frac{\tau_{10}^*}{\tau_{20}^*}) + 2W_{11}^{(1)}(-\frac{\tau_{10}^*}{\tau_{20}^*})q^{(2)}(-\frac{\tau_{10}^*}{\tau_{20}^*})\Big]$$
$$+ 6\bar{q}_2^* b_{24}[q^{(2)}(0)]^2\bar{q}^{(2)}(0) + \bar{q}_3^* c_{31}\Big[4W_{11}^{(2)}(-1)q^{(2)}(-1)$$
$$+ 2\bar{q}^{(2)}(-1)W_{20}^{(2)}(-1)\Big] + \bar{q}_3^* c_{32}\Big[2\bar{q}^{(3)}(-1)[q^{(2)}(-1)]^2$$
$$+ 4q^{(2)}(-1)\bar{q}^{(2)}(-1)q^{(3)}(-1)\Big] + 6c_{33}\bar{q}_3^*[q^{(2)}(-1)]^2\bar{q}^{(2)}(-1)\Big\}, \quad (43)$$

where

$$W_{20}(\theta) = \frac{ig_{20}q(0)}{\omega_2^*\tau_{20}^*}e^{i\omega_2^*\tau_{20}^*\theta} + \frac{i\bar{g}_{02}\bar{q}(0)}{3\omega_2^*\tau_{20}^*}e^{-i\omega_2^*\tau_{20}^*\theta} + J_1 e^{2i\omega_2^*\tau_{20}^*\theta},$$
$$W_{11}(\theta) = -\frac{ig_{11}q(0)}{\omega_2^*\tau_{20}^*}e^{i\omega_2^*\tau_{20}^*\theta} + \frac{i\bar{g}_{11}\bar{q}(0)}{\omega_2^*\tau_{20}^*}e^{-i\omega_2^*\tau_{20}^*\theta} + J_2. \quad (44)$$

J_1 and J_2 will be computed using the following relations

$$\begin{pmatrix} 2i\omega_2^* - A_1 & -A_2 & 0 \\ -B_1 e^{-2i\omega_2^*\tau_{20}^*} & -A_3 - B_2 e^{-2i\omega_2^*\tau_{20}^*} + 2i\omega_2^* & -A_4 \\ 0 & -C_1 e^{-2i\omega_2^*\tau_{20}^*} & -A_5 - C_2 e^{-2i\omega_2^*\tau_{20}^*} \end{pmatrix} J_1$$

$$= 2 \begin{pmatrix} J_1^{(1)} \\ J_1^{(2)} \\ J_1^{(3)} \end{pmatrix} \tag{45}$$

and

$$\begin{pmatrix} A_1 & A_2 & 0 \\ B_1 & A_3 + B_2 & A_4 \\ 0 & C_1 & A_5 + C_2 \end{pmatrix} J_2 = - \begin{pmatrix} J_2^{(1)} \\ J_2^{(2)} \\ J_2^{(3)} \end{pmatrix} \tag{46}$$

with
$J_1^{(1)} = a_{11} + q^{(2)}(0)$, $J_1^{(2)} = [q^{(2)}(0)]^2 + b_{23} q^{(1)} \left(-\frac{\tau_{10}^*}{\tau_{20}^*}\right) q^{(2)} \left(-\frac{\tau_{10}^*}{\tau_{20}^*}\right)$, $J_1^{(3)} = c_{31}[q^{(2)}(-1)]^2$, $J_2^{(1)} = 2a_{11} + \bar{q}^{(2)}(0) + q^{(2)}(0)$, $J_2^{(2)} = 2q^{(2)}(0)\bar{q}^{(2)}(0) + b_{23} \left[q^{(1)} \left(-\frac{\tau_{10}^*}{\tau_{20}^*}\right) \bar{q}^{(2)} \left(-\frac{\tau_{10}^*}{\tau_{20}^*}\right) + \bar{q}^{(1)} \left(-\frac{\tau_{10}^*}{\tau_{20}^*}\right) q^{(2)} \left(-\frac{\tau_{10}^*}{\tau_{20}^*}\right) \right]$,
$J_2^{(3)} = 2c_{31} q^{(2)}(-1)\bar{q}^{(2)}(-1)$.

One can see that each g_{ij} in (43) is determined by the system parameters and delays. Therefore, we can compute the following quantities:

$$\begin{aligned} c_1(0) &= \frac{1}{2\omega_2^*\tau_{20}^*}\left(g_{11}g_{20} - 2|g_{11}|^2 - \frac{|g_{02}|^2}{3}\right) + \frac{g_{21}}{2}, \\ \psi_2 &= -\frac{Re\{c_1(0)\}}{Re\{\xi'(\tau_{20}^*)\}}, \\ \beta_2 &= 2Re\{c_1(0)\}, \\ T_2 &= -\frac{Img\{c_1(0)\} + \mu_2\, Img\{\xi'(\tau_{20}^*)\}}{\omega_2^*\tau_{20}^*}. \end{aligned} \tag{47}$$

We summarize these results in the following theorem.

Theorem 5.1. *For system (3),*

(i) ψ_2 *determines the direction of Hopf bifurcation. If* $\psi_2 > 0$ ($\psi_2 < 0$), *the Hopf bifurcation is supercritical (subcritical).*

(ii) β_2 *determines the stability of the bifurcating periodic solutions. If* $\beta_2 < 0$ ($\beta_2 > 0$), *the bifurcating periodic solutions are stable (unstable).*

(iii) T_2 *determines the period of the bifurcating periodic solutions. If* $T_2 > 0$ ($T_2 < 0$), *then the period of the bifurcating periodic solutions increases (decreases).*

6. Numerical simulation

To illustrate our theoretical results, we consider parameter values $r = 3,\ K = 40,\ a = 15,\ \alpha = 0.35,\ \mu = 0.28,\ m = 0.52,\ \lambda = 0.053,\ d = 0.09$ as in [10]. For these parameter values one can easily calculate that $\frac{\mu}{K} = 0.007$, $\underline{m} = 0.4468$ and $\overline{m} = 0.6364$. Clearly $\lambda(= 0.053) > \frac{\mu}{K}$ and $\underline{m} < m(= 0.52) < \overline{m}$ are satisfied. Therefore, following Theorem 4.1, the coexistence equilibrium E^* is locally asymptotically stable in absence of delays (Fig. 1).

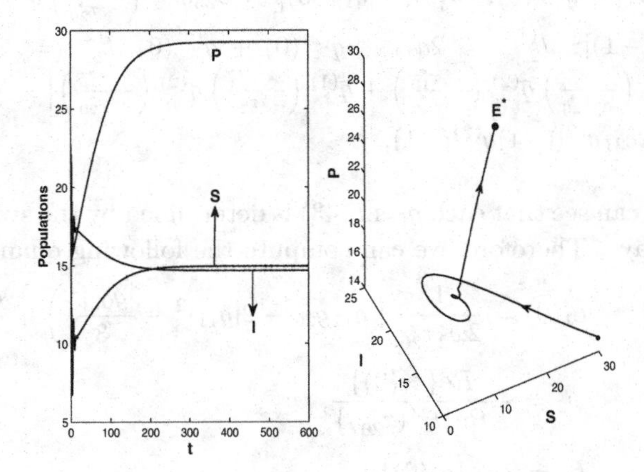

Figure 1. Left figure shows the time series evolution of system (3) in absence of delays and the right figure shows the corresponding phase plane. Parameters are $r = 3,\ K = 40,\ a = 15,\ \alpha = 0.35,\ \mu = 0.28,\ m = 0.52,\ \lambda = 0.053,\ d = 0.09$.

We now consider the case where $\tau_1 > 0$, $\tau_2 = 0$. Eq. (23) in this case has two positive roots namely, $h_{+,1} = 0.0365$ and $h_{+,2} = 0.0201$. Hence, we get two positive ω_1, viz. $\omega_{1,1} = \sqrt{h_{+,1}} = 0.1911$ and $\omega_{1,2} = \sqrt{h_{+,2}} = 0.1419$. Corresponding these values of ω_1, we get two critical values of τ_1 given by $\tau_{1,1}^* = 1.92$ and $\tau_{1,2}^* = 215.5267$. The transversality conditions are given by $\left[Re\left(\frac{d\xi}{d\tau_1}\right)\right]^{-1}_{\xi=i\omega_{1,1},\ \tau_1=\tau_{1,1}^*} = 1.1342 > 0$ and $\left[Re\left(\frac{d\xi}{d\tau_1}\right)\right]^{-1}_{\xi=i\omega_{1,2},\ \tau_1=\tau_{1,2}^*} = -2.0781 < 0$. Therefore, following Theorem 4.2, E^* is stable for $\tau_1 < 1.92$ and unstable for $\tau_1 > 1.92$. A Hopf bifurcation in forward direction occurs at $\tau_1 = 1.92$ (Fig. 2). Moreover, E^* again becomes stable for $\tau_1 > 215.5267$ through another Hopf bifurcation in the backward direction at $\tau_1 = 215.5267$. However, we ignore the later case as the critical value of $\tau_1 = 215.5267$ is biologically unrealistic.

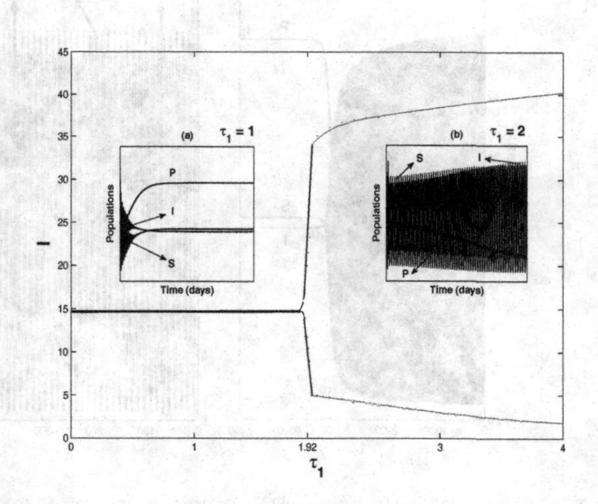

Figure 2. Bifurcation diagram of infected prey population (I) with τ_1 as the bifurcation parameter. This figure shows that the system is stable for $\tau_1 < 1.92$ and unstable for $\tau_1 > 1.92$. Inset figures show time series solutions of system (3) for some particular values of τ_1. Here $\tau_2 = 0$ and other parameters are as in Fig. 1.

To observe the effect of both delays, we fix $\tau_1 = 1.68$ from its stable range $(0, \tau_{1,1}^*)$ and treat τ_2 as a free parameter, where $\tau_{1,1}^* = 1.92$. From (31), we compute $(C+F)^2 - (J+M)^2 = -0.0485 < 0$. Therefore, one positive root of Eq. (30) is given by $\widehat{\omega}_1 = 0.0405$ and the corresponding τ_1−dependent critical value of τ_2 is given by $\tau_2^*(\tau_1) = 1.557$. The associated transversality condition is verified as $Re\left[\left(\frac{d\xi}{d\tau_2}\right)^{-1}\right]_{\omega=\widehat{\omega}_1,\ \tau_2=\tau_2^*(\tau_1)} =$ $0.0271(>0)$. Therefore, following Theorem 4.4, E^* is locally asymptotically stable for $\tau_2 < 1.557$, unstable for $\tau_2 > 1.557$ and a Hopf bifurcation in the forward direction occurs at $\tau_2 = 1.557$. From (47), we also calculate $\psi_2 = 47.5531(>0)$, $\beta_2 = -5.8211(<0)$ and $T_2 = 7.9276(>0)$. Following Theorem 5.1, the system undergoes a stable, supercritical Hopf bifurcation at E^* when τ_2 crosses $\tau_2^*(\tau_1)$. Also, the bifurcating periodic solution is stable and the amplitude of its period increases with τ_2, when τ_1 is fixed in a range where E^* is conditionally stable with $\tau_2 = 0$.

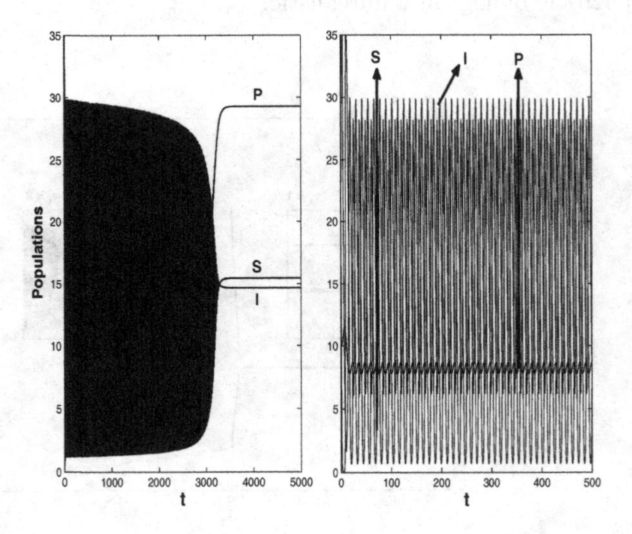

Figure 3. Left figure shows the time series evolution of system (3) with $\tau_2 = 1.4$ and the right figure shows the same with $\tau_2 = 1.6$. In both figures $\tau_1 = 1.68 \in (0, \tau_{1,1}^*)$ and other parameters are as in Fig. 1.

In Fig. 3, we have drawn the time series evolutions of system (3) for $\tau_2 = 1.4 < 1.557$ Fig. 3(left) and for $\tau_2 = 1.6 > 1.557$ Fig. 3(right), where $\tau_1 = 1.68 \in (0, \tau_{1,1}^*)$. These figures clearly show that E^* is locally asymptotically stable for $\tau_2 = 1.4$ and unstable for $\tau_2 = 1.6$. The bifurcation diagram (Fig. 4) also shows that the amplitude of the bifurcating periodic solutions increases as τ_2 increases.

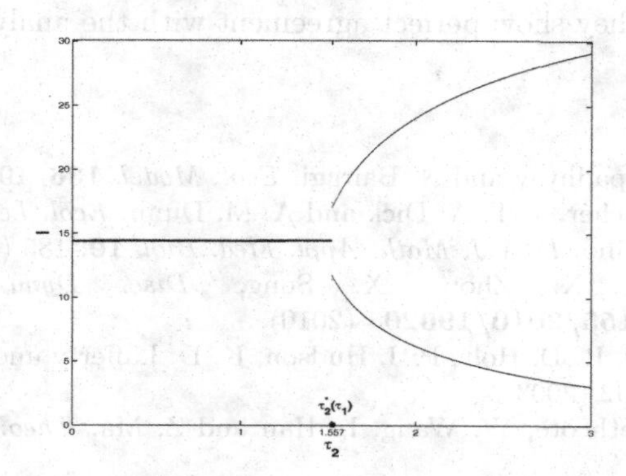

Figure 4. Bifurcation diagram of infected prey population (I) with τ_2 as the bifurcation parameter and $\tau_1(= 1.68)$ is fixed in the interval $(0, \tau_{1,1}^*)$, where $\tau_{1,1}^* = 1.92$. This figure shows that the system is stable for $\tau_2 < \tau_2^*(\tau_1)$ and unstable for $\tau_2 > \tau_2^*(\tau_1)$, where $\tau_2^*(\tau_1) = 1.557$. Other parameters are as in Fig. 1.

7. Discussion

In this paper we have analyzed an eco-epidemiological (or predator-prey-parasite) model where prey species is infected by some micro parasites. Predators consume only infected prey, following type-II response function, as they can be caught easily due to their lower fitness. Infection process is, however, not instantaneous but require some time to complete various intracellular processes, known as incubation delay. Reproduction of predator after consuming prey also requires times, known as reproduction delay. Therefore, to make the model biologically more realistic, we consider both the incubation delay and reproduction delay in the model system. For the coexistence equilibrium, we have derived conditions for linear stability and Hopf bifurcation in terms of system parameters for both single delay and multiple delays. We have also derived sufficient conditions on the parameters for which the delay-induced system is asymptotically stable around the coexistence equilibrium for all values of the delay parameter, i.e., the system is absolutely stable around the coexistence equilibrium. Under some other parametric conditions, there exists some critical value of the delay parameters below which the coexistence equilibrium is stable and above which it is unstable. Using normal form theory and the center manifold reduction, we have found the direction of Hopf bifurcation and derived the conditions for stability of bifurcating periodic solutions. Numerical simulations have been performed to illustrate different types of dynamics in the system and they show perfect agreement with the analytical findings.

References

1. J. Chattopadhyay and N. Bairagi, *Ecol. Model.* **136**, 103 (2001).
2. M. J. Hatcher, J. T. A. Dick and A. M. Dunn, *Ecol. Lett.* **9**, 1253 (2006).
3. E. Venturino, *IMA J. Math. Appl. Med. Biol.* **19**, 185 (2002).
4. X. Shi, X. Zhou, X. Song, *Disc. Dyna. Nat. and Soc.* **doi:10.1155/2010/196204** (2010).
5. C. Packer, R. D. Holt, P. J. Hudson, K. D. Lafferty and A. P. Dobson, *Eco. Lett.* **6**, 792 (2003).
6. H. W. Hethcote, W. Wang, L. Han and Z. Ma, *Theo. Popl. Biol.* **66**, 259 (2004).
7. N. Bairagi, P. K. Roy and J. Chattopadhyay, *J. Theo. Bio.* **248**, 10 (2007).
8. N. Bairagi and D. Adak, *Ecol. Complex.* **22**, 1 (2015).
9. D. Adak and N. Bairagi, *Chaos, Solitons & Frac.* **81**, 271–289 (2015).
10. N. Bairagi, R. R. Sarkar and J. Chattopadhyay, *Bull. of Math. Biol.* **70**, 2017 (2008).
11. J. Liu, *J. Appl. Math. Comput.* **50**, 557 (2016).
12. J. Wei and S. Ruan, *Physica D* **130**, 255 (1999).
13. X. Li, S. Ruan and J. Wei, *J. Math. Ana. Appl.* **236**, 254 (1999).
14. M. MacDonald, *Camb. Univ. Press Camb.* (1989).
15. C. Sun, Y. Lin and M. Han, *Chaos Solitons. & Frac.* **30**, 204 (2006).
16. N. Bairagi, *Int. J. Diff. Eq.* DOI: 10.1155/2011/978387 (2011).
17. J. K. Hale and S. M. Verduyn Lunel, *Springer Verlag* (1993).
18. J. K. Hale and P. Waltman, *SIAM J. Math. Anal.* **20**, 388 (1989).
19. H. Smith, *J. Diff. Eqns.* **66**, 420 (1987).
20. K. Gopalsamy, *Kluwer Academic, Dordrecht/Norwell, MA* (1992).
21. F. Brauer, *J. Diff. Eq.* **69**, 185 (1987).
22. Y. Kuang, *Academic Press, Boston, Mass. USA* **191** (1993).
23. B. D. Hassard, N. D. Kazarinoff and Y. H. Wan, *Cambridge Univ. Press, Cambridge, UK* (1981).
24. M. Ferrara, L. Guerrini and C. Bianca, *Appl. Math. & Infor. Sci.* **7**, 21 (2013).

PROPAGATION OF EXTRINSIC PERTURBATION IN MULTISTEP BIOCHEMICAL PATHWAYS

AATIRA G. NEDUNGADI
E-mail: aatira.nedungadi@gmail.com

SOMDATTA SINHA
Department of Biological Sciences
Indian Institute of Science Education and Research Mohali, India
E-mail: ssinha@iisermohali.ac.in

Biochemical pathways that underlie all cellular processes are inter-connected chemical reactions forming intricate networks of functional and physical interactions between molecular species in the cell. Much of regulation in these pathways are through negative and positive feedback processes. Many of the steps in a multi-step pathway are subjected to different environmental milieu as they take place at different cellular compartments. It is thus important to study, how such an interacting dynamical system faithfully transmits signals in spite of different types of perturbations act at different steps of the pathway. To study this we consider a minimal model of a three-step biochemical pathway comprising of two feedback loops — one negative and one positive — in the form of common cellular control processes in cells. This activator-inhibitor pathway is known to exhibit different types of simple and complex dynamics (equilibrium, periodic, biorhythmic, chaos) on variations in parameters. We delineate the qualitative and quantitative changes in the pathway dynamics for constant (*bias*) and *random* external perturbations acting on the pathway steps locally, or globally to all steps. We show that random fluctuations merely cause quantitative variations in the concentrations of the different variables, but *bias* induces qualitative change in dynamics of the pathway. These perturbations also are transmitted differentially when applied at different steps of the pathway. Thus, the dynamic response of multi-step biochemical pathways to external perturbations depends on their stoichiometry, network topology and, most non-intuitively, on the pathway dynamics.

1. Introduction

A variety of chemical reactions are carried out in the cells of living organisms. These reactions are catalyzed by many enzymes that are present inside the cell. In these reactions a substrate is converted into a product, which can then become the substrate for another reaction. In this way a series of reaction take place to get the final end-product molecule that is useful for cellular processes. Such a series of reactions is called as a biochemical pathway. Cells sense, transmit intra and extracellular signals, and allow cross talk among different processes through these pathways in different parts of the cell. Biochemical pathways are mostly multistep pathways. These pathways are generally regulated by feedback mechanisms. It can be activatory and inhibitory feedback or feed forward processes. In a regulated, multistep biochemical pathway, reactions can take place in different parts of the cell (e.g., inside different organelles) at different environmental conditions. Existence of noise is ubiquitous in intra-cellular environment due to the presence of small number of different regulatory molecules. Therefore, different parts of the cell may experience different types and strengths of noise.

Noise is known to affect the function and performance of biochemical networks, both constructively and destructively, as has been shown for noise in protein expression[26]. Stochasticity in gene expression and their regulation is a widely studied subject[8 12 14 3 15]. The co-ordinated transcriptional micro-RNA mediated regulation is suggested to be a recurrent motif to enhance the robustness of gene regulation in mammalian genomes[15]. Theoretical studies suggest that cells are able to achieve a higher net fitness in a stochastic environment than under periodic oscillations[22]. Noise in the transcription of a regulatory protein can lead to increased cell-cell variability in the target gene output, resulting in prolonged bistable expression states[27]. Intrinsic noise can be utilized as an advantage[24], but when propagated through a genetic cascade can lead to inefficient translation rates[20]. Attenuation of noise in an input signal can also be achieved in case of fluctuation bounded cascades[21]. It was shown, using a model pathway, that cellular dynamics can exhibit both robust and non robust behavior in noisy environment[5]. It is known that fluctuations in local enzyme levels need not affect global function[7].

The study of various causes and effects of noise in signaling pathways promotes a probabilistic view of signal propagation and explains the importance of presence of noise[11]. Regulation in biochemical pathways involve multiple levels of positive and negative feedback loops in the form of complex chemical reactions. Thus, to understand the working of any regulated multi-step biochemical pathway, one needs to know the reaction kinetics of the participating molecular species, the connectivity patterns of the feedback processes, and how the system adapts to the various environmental changes. Therefore it is interesting and important to study the functional dynamics of the pathway under the influence of perturbation at the different steps. There are studies which suggest that negative feedback efficiently decreases system noise[20] [9], while noise suppressed by positive or negative feedback loops cannot be reduced without limitation even in the case of slow transcritption[16]. The positive feedback can be used to improve sensitivity without a compromise in the ability to buffer propagated noise[10]. It has been shown that the stability of a network is significantly enhanced by the allosteric enzyme regulation[25]. Theoretical studies show that cellular signaling networks, in some cases, can act as filters to suppress noise[18]. These studies show that stochasticity can influence the reaction pathway at different steps in a variety of ways. Thus, the effect of noise on the end product dynamics, in presence of different types of regulation, needs to be explored. In an earlier study, the case of a model reaction pathway with a single negative feedback loop was studied, and the effect of stochasticity and its transmission through different steps in the pathway studied[23]. It was shown that constant perturbation can induce qualitative changes whereas *random* perturbation can cause significant quantitative variations in the concentrations of the different variables in the pathway dynamics in a negatively auto regulated pathway[23].

Since both negative and positive feedback processes are common in biochemical pathways, the focus in this paper is to study the propagation of noise in a pathway having both positive as well as negative feedback loops, so as to determine the robustness or sensitivity of the pathway to environmental noise. A minimal "activator-inhibitor" model with one positive feedback loop and negative feedback loop is used for this study, which has been shown to exhibit a variety of dynamics (stable, oscillatory, chaotic), which are observed in cellular functions[4]. Like the negatively auto-regulated pathway[23], the transmission of noise through the "activator-inhibitor" network is studied. The influence of inherent dynamical state and the nonlinear reaction chemistry (i.e., stoichiometry of the reactions) on the transmission of noise at different steps in presence of both negative and positive feedback loop is analyzed. We have shown that the noise sensitivity of the multi-step pathway is dependent on how the opposite feedback processes interact in the pathway, nonlinearity involved in reactions, and the dynamical states of the pathway.

2. Models and Methods

The model biochemical pathway considered here is shown in Fig. 1. It is a three step reaction sequence where the substrate S1 is converted to S2, which is then converted to the product S3 through an enzyme E[4]. This linear chemical reaction is regulated through two feedback loops. S1 is inhibited by the end product S3 (a negative feedback loop). The S3-S1 interaction follows co-operative kinetics. There is also a positive feedback loop due to the autocatalytic production of S3 from S2, through the allosteric property of the enzyme E catalysing this step.

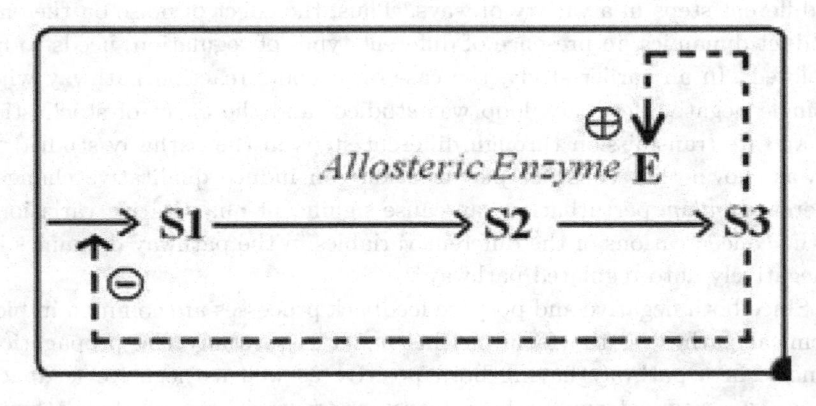

Figure 1. The minimal biochemical pathway: a three step reaction sequence with inhibition of S1 by end product S3 and activation of S2 through the allosteric enzyme E by S3.

The time evolution of this multistep biochemical pathway is given by[4]

$$\begin{aligned} \frac{dx}{dt} &= F(z) - kx, \\ \frac{dy}{dt} &= x - G(y,z), \\ \frac{dz}{dt} &= G(y,z) - qz, \end{aligned} \tag{1}$$

where

$$F(z) = \frac{1}{1+z^n},$$

$$G(y,z) = \frac{Ty(1+y)(1+z)^2}{L+(1+y)^2(1+z)^2}.$$

Here x, y and z are the normalized concentrations of the substrates S1, S2 and S3. The parameters k and q are the rates of degradation of S1 and S3. The functions $F(z)$ and $G(y,z)$ represent the negative and positive feedback terms respectively where the parameters T and L are maximum velocity and allosteric constant of the enzyme E[4]. To study the effect of perturbations acting at each step locally and globally to all steps of the pathway, the following system of equations are modelled[23]:

$$\frac{dx}{dt} = F(z) - kx + L1,$$
$$\frac{dy}{dt} = x - G(y,z) + L2, \tag{2}$$
$$\frac{dz}{dt} = G(y,z) - qz + L3,$$

where $L1$, $L2$ and $L3$ are the *local perturbations* to the evolution equations of x, y and z respectively. For *global perturbation*, we apply $L = L1 = L2 = L3$ to all the steps of the pathway.

There are two types of perturbations considered — *bias* and *random perturbation.* Constant environmental perturbations, such as, the presence of inducer, chemotactic agent, temperature, pH, or morphogen that can act at particular steps of the pathway or may also have global effects, is modelled using *bias. Bias* is modelled by adding a constant amount at each time point, whereas the random variations of various origins, as observed in real systems, are modelled using *random perturbation* with zero mean and uniform distribution. The strength of perturbations varied between 10-100% of the steady state value of the particular variable. For example, *random* perturbations of strength 20% to the variable x is given by

$$L1 = x_s \pm 0.2x_s \tag{3}$$

where x_s is the steady state (stable or unstable) value of x. Similarly, we apply *random* perturbation to y and z. The strength of perturbation is arbitrary in case of global perturbation. Since the steady state values of the three variables can be orders of magnitude different, this method of perturbation keeps a parity among perturbations applied to different steps vis-a-vis the concentration profile shown by the model pathway. The *global* perturbations, on the other hand, are applied at arbitrary strengths.

Three dynamical states of the pathway were studied for *bias* analyses: (1) Equilibrium state; (2) Oscillatory state; and (3) Chaotic state. Equilibrium and oscillatory pathways were also studied for *random* perturbations.

In case of *random* perturbations, the relative change in the variables at steady state was quantified using coefficient of variation given by

$$CV = \frac{standard - deviation}{mean} \times 100. \tag{4}$$

The basal parameters are kept at $n = 4$, $L = 10^6$, and $T = 10^4$. Numerical simulations were carried out in XPPAUT (www.math.pitt.edu/$\tilde{}$bard/xpp/xpp.html).

3. Results and Discussion

The dynamics of the pathway, shown in Fig. 1 and modelled by Eq. (2), is depicted in Fig. 2 by the bifurcation diagram at $q = 0.1$, for increasing k, which is the parameter controlling the rates of degradation of S1. Other parameters are kept at their basal values.

It can be observed that for small values of k (around 10^{-5}) the system dynamics is in equilibrium. The pathway also returns to equilibrium for large value of k. But for a large range of intermediate values of k, the pathway shows a variety of dynamics — oscillatory dynamics of small and large amplitudes, birhythmicity, complex oscillations,and chaotic behaviour[4]. During birhythmic dynamics, the pathway shows coexistence two different attractors for the same parameter values for different initial conditions. We now enumerate the analysis of the pathway when perturbed by local and global perturbations.

3.1. *Stability analysis of the pathway with bias*

We considered the system Eq. (2) with $n = 4$ for stability analysis. Equating the time derivatives to zero, we obtained the fixed points as

$$\begin{aligned} x_0 &= qz_0 - L2 - L3, \\ y_0 &= \frac{(2qz_0 - T - 2L3)}{(qz_0 - T - L3)} \\ &\quad + \frac{\sqrt{(T-2qz_0+2L3)^2(1+z_0)^2+4(T+L3-qz_0)(L+(1+z_0)^2)(qz_0-L3)}}{2(qz_0-T-L3)(1+z_0)}, \\ &z_0, \end{aligned} \tag{5}$$

where, z_0 is the only positive root (as concentrations are always positive and real) of the polynomial

$$kqz^5 - (L1 + kL2 + kL3)z^4 + kqz - 1 - L1 - kL2 - kL3.$$

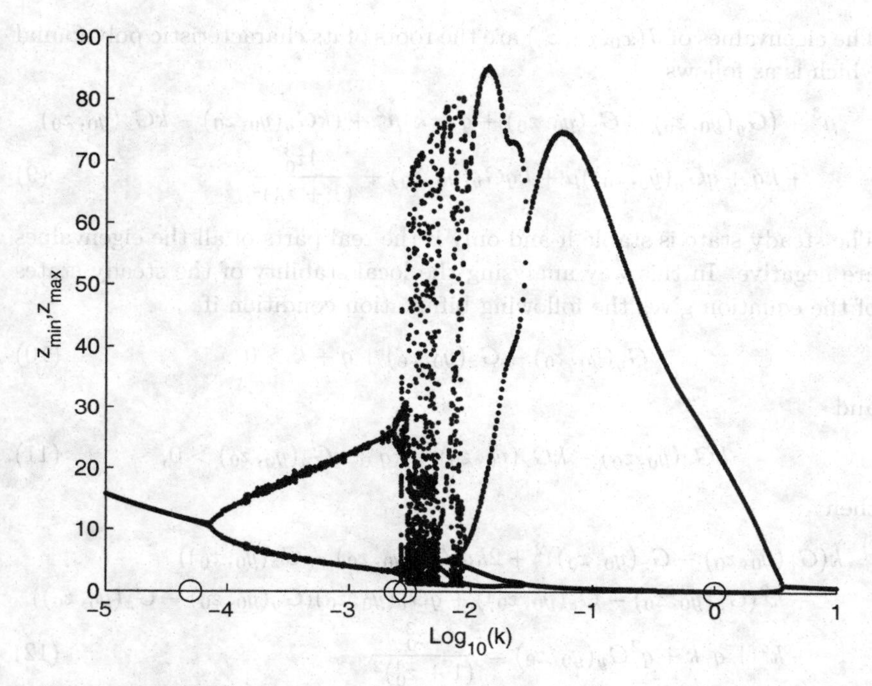

Figure 2. Bifurcation diagram of z for increasing k, showing the different dynamic regimes exhibited by the pathway. $\log(k)$ is plotted on the X-axis, and the maxima and minima of z are plotted on the Y-axis. The circles in the X-axis are parameters chosen for the perturbation studies.

The local stability of the fixed point (x_0, y_0, z_0) can be analysed by determining their respective eigenvalues and eigenvectors. Beginning with the Jacobian of the linearised system:

$$J(x_0, y_0, z_0) = \begin{pmatrix} -k & 0 & \frac{-4z_0^3}{(1+z_0^4)^2} \\ 1 & -G_y(y_0, z_0) & -G_z(y_0, z_0) \\ 0 & G_y(y_0, z_0) & G_z(y_0, z_0) - q \end{pmatrix} \tag{6}$$

where

$$G_y(y_0, z_0) = \frac{T(1+z_0)^2(1+2y_0)}{L+(1+y_0)^2(1+z_0)^2} - \frac{2Ty_0(1+y_0)^2(1+z_0)^4}{(L+(1+y_0)^2(1+z_0)^2)^2} \tag{7}$$

$$G_z(y_0, z_0) = \frac{2Ty_0(1+y_0)(1+z_0)}{L+(1+y_0)^2(1+z_0)^2} - \frac{2Ty_0(1+y_0)^3(1+z_0)^3}{(L+(1+y_0)^2(1+z_0)^2)^2} \tag{8}$$

The eigenvalues of $J(x_0, y_0, z_0)$ are the roots of its characteristic polynomial which is as follows:

$$\mu^3 + (G_y(y_0, z_0) - G_z(y_0, z_0) + q + k)\mu^2 + (kG_y(y_0, z_0) - kG_z(y_0, z_0) + kq + qG_y(y_0, z_0))\mu + kqG_y(y_0, z_0) + \frac{4z_0^3}{(1+z_0^4)^2} \quad (9)$$

The steady state is stable if and only if the real parts of all the eigenvalues are negative. In this way analysing the local stability of the steady states of the equation gives the following bifurcation condition if

$$G_y(y_0, z_0) - G_z(y_0, z_0) + q + k > 0 \quad (10)$$

and

$$kG_y(y_0, z_0) - kG_z(y_0, z_0) + kq + qG_y(y_0, z_0) > 0, \quad (11)$$

then

$$k(G_y(y_0, z_0) - G_z(y_0, z_0))^2 + 2kq(G_y(y_0, z_0) - G_z(y_0, z_0)) + k^2(G_y(y_0, z_0) - G_z(y_0, z_0)) + qG_y(y_0, z_0)(G_y(y_0, z_0) - G_z(y_0, z_0)) + k^q + q^2k + q^2G_y(y_0, z_0) = \frac{4z_0^3}{(1+z_0^4)^2} \quad (12)$$

Note in both expression (9) and (12), $G_y(y_0, z_0)$ and $G_z(y_0, z_0)$ is defined as in (6). If the L.H.S is greater than the R.H.S., then the corresponding fixed point is locally stable, else it is unstable.

3.2. *Dynamics of the pathway with bias*

The quantitative changes in the concentrations of the variables and their dynamics, in response to constant perturbation, are presented in this section. It may be noted that addition of *bias* to the system leads to change only in the steady state values (indicated by x_s, y_s, z_s), but the Hopf bifurcation remains the same. The dynamic regimes of the pathway at which *bias* have been applied are — (1) chaotic state ($k = 0.0029$), (2) oscillatory state ($k = 1$), and (3) equilibrium ($k = 0.00005$). These are indicated by circles in Fig. 2.

Pathway at chaotic state ($k = 0.0029$): The fixed point was obtained at $x_s = 0.5098$, $y_s = 37.5387$, $z_s = 5.0979$. The effect of positive and negative local *bias* ($L3$) at the third step of the pathway at chaotic state is depicted through examples of phase portraits of the chaotic attractor in Fig. 3. It can be observed that by addition of positive *bias* to z in the chaotic pathway (Fig. 3(d))leads to stability (Figs. 3(a)–(c)). But addition of negative *bias* leads to stable oscillatory states ((Figs. 3(e)–(g)). The dynamic response of the end-product of the pathway, z, for different strengths of positive *bias* at different steps of the reaction is summarised in Fig. 4. The X-axis shows the strength of positive perturbation and the Y-axis shows the dynamics of the system corresponding to local or global perturbations. In the figure, $L1$ depicts the effect of *bias* when the *bias* is applied only to the time evolution equation of x in Eq. (2). Similarly $L2$ and $L3$ represent the effect of *bias* when they are applied only to the time evolution equation of y and z respectively, in Eq. (2). The effect of global *bias* is represented by L. Black portion represents values of *bias* for which the pathway dynamics is stable, and grey represents unstable dynamics (chaotic, oscillatory).

It is clear from the figure that the chaotic pathway attains stability for both local and global *bias* at all steps, but the change in stability (bifurcations) occur for $L1 = 0.002$, $L2 = 1.024$, $L3 = 0.156$ and $L = 0.002$. Clearly low values of perturbation to x leads to stability of the pathway, while much larger perturbation is required on y to stabilise the system. This is because positive perturbation effectively acts like decreasing the degradation rate of x. From Fig. 2, it is clear that for smaller values of k, the system is stable. Hence even small perturbations have large stabilising effect when applied on x. On the other hand, larger positive constant perturbations on y or z are required (because of the interaction between the positive and the negative feedback loops) to effectively result in inducing stability to the pathway. Hence y and z are more robust to *bias* in this case. Thus it is shown that a fixed amount of perturbation (*bias*) can change the dynamics of the biochemical pathway from chaotic to oscillatory and oscillatory to stable steady states, and different steps have differential sensitivity to perturbation for stabilisation of the pathway, with x being the most sensitive and y being most robust to the application of *bias*.

Pathway at oscillatory state ($k = 1$): Like in the case of chaotic state, addition of positive *bias* stabilises the pathway. Figure 5 depicts the consolidated description of the dynamics of the pathway for different strengths of positive *bias* at different steps of the reaction pathway. In contrast to the chaotic pathway, here the bifurcations occur at $L1 = 1.048$, $L2 = 1.048$, $L3 = 0.021$ and $L = 0.029$ — same values of local *bias* in the x and y equations. A much smaller amount of *bias* to z stabilises the pathway.

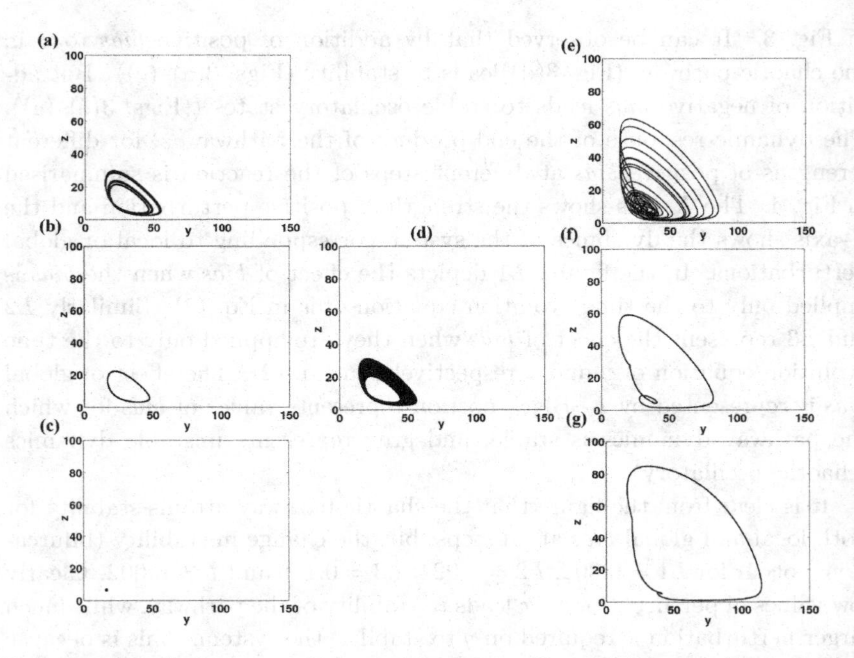

Figure 3. $(x - y)$ phase portrait for chaotic pathway ($k = 0.0029$) when *bias* is applied to the third variable in (Eq. (2), for: (a) $L3 = 0.005$, (b) $L3 = 0.05$, (c) $L3 = 0.5$, (d) $L3 = 0$, (e) $L3 = -0.001$, (f) $L3 = -0.01$, (g) $L3 = -0.1$

Thus, in this case of oscillatory pathway, x and y are more robust to fixed perturbation compared to that in z. This can be explained by the fact that larger positive *bias* to x is required to effectively reduce higher values of k (($k = 1$)), thus making x more robust.

Pathway at equilibrium state ($k = 0.00005$): Unlike the oscillatory pathways, the bifurcation condition here is not satisfied for positive *bias*, but negative *bias* leads to instability. Bifurcation occurs for $L1 = -0.00002$, $L2 = -0.5745$, $L3 = -0.0648$, and $L = -0.00002$. Figure 6 gives the summary result of constant negative perturbation strengths required to destabilise the pathway for local and global negative *bias*. As before it can be argued that addition of negative *bias* effectively acts to increase the k value, thereby pushing the dynamics to the left in Fig. 2. Hence a small magnitude of negative perturbation on x will destabilise the system. However, if small negative perturbations are applied on y or z, the resultant perturbation on x will be too small (because of the negative feedback loop) to destabilise

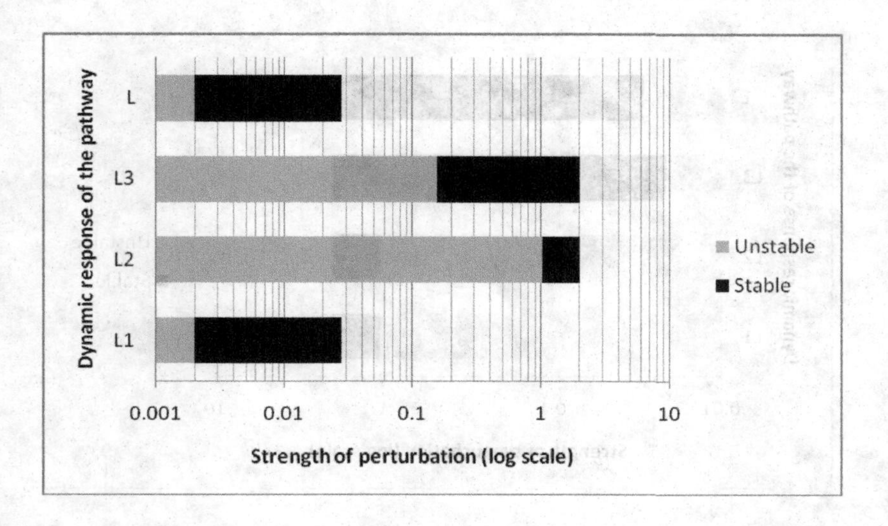

Figure 4. Pathway dynamics (chaotic at $k = 0.0029$) for different strengths of positive *bias* applied at different steps of the pathway. X-axis shows perturbation strength, and Y-axis shows the dynamics of the system for the particular steps (variable equations) perturbed. Black represents stable and grey represents unstable dynamics.

the system. Hence y and z are more robust to negative *bias* to induce instability when the pathway is at equilibrium.

Thus, it is shown that addition of constant perturbation (*bias*), locally or globally, can change the dynamics of the model biochemical pathway in a non-intuitive manner, as it depends on various factors, such as the step at which it is applied, presence of various regulatory feedback processes and their interactions, and the inherent dynamical state of the pathway.

3.3. *Effect of random perturbation*

In this section, the effects of *random* perturbation applied locally and globally to the multistep pathway is studied when the pathway is at two dynamic states — equilibrium and oscillatory. Interestingly, this system does not show any change in its dynamics under noise. Noise only induces quantitative variations in the concentrations of variables. Therefore, we used Coefficient of Variation (CV) in case of equilibrium state and Mean Square Deviation (MSD) in case of oscillatory state of the variables, for quantifying the changes under *random* perturbations.

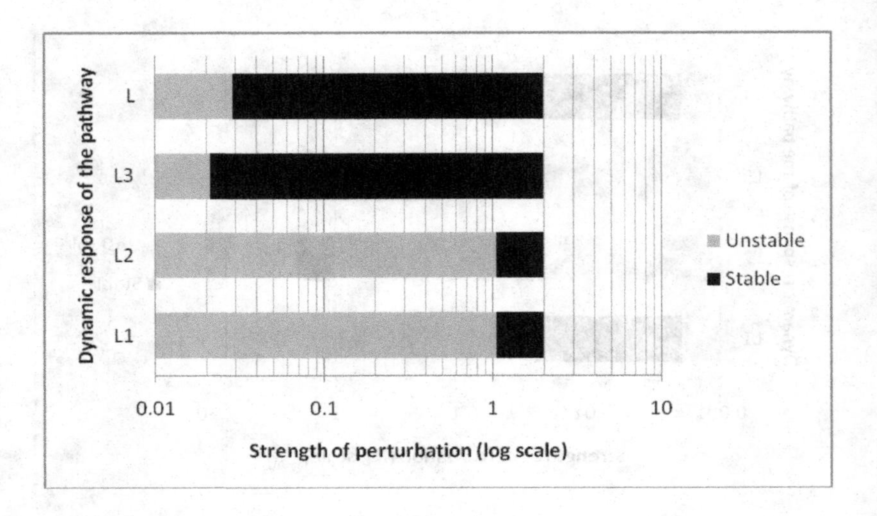

Figure 5. Pathway dynamics (oscillatory at $k = 1$) for different strengths of positive *bias* applied at different steps of the pathway. The X-axis shows positive perturbation strength, and the Y-axis shows the dynamics of the pathway for the variables perturbed. Black represents stable and grey represents unstable.

Pathway at equilibrium (k = 0.00005) Here the stable steady state was obtained at $x_s = 1.1487$, $y_s = 28.42027$, $z_s = 11.4869$. The local and global *random* perturbations was applied as discussed earlier. The CV for different strengths of perturbation, applied individually on x, y, z and globally on all variables simultaneously, were computed. The results are shown in the Fig. 7. The X-axis shows the percentage of perturbation (to the steady state), while the Y-axis shows the CV from the steady state value in logarithmic scale. It may be noted that even though the x-axis is same for Fig. 7(a)–(c), the actual values are quite different because of the difference in the numerical values of the steady states. As can be seen from Fig. 7, the effect of perturbation is not always the maximum for the variable on which the perturbation is applied. It is because in a regulated pathway with nonlinear processes, the random perturbation is not transmitted linearly always. The regulatory controls play an important role in suppressing or enhancing noise.

Fig. 7(a) shows that *random* perturbation on x causes similar variations in y and z. The noise from x is transmitted linearly to y, but from y it is amplified and transmitted to z due to the positive feedback. Therefore even though the perturbation is small as compared to the steady state value of

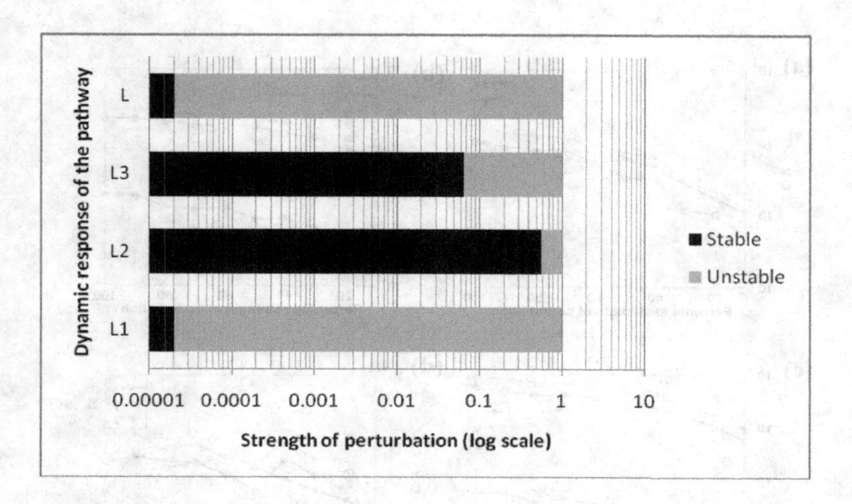

Figure 6. Pathway dynamics (equilibrium at $k = 0.00005$) for different strengths of negative *bias* applied at different steps of the pathway. The X-axis shows negative perturbation strength, and the Y-axis shows the dynamics of the pathway for the variables perturbed. Black represents stable and grey represents unstable pathway dynamics.

z, and it also not applied directly to z, still it causes more variations in z compared to the variable perturbed (i.e., x). Similarly, in Fig. 7(b), when y is perturbed, the variation in z is more than y due to the positive feedback. But when transmitted from z to x, the magnitude gets reduced considerably because of the negative feedback loop. Therefore the variation in x is very small as compared to y and z. On the same lines, the variation in z is maximum when z is perturbed as is seen in Fig. 7(c), and similar variations are obtained for y because of the positive feedback loop from z, and negative feedback loop causes smaller much variation in x. In Fig. 7(d), CV in x, y and z is shown when the steps of the pathway are subjected to global *random* perturbations. It can be observed that all three variables show variation of almost same order where z has the maximum variation which is expected. Thus in this case, x seems to be the least vulnerable for *random* perturbation unlike the *bias* case.

Pathway at oscillatory state ($k = 1$): The unstable steady state is $x_s = 0.1533$, $y_s = 48.7699$, $z_s = 1.533$. Figures 8(a)–(c) show the time evolution of x, y and z when unperturbed. Figures 8(d)–(f), 8(g)–(i), and 8(j)–(l) show that of x, y and z when *random* perturbation of strength 80% (of the respective steady states) are applied only on x, or y or z,

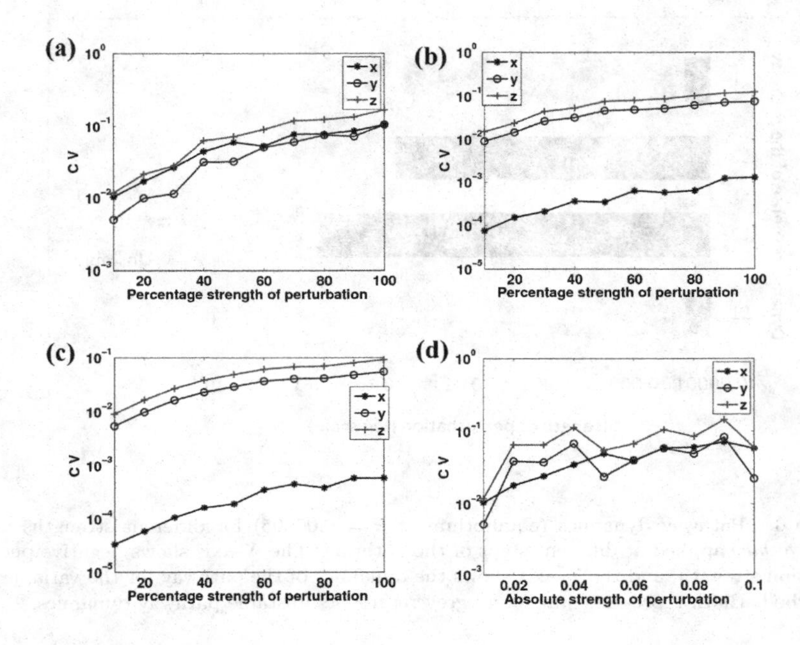

Figure 7. Coefficient of variation (CV) of the pathway variables for *random* perturbation on (a) x, (b) y, (c) z and (d) all variables simultaneously. X-axis shows percentage of strength of perturbation to the respective steady states, except for (d) where it is the absolute value. Y-axis shows the CV of $x(***)$, $y(ooo)$, $z(+++)$.

respectively.

In the unperturbed case, x, y and z oscillate with different amplitudes. Here x_max is 2.9×10^5 times the x_min, y_max is 6.8 times the y_min, and the z_max is 1.278×10^2 times the z_min. Thus a large amplitude variation in x will be suppressed to smaller amplitude variations in y and z, even though, due to the positive feedback between y and z, there is some increase in the amplitude of variation in z. It is also clear from the shape of the oscillations in Figs. 8(a)–(c) that x, during a period of oscillation, is either in the maximum phase or in the minimum phase mostly, whereas the rise and fall of y is gradual, and z remains mostly at low, except for a small fraction of time where it rises to its maxima and again falls down rapidly. This indicates that the *random* perturbation will affect x primarily at its maxima or minima, and z mostly at its minima. The effect of *random* perturbation on y, on the other hand, will be at all phases during the period. Figures 8(d)–(f) shows that even high perturbation in x have very

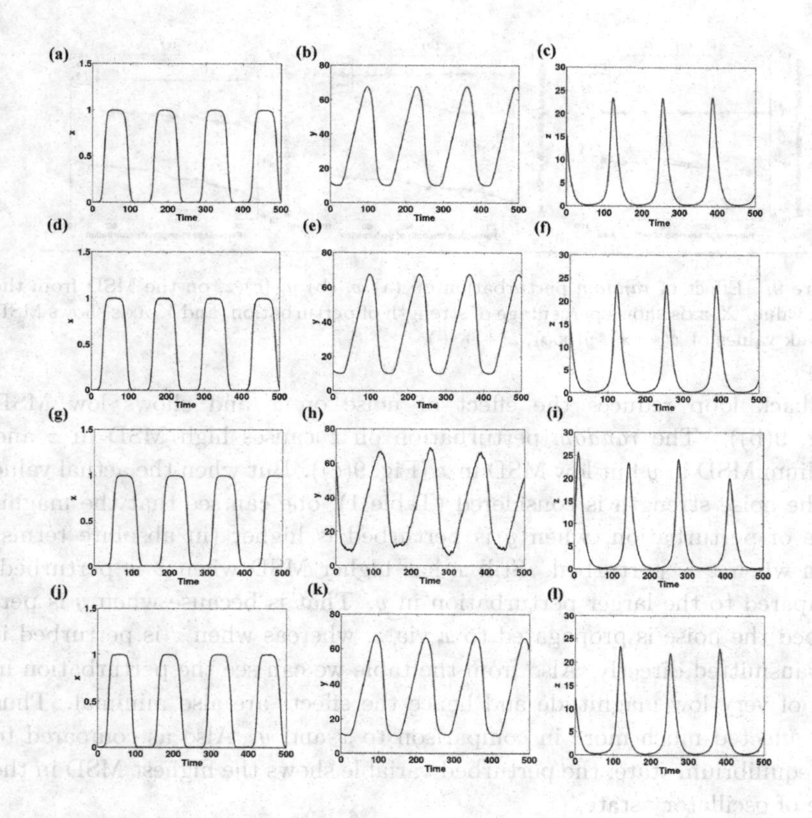

Figure 8. Time evolution of the three variables in the oscillatory pathway: (a) without *random* perturbation, and with local *random* perturbation on — (b) x, (c) y, (d) z, with noise strength 80% of the respective unstable steady state values.

negligible effect on y and z, primarily because the absolute value of noise is very small due to low x_s value. As expected, perturbation on y have the maximum effect both on y and z (Figs. 8(g)–(i)), and perturbation on z, have its maximum effect on its peak value, but suppressed in x and y due to the negative feedback (Figs. 8(j)–(l)).

Figure 9 shows the comparison of Mean Square Deviation (MSD) from the peak value between x, y and z when they are randomly perturbed individually. As expected and explained earlier, the effect on y and z is very low when x is perturbed (Fig. 9(a)) since even large noise on x gets suppressed due to the negative feedback. Similarly when y is perturbed, z also shows significant MSD as the noise gets amplified. But presence of negative

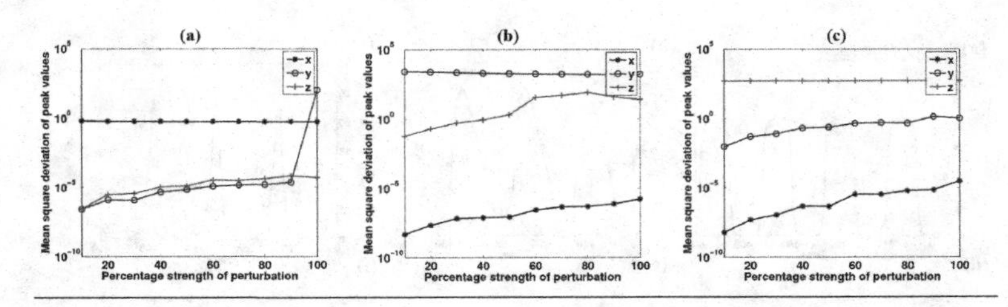

Figure 9. Effect of *random* perturbation on (a) x, (b) y, (c) z, on the MSD from the peak value. X-axis shows percentage of strength of perturbation, and Y-axis shows MSD of peak values of $x(***)$, $y(ooo)$, $z(+++)$.

feedback loop reduces the effect of noise on x and shows low MSD (Fig. 9(b)). The *random* perturbation on z causes high MSD in z and medium MSD in y but low MSD in x (Fig. 9(c)). But when the actual value of the noise strength is considered (Table 1), one can see that the magnitude of perturbation, when y is perturbed is higher (in absolute terms) than when z is perturbed. Still x has higher MSD when z is perturbed, compared to the larger perturbation in y. That is because when y is perturbed the noise is propagated to x via z, whereas when z is perturbed it is transmitted directly. Also from the table we can see the perturbation in x is of very low magnitude and hence the effects are also minimal. Thus z is affected much more in comparison to x and y. Also as compared to the equilibrium state, the perturbed variable shows the highest MSD in the case of oscillatory state.

Table 1. The effect of *random* perturbation (80%) on different variables.

Variable perturbed Absolute strength of perturbation	MSD of x	MSD of y	MSD of z
x (0.1533 $\pm$ 0.12264)	0.5570	1.5823e-005	4.3287e-005
y(48.7699 $\pm$ 39.01592)	4.9363e-007	1.6605e+003	81.3344
z(1.533$\pm$ 1.2264)	6.1887e-006	0.4673	529.59343

Pathway at birhythmic state ($k = 0.0024$): The steady state value is $x_s = 0.5295$, $y_s = 37.0435$, $z_s = 5.2948$. For this set of parameter values, the pathway exhibits coexistence of two different attractor depending upon initial conditions, for the same parameter values[5]. Figure 10 shows both the attractors.

Figure 11 and Fig. 12 show that when *random* perturbation of strength

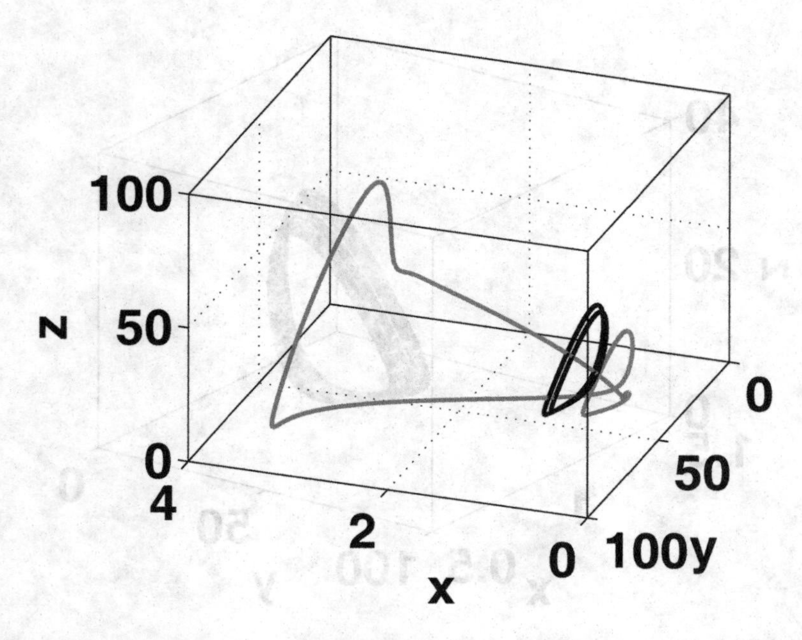

Figure 10. Three dimensional phase plot of the coexisting attractors at $k = 0.0024$. Type I attractor is in black and type II attractor is in grey.

10% was applied to x after getting Type 1 or Type 2 attractors, the resultant attractors was always of Type I for 10/10 cases. For the Type II attractor *random* perturbation to x was applied for a long time (30000 time points), but the resultant Type 1 attractor was robust under noise as shown in Fig. 12. This shows type I attractor is more robust to noise in x where as there is a very less chance of obtaining Type II attractor with *random* perturbation in x. A detailed analysis of the stability of these attractors have been done earlier[5].

4. Conclusions

The main objective of this analysis is to address a few questions, using a model biochemical pathway, as to how more complex regulated, multi-step reaction pathways common in nonlinear cellular processes respond to constant and random local and global perturbations. It would be interesting to know how the feedback and regulatory controls in highly networked pathways work and it adapts to various environmental changes. It would also be

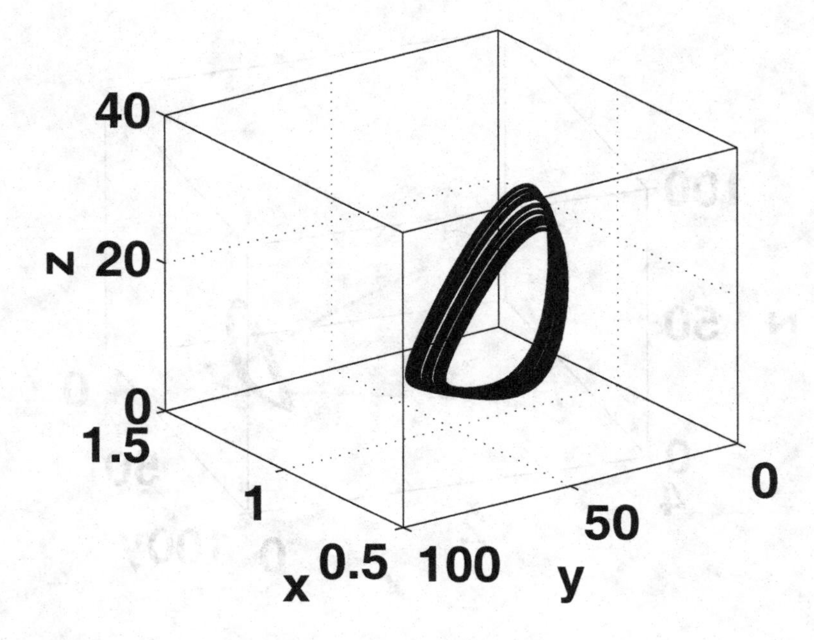

Figure 11. Three dimensional phase plot of the Type 1 attractor when x was randomly perturbed with 10% strength.

interesting to know about the features that make the pathway more or less robust, and the role of noise in decision making of the system. A detailed and comprehensive study of the similarities and differences in dynamics of a three step negatively auto-regulated model biochemical pathway showed that the noise in the intermediate step or the output step is lost due to the negative feed back of the end-product at the input level — a common design of regulation in intracellular pathways[23]. It was also shown that the presence of positive feedback can amplify noise[1 2 17]. In this study we chose a model pathway with a biochemically relevant positive feedback (through an autocatalytic enzyme) coupled to the commonly occurring design of negative feedback (end-product inhibition) in a three-step pathway.

The role of noise in decision making and heterogeneity in populations has been well studied[13 6 19]. Here we studied the role of constant *bias* that mimics the effect of constant environmental perturbations like the presence of inducer, chemotactic agent, temperature, pH, or morphogen that can act at particular steps of the pathway or have global effect. We show that

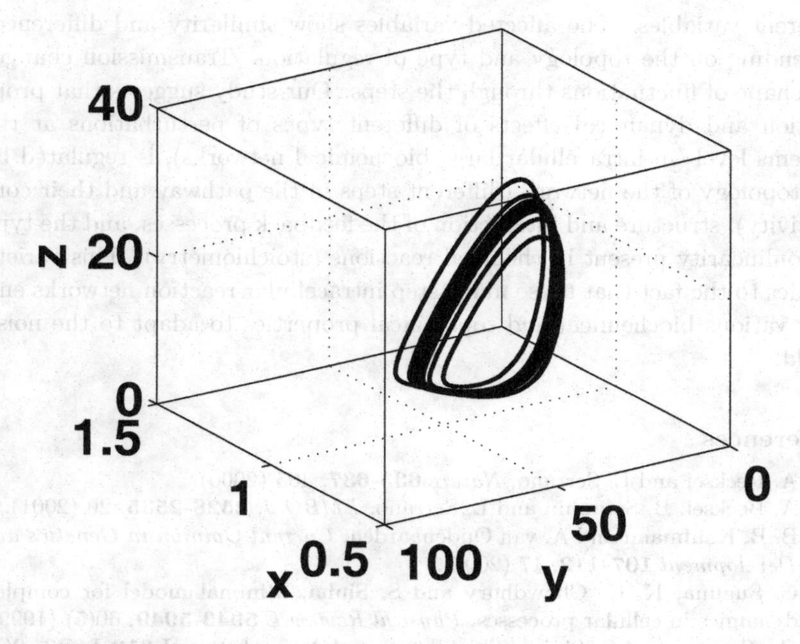

Figure 12. Three dimensional phase plot of the Type 2 attractor when x was randomly perturbed with 10% strength.

such constant perturbations, with appropriate magnitude, can change the dynamics of the pathway from unstable to stable or vice versa depending on the type and (i.e., strength and sign) and point of action (steps of the pathway). Thus *bias* can be considered as a tool to change the functional dynamics of the reaction pathway, and thus take part in the decision making of the system. On the other hand, we find that random noise does not cause any qualitative change in dynamics of pathways and transmits through the pathway steps depending on the topology and type of the feedback processes, indicating the robustness of pathways to small variations in the environment. thus, environmental and systematic stochasticity in terms of *bias* can expose and/or conceal the dynamical properties of the pathway in otherwise stable (fixed point or oscillatory) dynamical state. It can be used for inducing temporary phenotypic changes in pathway functional states. Thus a qualitative change in the system dynamics can be induced by addition of constant perturbation to the system. *Random* perturbations can only cause significant quantitative variations in the concentrations of the different variables. The affected variables show similarity and differences depending on the topology and type of regulation. Transmission changes the shape of fluctuations through the steps. Our study suggests that propagation and dynamical effects of different types of perturbations at the systems level (in intracellular large biochemical networks), is regulated by the topology of the network (different steps in the pathway and their connectivity), structure and interaction of the feedback processes, and the type of nonlinearity present in chemical reactions (stoichiometry). This variety alludes to the fact that these multi-step intracellular reaction networks employ various biochemical and topological properties to adapt to the noisy world.

References

1. A. Becksei and L. Serrano, *Nature* **633-637**, 405 (2000).
2. A. Becksei, B. Seraphin and L. Serrano, *EMBO J.* **2528-2535**, 20 (2001).
3. B. B. Kaufmann and A. van Oudenaarden, *Current Opinion in Genetics and Development* **107-112**, 17 (2007).
4. C. Suguna, K. K. Chowdhury and S. Sinha, Minimal model for complex dynamics in cellular processes, *Physical Review E* **5943-5949**, 60(5) (1999).
5. C . Suguna and S. Sinha, *Fluctuation and Noise Letters* **L313-L326**, 2(4) (2002).
6. C. V. Rao, D. M. Wolf and A. P. Arkin, *Nat. Genet.* **231-237**, 420 (2002).
7. E. Levine and T. Hwa, *PNAS* **9224-9229**, 104(22) (2007).
8. E. M. Ozbudak, M. Thattai, I. Kurtser, A. D. Grossman and A. van Oudenaarden, *Nature Genetics* **69-73**, 31 (2002).
9. F. J. Bruggeman, N. Bluthgen, H. V. Westerhoff, *PLoS Comput Biol* **e1000506**, 5(9) (2009).
10. G. Hornung, N. Barkai, *PLoS Comput Biol* **e8**, 4(1) (2008).
11. J. E. Ladbury and S. T. Arold, *Trends Biochem Sci* **173-178**, 37(5) (2012).
12. J. N Pedraza, A. van Oudenaarden, *Science* **1965-1968**, 307 (2005).
13. J. Spudich and D. E. Jr. Koshland, *Nature* **467-471**, 262 (1976).
14. J. T. Mettetal, D. Muzzey, J. M. Pedraza, E. M. Ozbudak, A. van Oudenaarden, *PNAS* **7304-7309**, 103(19) (2006).
15. J. Tsang, J. Zhu, A. van Oudenaarden, *Molecular Cell* **753-767**, 26 (2007).
16. J. Zhang, Z. Yuan and T. Zhou, *Phys. Biol.* **046009**, 6 (2009).
17. L. S. Weinberger, J. C. Burnett, J. E. Toettcher, A. P. Arkin and D. V. Schaffer, *Cell* **169-182**, 122 (2005).
18. M. Hinczewski and D. Thirumalai, *Physical Review X* **041017** 4 (2014).
19. M. Kaern, T. C Elston, W. Blake and J. J. Collins, *Nat. Rev. Genet.* **451-464**, 6 (2005).
20. M. Thattai, A. van Oudenaarden, *PNAS* **8614-8619**, 98(15) (2001).
21. M. Thattai, A. van Oudenaarden, *Biophysical Journal* **2943-2950**, 82 (2002).
22. M. Thattai, A. van Oudenaarden, *Genetics* **523-530**, 167 (2004).

23. R. Maithreye and S. Sinha, *Physical Biology* **48-59**, 4 (2007).
24. R. Steuer, C. Zhou, J. Kurths, *Biosystems* **241-251**, 72(3) (2003).
25. S. Grimbs, J. Selbig, S. Bulik, H. G. Holzhutter, R. Steuer, *Molecular Systems Biology* **146**, 3 (2007).
26. V. Shahrezaei, J. F. Ollivier and P. S. Swain, *Molecular Systems Biology* **196**, 4 (2008).
27. W. J. Blake, M. Kaern, C. R. Cantor and J. J. Collins, *Nature* **633-637**, 422(6932) (2003).

CATASTROPHIC TRANSITIONS IN CORAL REEF BIOME UNDER INVASION AND OVERFISHING

SAMARES PAL

Department of Mathematics
University of Kalyani, Kalyani 741235, India
E-mail: samaresp@gmail.com

JOYDEB BHATTACHARYYA

Department of Wildlife and Fisheries Sciences
Texas A&M University, Texas 77840, USA
E-mail: b.joydeb@gmail.com

Macroalgae and corals compete for the available algal turf in coral reef ecosystem. While herbivorous *Parrotfish* play a beneficial role in decreasing the growth of macroalgae, in presence of invasive lionfish (*Pterois volitans*), there is a reduction in herbivory, leading to the proliferation of macroalgae in coral reef ecosystem. Abundance of macroalgae changes the community structure form coral-dominated to macroalgae-dominated reef ecosystem. We investigate coral-macroalgal phase shift due to the effects of invasion and overfishing by means of a continuous time model in a food chain. Conditions for local asymptotic stability of steady states are derived. It is observed that the system undergoes a Hopf bifurcation when the maximal harvesting rate of *Pterois volitans* crosses certain threshold. Further, we observe that overfishing of *Parrotfish* and invasiveness of predatory *Pterois volitans* play a critical role in sudden change of transition from coral-dominated regime to macroalgae-dominated regime followed by hysteresis. Computer simulations have been carried out to illustrate different analytical results.

Keywords: Invasion; phase shifts; hysteresis; harvesting; Hopf bifurcation.

1. Introduction

Macroalgae play significant roles in the ecology of coral reefs. Despite their importance in coral reefs, macroalgal cover on reefs is increasingly related to the decline of coral reefs[1,2]. Coral reef ecosystems under natural and anthropogenic stresses exhibit two common alternative states, one has a high level of coral cover and the other has a degraded coral-depleted state corresponding to high level of macroalgae[3]. Degradation of coral reefs often involves sudden change in transition in community structure for which corals decline with an increase in abundance of macroalgae[4]. Phase shifts in coral reefs are mostly driven partly by the competition for light and space between corals and algae. Macroalgae may compete with corals by basal encroachment, shading or abrasion and allelopathic chemical defenses[5]. Also, macroalgae can dominate corals through space pre-emption by reducing available space for the successful settlement of coral larvae[6]. Coral can inhibit algae growth by shading, stinging, allelopathic chemical defenses, the occupation of space, mucus secretion, and overgrowth[7]. Although the growth rate of corals is very slow[8], the collective growth of many colonies across a large area can return the area to coral dominance within a few years.

Coral reefs throughout the world have experienced sustained declines in coral cover over due to rapid loss of herbivores[9]. Under conditions of reduced herbivory in coral reefs, the faster growing macroalgae often overgrow corals, depriving them of essential sunlight and causing their decline. The grazing of macroalgae by herbivores contribute to the resilience of the coral-dominated reef[10]. *Pterois volitans* is one of the top levels of the food web in many coral reefs[11]. There is substantial evidence that herbivore removal by the invasion of *Pterois volitans* has resulted in increased algal growth on coral reefs at the expense of living coral cover[12,13]. This can resulted in a permanent shift in state in which macroalgae, once dominant, inhibit coral settlement for ever[14]. This leads to the disruption of symbiosis between coral polyps and zooxanthellae, resulting in the expulsion of zooxanthellae and loss of photosynthetic pigments. A shift of regime occurs when the grazing rate of *Parrotfish* or the rate of invasion of adult *Pterois volitans* cross some critical thresholds. As different sets of feedbacks stabilize each alternative regime, returning the system from macroalgae-dominated regime to coral-dominated regime becomes difficult. Consequently, the threshold for a shift of regime and for its reversal can become different, a phenomenon known as hysteresis. If stresses continue for long enough, corals can suffer extensive mortality, and eventually bleaches[15]. According to NOAA (National Oceanic and Atmospheric Administration), commercial harvesting of adult *Pterois volitans* is necessary to mitigate their impact on coral reef ecosystems[16,17].

We study a model in which macroalgae and corals compete for occupying turf algae in presence of herbivorous *Parrotfish* and invasive *Pterois volitans*. *Parrotfish* are harvested with a non-constant harvesting policy[18]. In

particular, for a more realistic approach, a rational harvesting function has been considered in the model which provides diminishing marginal returns of the harvesting organization.
In the present paper the main emphasis will be put in studying the the dynamic behaviour of the system and role of invasive *Pterois volitans* in coral-macroalgal competition for space. We have studied the model analytically as well as numerically; proofs are all relegated to the Appendix.

2. The Basic Model

We begin with an analytic model in terms of the fraction of seabed in a given area, available for the growth of three benthic groups - corals, turf algae and macroalgae. Let $C(t), M(t)$ and $T(t)$ represent the cover of corals, macroalgae and turf algae respectively as a fraction of available seabed so that $M(t) + C(t) + T(t) = c$ (constant) at any instant t. Let $P(t)$ and $L(t)$ represent the concentrations of *Parrotfish* and *Pterois volitans* respectively at any instant t.
We make the following assumptions in formulating the mathematical model:
(H_1) The space freed due to the mortality of corals and macroalgae is recolonized by algal turfs.
(H_2) Corals are overgrown by macroalgae at a rate α.
(H_3) Macroalgae spread vegetatively over algal turfs at a rate β.
(H_4) Colonization rate of newly immigrated macroalgae on algal turf is γ.
(H_5) Corals recruit to and overgrow algal turfs at a rate r.
(H_6) Macroalgae and corals have natural mortality rates d_1 and d_2 respectively.
(H_7) In absence of *Pterois volitans*, *Parrotfish* follow logistic growth with intrinsic growth rate s and carrying capacity $k(M + T)$ as a function of macroalgae and turf algae cover. In absence of corals, we have $M + T = 1$ and consequently, the carrying capacity reaches its maximum, k.
(H_8) Grazing on algal turfs prevents macroalgal succession from the turf. In absence of harvesting of *Parrotfish*, the grazing intensity of *Parrotfish* arrives at its maximum, g.
(H_9) The grazing intensity $\frac{gP}{k}$ of *Parrotfish* is proportional to the abundance of *Parrotfish* relative to its maximum carrying capacity with maximal grazing rate, g. The loss of macroalgal cover and subsequent recolonization of algal turfs due to grazing is at a rate $\frac{gMP}{k(M+T)}$.
(H_{10}) The predation form of *Pterois volitans* is assumed to follow Holling-II functional response with maximal growth rate m, half saturation constant

a and growth efficiency e $(0 < e < 1)$. *Pterois volitans* have natural mortality rate d_3.
(H_{11}) *Parrotfish* and *Pterois volitans* are harvested with maximal harvesting rates h_i. Also, c_i represent the concentration of the fishes for which the rate of harvesting is half its maximal $(i = 1, 2)$.
The equations representing reef-dynamics in presence of grazing and invasion are given by:

$$\begin{aligned}
\frac{dM}{dt} &= M\left\{\alpha C - d_1 - \frac{gP}{k(M+T)}\right\} + (\beta M + \gamma)T \\
\frac{dC}{dt} &= C(rT - \alpha M - d_2) \\
\frac{dT}{dt} &= M\left\{\frac{gP}{k(M+T)} + d_1\right\} + d_2 C - (\beta M + rC + \gamma)T \\
\frac{dP}{dt} &= P\left[s\left\{1 - \frac{P}{k(M+T)}\right\} - \frac{mL}{a+P} - \frac{h_1}{c_1+P}\right] \\
\frac{dL}{dt} &= L\left(\frac{emP}{a+P} - \frac{h_2}{c_2+L} - d_3\right)
\end{aligned} \tag{1}$$

where $M(0) \geq 0, C(0) \geq 0, T(0) \geq 0, P(0) \geq 0$ and $L(0) \geq 0$.
Without any loss of generality we assume that $c = 1$. Then the system (1) reduces to

$$\begin{aligned}
\frac{dM}{dt} &= M\left\{\alpha C - d_1 - \frac{gP}{k(1-C)}\right\} + (\beta M + \gamma)(1 - M - C) \equiv F^1 \\
\frac{dC}{dt} &= C\{r(1 - M - C) - \alpha M - d_2\} \equiv F^2 \\
\frac{dP}{dt} &= P\left[s\left\{1 - \frac{P}{k(1-C)}\right\} - \frac{mL}{a+P} - \frac{h_1}{c_1+P}\right] \equiv F^3 \\
\frac{dL}{dt} &= L\left(\frac{emP}{a+P} - \frac{h_2}{c_2+L} - d_3\right) \equiv F^4
\end{aligned} \tag{2}$$

where $M(0) \geq 0, C(0) \geq 0, P(0) \geq 0$ and $L(0) \geq 0$.
Obviously the right hand sides of system (2) are continuous smooth functions on $\mathbf{R}^4_+ = \{(M, C, P, L) : M, C, P, L \geq 0\}$. Indeed, they are Lipschitzian on $\mathbf{R}^4_+$ and so the solution of the system (2) exists and is unique. Therefore, the interior of the positive octant of $\mathbf{R}^4$ is an invariant region.

3. Boundedness and Permanence of the System

Theorem 3.1.
For all $\epsilon > 0$, there exists $t_\epsilon > 0$ such that all the solutions of (2) *enter*

into the set $\left\{(M,C,P,L) \in \boldsymbol{R}^4 : M(t)+C(t)+P(t)+L(t) \le 1+\frac{s}{d_0}+\epsilon\right\}$ *whenever* $t \ge t_\epsilon$, *where* $d_0 = \min\{s, d_3\}$.

The above theorem states that, with non-negative initial values and other parameter values in the ecosystem the population of the organisms in the system will always remain finite and bounded.

Let us define $\lambda_1 = \frac{ad_3}{em-d_3}$, for $m > \frac{d_3}{e}$. Then λ_1 represents break-even concentration, the concentration of *Parrotfish* for which *Pterois volitans* have a constant growth.

The following theorem states the condition under which *Pterois volitans* cannot survive even, in the absence of harvesting:

Theorem 3.2.

(i) If $m \le \frac{d_3}{e}$, *then* $lim_{t\to\infty}L(t) = 0$.

(ii) If $m > \frac{d_3}{e}$ *and* $\lambda_1 > 1$ *then* $lim_{t\to\infty}L(t) = 0$.

Under the hypothesis of theorem (3.2), it follows that

(*i*) if the maximal growth rate of *Pterois volitans* is less than or equal to $\frac{d_3}{e}$ then *Pterois volitans* will not survive in the system.

(*ii*) If the maximal growth rate of *Pterois volitans* exceeds $\frac{d_3}{e}$ and the break-even concentration is greater than unity, then *Pterois volitans* cannot survive.

Therefore, for the survival of *Pterois volitans* in the system, we must have $m > \frac{d_3}{e}$ and $\lambda_1 \le 1$.

By definition, the system (2) will be permanent if there exists $u_i, U_i \in (0,\infty)$ such that $u_i \le lim_{t\to\infty}x_i(t) \le U_i$, for each organism $x_i(t)$ in the system[19,20]. Permanence represents convergence at an interior attractor from any positive initial conditions and so it can be regarded as a strong form of coexistence[21]. From a biological point of view, permanence of a system ensures the survival of all the organisms in the long run.

Since $M(t)+C(t)+P(t)+L(t) \le 1+\frac{s}{d_0}$ as $t \to \infty$, it follows that there exists a positive number $U < 1+\frac{s}{d_0}$ such that $M(t), C(t), P(t), L(t) \le U$ for large values of t.

The following theorem rules out the possibility of extinction of any organism in the system (2):

Theorem 3.3.

For large t, *if* $g < \frac{k\alpha}{4}\left(1-\frac{d_1}{\alpha}\right)^2, h_1 < c_1\left(s-\frac{mU}{a}\right)$ *and* $h_2 > \frac{c_2(em-d_3)(p_0-\lambda_1)}{a+p_0}$, *there exists* $m_0 > 0$ *such that all the solutions of* (2) *enter into the set* $\{(M,C,P,L) : m_0 \le M(t) \le U, c_0 \le C(t) \le C_0, p_0 \le P(t) \le U, l_0 \le L(t) \le U\}$ *and will remain there forever, where* $c_0 = \frac{\alpha+d_1-\sqrt{(\alpha-d_1)^2-\frac{4\alpha g}{k}}}{2\alpha}, C_0 = \frac{\alpha+d_1+\sqrt{(\alpha-d_1)^2-\frac{4\alpha g}{k}}}{2\alpha}, p_0 = k(1-$

$C_0)\left(1-\frac{mc_1U+ah_1}{asc_1}\right)$ *and* $l_0=\frac{h_2(a+p_0)}{(em-d_3)(p_0-\lambda_1)}-c_2$.

Thus, restricted harvesting of *Parrotfish* and high rate of harvesting of *Pterois volitans* leads to the coexistence of all the organisms in the system.

4. Equilibria and Their Stability

In this section we determine biologically feasible equilibria of the model and investigate the dependence of their stability on several key parameters.

The system (2) possesses the following equilibria:

(*i*) Fish-free equilibrium $E_0(M_0,0,0,0)$ in absence of coral cover, where $M_0=\frac{\beta-\gamma-d_1+\sqrt{(\beta-\gamma-d_1)^2+4\beta\gamma}}{2\beta}$. The steady state E_0 is macroalgae-dominated and coral-depleted in absence of fish.

(*ii*) The macroalgae-dominated and coral-depleted steady state $E_1(M_1,0,P_1,0)$ in absence of *Pterois volitans*, where $M_1=\frac{\beta-\gamma-d_1-\frac{gP_1}{k}+\sqrt{\left(\beta-\gamma-d_1-\frac{gP_1}{k}\right)^2+4\beta\gamma}}{2\beta}$ and P_1 is a positive root of the equation $sP^2+sP(c_1-k)+k(h_1-sc_1)=0$.

(*iii*) The macroalgae-dominated and coral-depleted steady state $E_2(M_2,0,P_2,L_2)$ in presence of fish, where P_2 is a positive root of the equation

$$s\left(1-\frac{P}{k}\right)-\frac{mf(P)}{a+P}-\frac{h_1}{c_1+P}=0,\ M_2=\frac{\beta-\gamma-d_1-\frac{gP_2}{k}+\sqrt{\left(\beta-\gamma-d_1-\frac{gP_2}{k}\right)^2+4\beta\gamma}}{2\beta}$$

and $L_2=f(P_2)=\frac{h_2(a+P_2)-c_2\{emP_2-d_3(a+P_2)\}}{emP_2-d_3(a+P_2)}$.

(*iv*) Fish-free equilibrium $E_3(M_3,C_3,0,0)$ in presence of coral and macroalgae cover, where M_3 is a positive root of $\alpha(rq+\beta)M^2+M(r\alpha p+\alpha\gamma+\beta d_2-rd_1)+\gamma d_2=0$, $C_3=p+qM_3$, $p=\frac{r-d_2}{r}$ and $q=-\frac{r+\alpha}{r}$.

(*v*) *Pterois volitans*-free equilibrium $E_4(M_4,C_4,P_4,0)$ in presence of coral and macroalgae cover, where P_4 is a positive root of $\left(\frac{f_1(P)-p}{q}\right)\left\{\alpha f(P)-d_1-\frac{gP}{k(1-f_1(P))}\right\}+\left(\frac{\{\beta f_1(P)-p\}}{q}+\gamma\right)\left(1+\frac{p}{q}-\frac{(q+1)f_1(P)}{q}\right)=0$, $C_4=f_1(P_4)=\frac{k(sc_1+sP_1-h_1)-sc_1P_4-sP_4^2}{k(sc_1+sP_4-h_1)}$ and $M_4=\frac{C_4-p}{q}$.

(*vi*) The interior equilibrium $E^*(M^*,C^*,P^*,L^*)$, where P^* is a positive root of the equation

$$\left(\frac{h^*(P)-p}{q}\right)\left(\alpha h^*(P)-d_1-\frac{gP}{k(1-h^*(P))}\right)+\left(\gamma+\beta\frac{h^*(P)-p}{q}\right)\left(1-h^*(P)-\frac{h*(P)-p}{q}\right)=0,$$

$M^*=\frac{h^*(P^*)-p}{q}$, $C^*=h^*(P^*)=1-\frac{sP^*(a+P^*)(c_1+P^*)}{k\{sP^*(a+P^*)(c_1+P^*)-mf^*(P^*)(c_1+P^*)-h_1(a+P^*)\}}$,

and $L^*=f^*(P^*)=\frac{h_2(a+P^*)-c_2(em-d_3)(P^*-\lambda_1)}{(em-d_3)(P^*-\lambda_1)}$.

The criterion for existence of the stationary states of the system (2) are summarized in Table 1.

Equilibria	Macroalgae-dominated state
E_0	Always exist
E_1	$h_1 < sc_1 + s\left(\frac{1-c_1}{2}\right)^2$
E_2	$P_2 > 0, \lambda_1 > \frac{c_2 P_2(em-d_3)-h_2(a+P_2)}{c_2(em-d_3)}$
E_3	$0 < C_3 < \frac{r-d_2}{2\alpha+r}$
E_4	$0 < C_4 < \frac{r-d_2}{2\alpha+r}, P_4 > 0$
E^*	$0 < C^* < \frac{r-d_2}{2\alpha+r}, \lambda_1 > \frac{c_2 P^*(em-d_3)-h_2(a+P^*)}{c_2(em-d_3)}$
Equilibria	**Coral-dominated state**
E_0	–
E_1	–
E_2	–
E_3	$\frac{r-d_2}{2\alpha+r} < C_3 < \frac{r-d_2}{r}$
E_4	$\frac{r-d_2}{2\alpha+r} < C_4 < \frac{r-d_2}{r}, P_4 > 0$
E^*	$\frac{r-d_2}{2\alpha+r} < C^* < \frac{r-d_2}{r}, \lambda_1 > \frac{c_2 P^*(em-d_3)-h_2(a+P^*)}{c_2(em-d_3)}$

Table 1	**Criterion for existence of the stationary states.**

Macroalgae-dominated irreversible regime

Due to increasing frequencies of coral bleaching, there can be a permanent shift from coral-dominated ecosystems to macroalgae-dominated ones. We study macroalgae-dominated system following mass coral bleaching in which there is no possibility of reversal of the regime.

Lemma 4.1.

Fish-free equilibrium $E_0(M_0, 0, 0, 0)$ in absence of coral cover is locally asymptotically stable if $h_1 > sc_1$ and $\alpha > \frac{r(1-M_0)-d_2}{M_0}$.

Thus, with high macroalgal overgrowth rate on corals and high rate of harvesting of *Parrotfish*, the system becomes macroalgae-dominated and stable at E_0 with complete elimination of corals and fishes.

Lemma 4.2.

Pterois volitans-free equilibrium $E_1(M_1, 0, P_1, 0)$ in absence of coral cover is locally asymptotically stable if $\alpha > \frac{r(1-M_1)-d_2}{M_1}, h_1 < \frac{s(c_1+P_1)^2}{k}$ and $h_2 > \frac{c_2(em-d_3)(P_1-\lambda_1)}{a+P_1}$.

Thus, with high macroalgal overgrowth rate on corals, limited harvesting of *Parrotfish* and high harvesting of *Pterois volitans* stabilizes the system at E_1 with complete elimination of corals and *Pterois volitans.*

Lemma 4.3.

Equilibrium of coexistence of fishes $E_2(M_2, 0, P_2, L_2)$ in absence of coral cover is locally asymptotically stable if $\alpha > \frac{r(1-M_2)-d_2}{M_2}$, $h_1 < (c_1 + P_2)^2 \left\{\frac{s}{k} - \frac{mL_2}{(a+P_2)^2} - \frac{h_2 L_2}{P_2(c_2+L_2)^2}\right\}$ *and* $h_2 < \frac{em^2 a(c_2+L_2)^2}{(a+P_2)^3}\left\{\frac{s}{k} - \frac{mL_2}{(a+P_2)^2} - \frac{h_1}{(c_1+P_2)^2}\right\}^{-1}$.

Thus, with high macroalgal overgrowth rate on corals and restricted harvesting of *Parrotfish* and *Pterois volitans* stabilizes the system at E_2 with complete elimination of corals.

Reversible regimes

Corals and macroalgae can coexist under two alternate stable states, one being dominated by corals and the other by macroalgae. It is observed that the equilibria E_3, E_4 and E^* are macroalgae-dominated if $0 < C_4 < \frac{r-d_2}{2r+\alpha}$ and are coral-dominated if $\frac{r-d_2}{2r+\alpha} < C_4 < \frac{r-d_2}{r}$.

We now analyze the stability criterion of coexistence steady states of macroalgae and corals.

Lemma 4.4.

Fish-free equilibrium $E_3(M_3, C_3, 0, , 0)$ in presence of macroalgal and coral cover is locally asymptotically stable if $h_1 > sc_1$ and $\beta < r + \alpha - \frac{\gamma}{M_3} + \frac{r\gamma(1-M_3-C_3)}{\alpha M_3^2}$.

Thus, with low vegetative growth of macroalgae on algal turf and high rate of harvesting of *Parrotfish*, the system becomes stable at E_3.

Lemma 4.5. *Pterois volitans*-free equilibrium $E_4(M_4, C_4, P_4, 0)$ in presence of macroalgal and coral cover is locally asymptotically stable if $g > g_0$, $h_1 < h_0$, $h_2 > \frac{c_2(em-d_3)(P_4-\lambda_1)}{a+P_4}$ and $\eta > 0$, where

$$g_0 = \frac{k(1-C_4)^2(c_1+P_4)^2}{h_1 M_4 P_4 (r+\alpha)}\left\{\frac{h_1}{(c_1+P_4)^2} - \frac{s}{k(1-C_4)}\right\}\left\{\frac{r\gamma}{M_4}(1 - M_4 - C_4) + \alpha M_4(r + \alpha - \beta) - \alpha\gamma\right\},$$

$$h_0 = \frac{(c_1 + P_4)^2}{P_4}\left\{\frac{sP_4}{k(1 - C_4)} + \frac{\gamma}{M_4}(1 - C_4) + rC_4 + \beta M_4\right\} \text{ and}$$

$$\eta = (rC_4 - F^1_M|_{E_4} - F^3_P|_{E_4})\left\{F^1_M|_{E_4}(F^3_P|_{E_4} - rC_4) - rC_4F^3_P|_{E_4} - (r + \alpha)C_4F^1_C|_{E_4}\right\}$$
$$- rC_4F^1_M|_{E_4}F^3_P|_{E_4} - C_4(r + \alpha)(F^1_C|_{E_4}F^3_P|_{E_4} - F^3_C|_{E_4}F^1_P|_{E_4})$$

Thus, with high grazing rate and low harvesting rate of *Parrotfish* together with high harvesting rate of *Pterois volitans*, the system becomes stable at E_4.

Lemma 4.6.
Equilibrium of coexistence E^* is locally asymptotically stable if $\eta_1^* > 0, \eta_1^*\eta_2^* > \eta_3^*$ and $\eta_1^*\eta_2^*\eta_3^* > \eta_3^{*2} + \eta_1^{*2}\eta_4^*$, where the values of η_i^* $(i = 1, \ldots, 4)$ are given in the Appendix.

5. Hopf Bifurcation

The linearized system of (2) about E^* is given by $\frac{dX}{dt} = J(E^*)X$, where $J(E^*)$ is the Jacobian matrix of the system (2) evaluated at E^*.
The characteristic equation of $J(E^*)$ is $u^4 + \eta_1^* u^3 + \eta_2^* u^2 + \eta_3^* u + \eta_4^* = 0$.
The necessary and sufficient conditions for Hopf bifurcation to occur at $\nu = \nu_{cr}$ are: (i) $f_1(\nu_{cr}) = f_2(\nu_{cr})$ and (ii) $Re\left[\frac{d\lambda_{1_j}}{d\nu}\right]_{\nu=\nu_{cr}} \neq 0, (j = 1, \ldots, 4)$
where $f_1(\nu) = \eta_1^*(\nu)\eta_2^*(\nu)\eta_3^*(\nu)$ and $f_2(\nu) = \eta_3^{*2}(\nu) + \eta_1^{*2}(\nu)\eta_4^*(\nu)$.
Let $g : (0, \infty) \to \mathbf{R}$ is a continuously differentiable function of ν defined by $g(\nu) = f_1(\nu) - f_2(\nu)$. The existence of ν_{cr} is ensured by solving the equation $g(\nu_{cr}) = 0$.
At $\nu = \nu_{cr}$, the characteristic equation is $\left(\lambda^2 + \frac{\eta_3^*(\nu_{cr})}{\eta_1^*(\nu_{cr})}\right)\left(\lambda^2 + \lambda\eta_1^*(\nu_{cr}) + \frac{\eta_1^*(\nu_{cr})\eta_4^*(\nu_{cr})}{\eta_3^*(\nu_{cr})}\right) = 0$.
For $\nu = \nu_{cr}$, let $\lambda_{1_i} (i = 1, \ldots, 4)$ be the roots of the characteristic equation with the pair of purely imaginary roots as $\lambda_{1_1} = i\omega_0$ and $\lambda_{1_2} = \bar{\lambda}_1$, where $\omega_0 = \sqrt{\frac{\eta_3^*(\nu_{cr})}{\eta_1^*(\nu_{cr})}}$.
Now, if λ_{1_3} and λ_{1_4} are not real, then $Re\lambda_{1_3} = \frac{1}{2}(\lambda_{1_3} + \lambda_{1_4}) = -\frac{1}{2}\eta_1^*(\nu_{cr}) < 0$.
If λ_{1_3} and λ_{1_4} are real roots, then $\lambda_{1_3} + \lambda_{1_4} < 0$ and $\lambda_{1_3}\lambda_{1_4} = \frac{\eta_4^*(\nu_{cr})}{\omega_0^2} > 0$ implies $\lambda_{1_3}, \lambda_{1_4} < 0$.
Since g is a continuously differentiable function of ν, there exists an open interval $(\nu_{cr} - \epsilon, \nu_{cr} + \epsilon)$ such that $\lambda_{1_1}(\nu) = \beta_1(\nu) + i\beta_2(\nu)$ and $\lambda_{1_2}(\nu) = \beta_1(\nu) - i\beta_2(\nu), \forall\nu \in (\nu_{cr} - \epsilon, \nu_{cr} + \epsilon)$.
Thus $\forall\nu \in (\nu_{cr} - \epsilon, \nu_{cr} + \epsilon)$, $\frac{d}{d\nu}(\lambda^4 + \eta_1^*\lambda^3 + \eta_2^*\lambda^2 + \eta_3^*\lambda + \eta_4^*) = 0$
$\Rightarrow (K(\nu) + iL(\nu))\frac{d\lambda}{d\nu} + (M(\nu) + iN(\nu)) = 0$, where
$K(\nu) = 4\beta_1^3(\nu) - 12\beta_1(\nu)\beta_2^2(\nu) + 3\left(\beta_1^2(\nu) - \beta_2^2(\nu)\right)\eta_1^*(\nu) + 2\beta_1(\nu)\eta_2^*(\nu) + \eta_3^*(\nu)$,
$L(\nu) = 12\beta_1^2(\nu)\beta_2(\nu) - 4\beta_2^3(\nu) + 6\beta_1(\nu)\beta_2(\nu)\eta_1^*(\nu) + 2\beta_2(\nu)\eta_2^*(\nu)$,
$M(\nu) = \beta_1^3(\nu)\eta_1^{*'}(\nu) - 3\beta_1(\nu)\beta_2^2(\nu)\eta_1^{*'}(\nu) + \left(\beta_1^2(\nu) - \beta_2^2(\nu)\right)\eta_2^{*'}(\nu) + \beta_1(\nu)\eta_3^{*'}(\nu)$,
$N(\nu) = 3\beta_1^2(\nu)\beta_2(\nu)\eta_1^{*'}(\nu) - \beta_2^3\eta_1^{*'}(\nu) + 2\beta_1(\nu)\beta_2(\nu)\eta_2^{*'}(\nu) + \beta_2(\nu)\eta_3^{*'}(\nu)$.
Therefore, $\frac{d\lambda}{d\nu} = -\frac{\{M(\nu)K(\nu) - N(\nu)L(\nu)\} + i\{N(\nu)K(\nu) - M(\nu)L(\nu)\}}{K^2(\nu) + L^2(\nu)}$
If $[M(\nu)K(\nu) - N(\nu)L(\nu)]_{\nu=\nu_{cr}} \neq 0$, then $Re\left[\frac{d\lambda_j}{d\nu}\right]_{\nu=\nu_{cr}} \neq 0$.

Thus, if (a) $f_1(\nu_{cr}) = f_2(\nu_{cr})$ and (b) $[M(\nu)K(\nu) - N(\nu)L(\nu)]_{\nu=\nu_{cr}} \neq 0$ hold, then a Hopf bifurcation occurs at $\nu = \nu_{cr}$ and also it is non-degenerate.

6. Numerical Simulations

In this section, we investigate numerically the effect of the various parameters on the qualitative behavior of the system using parameter values given in Table 2 throughout, unless otherwise stated.

Parameters	Description of Parameters	Value
α	Rate of macroalgal direct overgrowth over coral	0.45
r	Rate of coral recruitment to algal turfs	4.5
β	Rate of macroalgal vegetative growth on algal turfs	0.8
γ	Colonization rate of newly immigrated macroalgae on algal turf	0.05
d_1	Natural macroalgal mortality rate	0.13
d_2	Natural coral mortality rate	0.44
s	Intrinsic growth rate of *Parrotfish*	0.49
g	Maximal macroalgae-grazing rate of *Parrotfish*	1
k	Maximal carrying capacity of *Parrotfish*	1
m	Maximal growth rate of *Pterios volitans* on *Parrotfish*	1.5
a	Half saturation constant for uptake of P by L	0.5
h_1	Maximal rate of harvesting of *Parrotfish*	0
h_2	Maximal rate of harvesting of *Pterios volitans*	0.1
c_1	Half saturation constant on harvesting of *Parrotfish*	2
c_2	Half saturation constant on harvesting of *Pterios volitans*	2
e	Growth efficiency of *Pterios volitans*	0.75
d_3	Natural mortality rate of *Pterios volitans*	0.1

Table 2 **Parameter values used in the numerical analysis.**

By analyzing the system (2) we are now able to show that a sharp transition with hysteresis can be achieved by varying some of the parameter values.

Effect of invasion of *Pterois volitans*:

To identify the impact of *Pterois volitans*-invasion on coral cover, in Fig. 1(a), we plot the equilibrium values in the $C - m$ plane, yielding a bifurcation diagram. The region I represents monostability at E_4 for $m < 1.072$. In this region, the system will ultimately arrive at a coral-dominated state in absence of *Pterois volitans* corresponding to low levels of macroalgae. The bistable region is represented by II for $1.072 < m < 1.265$.

Hysteresis results in the region bounded by $A_1B_1C_1D_1$ with low invasion level ($m < 1.072$) followed by an increase in the *Pterois volitans*-invasion above a critical threshold $m = 1.265$. Any perturbation in the system that has resulted in a subsequent decline in *Pterois volitans*-invasion level below the threshold $m = 1.072$, returns the system to coral-dominated *Pterois volitans*-free stable state in region I. Once the intensity of invasion of *Pterois volitans* surpasses the threshold $m = 1.265$ but still less than $m = 1.7$, the system arrives at a coral-dominated stable coexistence state, represented by region III of monostability at E^*. For $m > 1.7$, the system becomes unstable as depicted by the region IV.

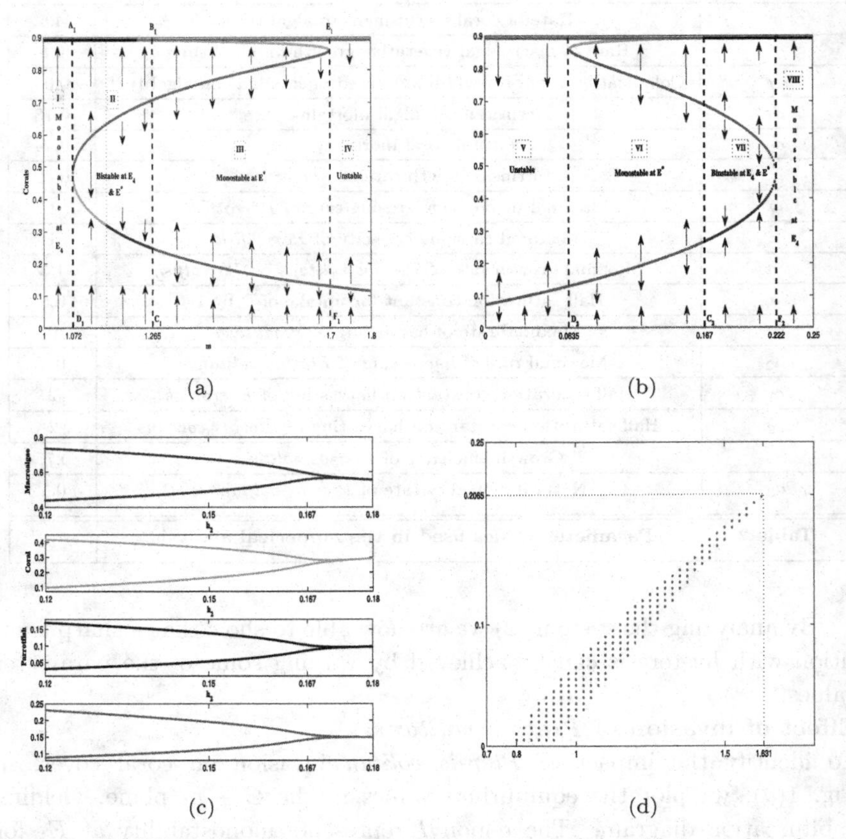

Figure 1. (a) Bifurcation diagram of invasion rate versus the equilibrium value of coral cover. (b) Bifurcation diagram of *Pterois volitans* harvesting rate versus the equilibrium value of coral cover. (c) The system undergoes a Hopf bifurcation as h_2 is crosses $h_2 = 0.167$. (d) Coexistence region in $m - h_2$ parameter space.

Effect of harvesting of *Pterois volitans*:
To identify the impact of harvesting-rate of *Pterois volitans* on coral cover, in Fig. 1(b), we plot the equilibrium values in the $C - h_2$ plane. The system is unstable in region V for $h_2 < h_{2_{cr_2}} = 0.0635$. The system is monostable in the region VI for $0.0635 < h_2 < 0.167$. The bistable region is represented by the region VII for $0.167 < h_2 < 0.222$. Hysteresis results in the region $A_2B_2C_2D_2$ with low harvesting level ($h_2 < 0.167$) followed by an increase of harvesting above a critical threshold $h_2 > 0.222$. For $h_2 > 0.222$, the system becomes monostable at the coral-dominated *Pterois volitans*-free equilibrium as depicted by the region VIII. From Fig. 1(c), it follows that the system undergoes a Hopf bifurcation when h_2 crosses $h_2 = 0.167$.
Combined effects of invasion and harvesting of *Pterois volitans*:
From Fig. 1(d) we can conclude that the stable coexistence is not possible with high m and low h_2 i.e. if the harvesting rate of *Pterois volitans* is low compared to the invasion rate of *Pterois volitans*, then coexistence of all the populations is not possible. More specifically, if $m < 0.8$ or $m > 1.63$, then coexistence of all the populations is not possible by changing the harvesting rate of *Pterois volitans*.
Effect of harvesting of *Parrotfish*:
To identify the impact of harvesting of *Parrotfish* on coral cover, in Fig. 2, we plot the equilibrium values in the $C - h_1$ plane, yielding a bifurcation diagram. The region IX represents monostability at the coral-dominated equilibrium E^* for $h_1 < 0.282$. In this region, the system will ultimately arrive at a coral-dominated state corresponding to low levels of macroalgae. The bistable coexistence region is represented by X for $0.282 < h_1 < 0.34$. In this region, the system will arrive at either coral-dominated or macroalgae-dominated coexistence state depending upon initial conditions. For $0.34 < h_1 < 0.65$, the system becomes bistable at macroalgae-dominated equilibria E_4 and E^* as depicted by the region XI. The system becomes monostable at the macroalgae-dominated coexistence equilibrium for $0.65 < h_1 < 0.922$, represented by the region XII. For $h_1 > 0.922$, the system becomes monostable at *Pterois volitans*-free equilibrium E_4 and further increase of h_1 stabilizes the system at E_0.
Effect of macroalgal overgrowth on corals:
To identify the impact of macroalgal overgrowth on corals, in Fig. 3(a), we plot the equilibrium values in the $C - \alpha$ plane, yielding a bifurcation diagram. The regions XIII and XV represent monostability at the coral-dominated equilibrium E^* for $\alpha < 0.326$ and $0.365 < \alpha < 0.53$ respectively. The bistable coexistence region is represented by XIV for

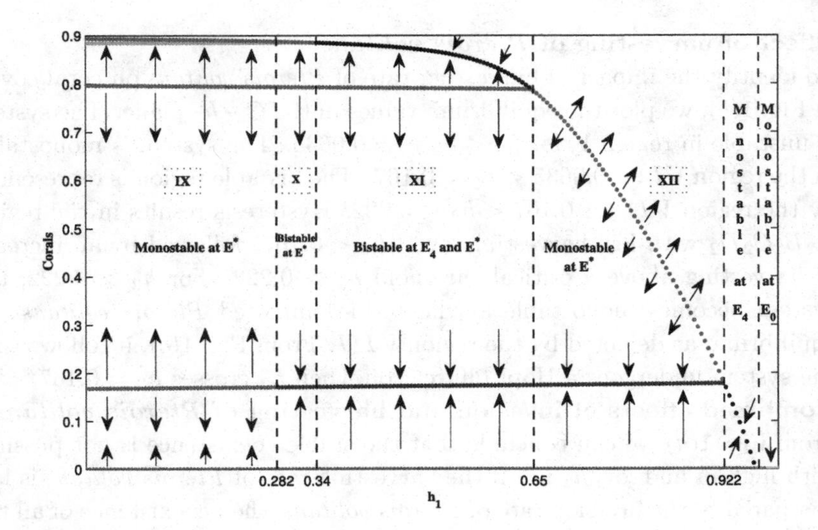

Figure 2. Bifurcation diagram of *Parrotfish*-harvesting rate versus the equilibrium value of coral cover.

$0.326 < \alpha < 0.365$. Hysteresis results in the region bounded by $A_3B_3C_3D_3$ with low macroalgal overgrowth on corals ($\alpha < 0.326$) followed by an increase in the macroalgal overgrowth on corals above a critical threshold ($\alpha > 0.365$). For $\alpha > 0.53$, the system becomes unstable as depicted by the region XVI.

Combined effects of macroalgal overgrowth on corals and harvesting of *Parrotfish*:

From Fig. 3(b) we can conclude that the stable coexistence is not possible with high α (viz. $\alpha > 0.7$). With low α (viz. $\alpha < 0.35$), coral-dominated stable coexistence occurs with *Parrotfish*-harvesting below a certain threshold (viz. $h_1 < 0.26$).

Combined effects of macroalgal overgrowth on corals and harvesting of *Pterois volitans*:

From Fig. 3(c) we can conclude that the stable coexistence is not possible with high α (viz. $\alpha > 0.5$). With low α (viz. $\alpha < 0.4$), coral-dominated stable coexistence occurs with *Pterois volitans*-harvesting below a certain threshold (viz. $h_2 < 0.16$).

Effect of macroalgal growth on turf algae:

To identify the impact of macroalgal overgrowth on turf algae, in Fig. 4(a), we plot the equilibrium values in the $C - \beta$ plane, yielding a bifurcation

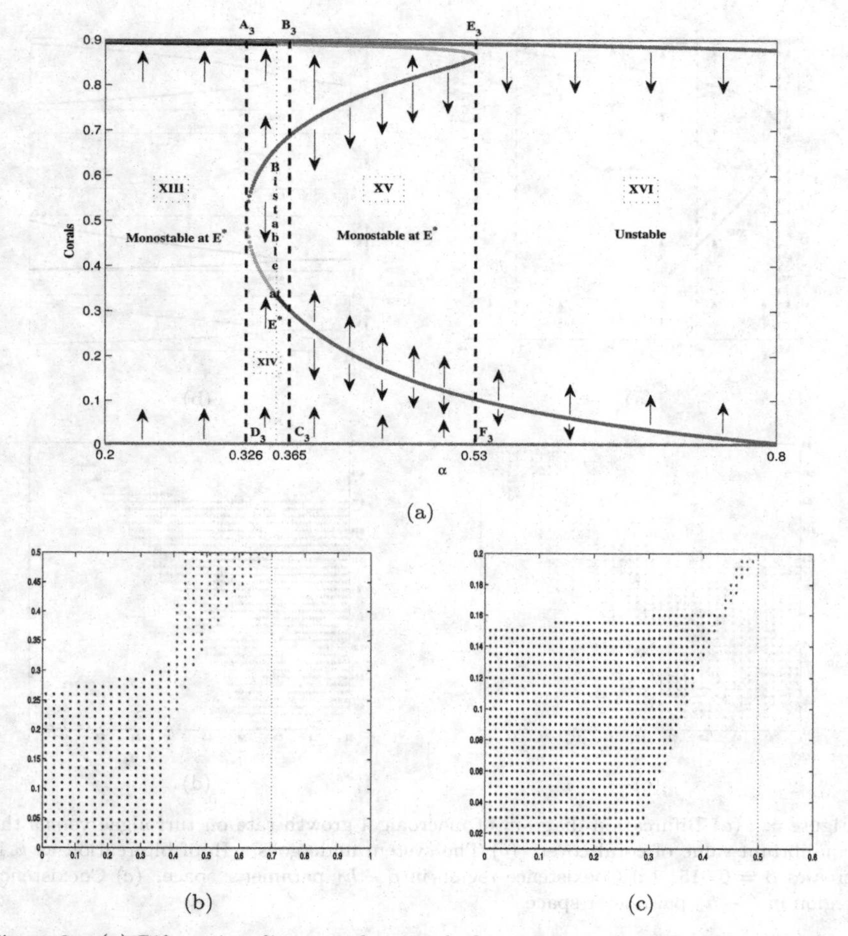

Figure 3. (*a*) Bifurcation diagram of macroalgal overgrowth rate on corals versus the equilibrium value of coral cover. (*b*) Coexistence region in $\alpha - h_1$ parameter space. (*c*) Coexistence region in $\alpha - h_2$ parameter space.

diagram. The regions XVII and XIX represents monostability at the coral-dominated equilibrium E^* for $\beta < 0.2$. The bistable coexistence region is represented by XVIII for $0.2 < \beta < 0.615$. Hysteresis results in the region bounded by $A_4B_4C_4D_4$ with low macroalgal growth on turf algae ($\beta < 0.2$) followed by an increase in the macroalgal growth on algal turf above the critical threshold $\beta = 0.615$. For $\beta > 1.53$, the system becomes unstable as depicted by the region XX. From Fig. 4(b), it follows that the system undergoes a Hopf bifurcation when β crosses $\beta = 0.615$.

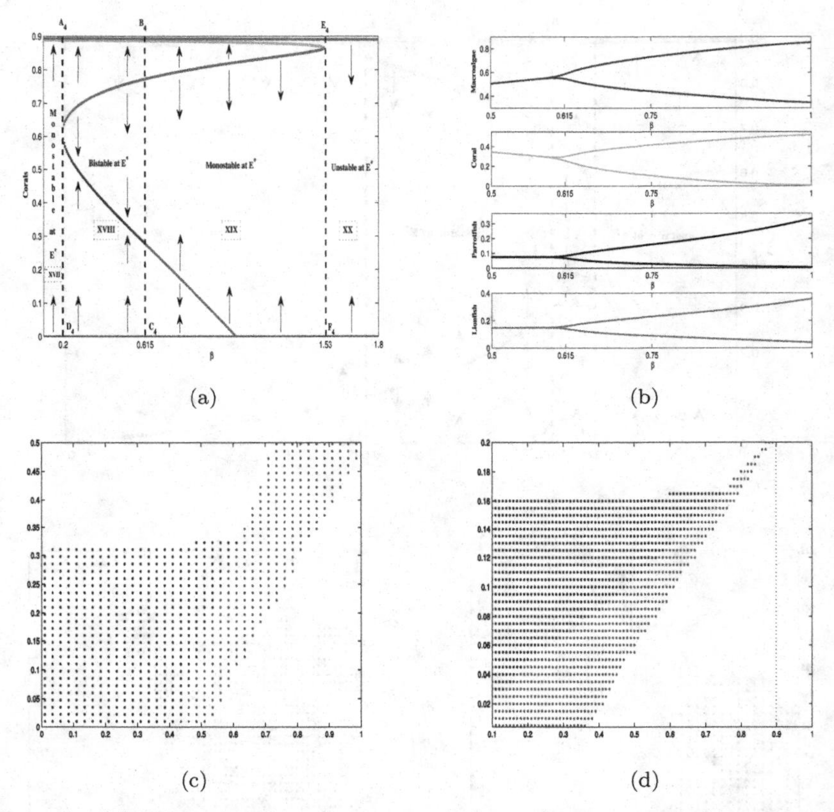

Figure 4. (*a*) Bifurcation diagram of macroalgal growth rate on turf algae versus the equilibrium value of coral cover. (*b*) The system undergoes a Hopf bifurcation as β is crosses $\beta = 0.615$. (*c*) Coexistence region in $\beta - h_1$ parameter space. (*c*) Coexistence region in $\beta - h_2$ parameter space.

Combined effects of macroalgal growth on turf algae and harvesting of *Parrotfish*:

From Fig. 4(*c*) we see that the stable coral-dominated coexistence is not possible with high harvesting rate of *Parrotfish* together with low macroalgal growth rate on turf algae. With moderately high macroalgal growth rate on turf algae, macroalgae-dominated stable coexistence occurs subject to low harvesting rate of *Parrotfish*.

Combined effects of macroalgal growth on turf algae and harvesting of *Pterois volitans*:

From Fig. 4(*d*) we see that the stable coexistence is not possible for $\beta > 0.9$. With low harvesting rate of *Pterois volitans*, macroalgae-dominated stable coexistence occurs subject to low macroalgal growth on turf algae. With low macroalgal growth on turf algae and high rate of harvesting of *Pterois volitans*, coral-dominated coexistence occurs.

Effect of grazing:

To identify the impact of grazing rate on corals, in Fig. 5(a), we plot the equilibrium values in the $C - g$ plane, yielding a bifurcation diagram. The region XXI represents instability for $g < 0.86$. The coral-dominated monostable coexistence regions are given by XXII and XXIV for $0.86 < g < 1.29$ and $g > 1.476$ respectively. The bistable coexistence region XXIII exists for $1.29 < \beta < 1.476$.

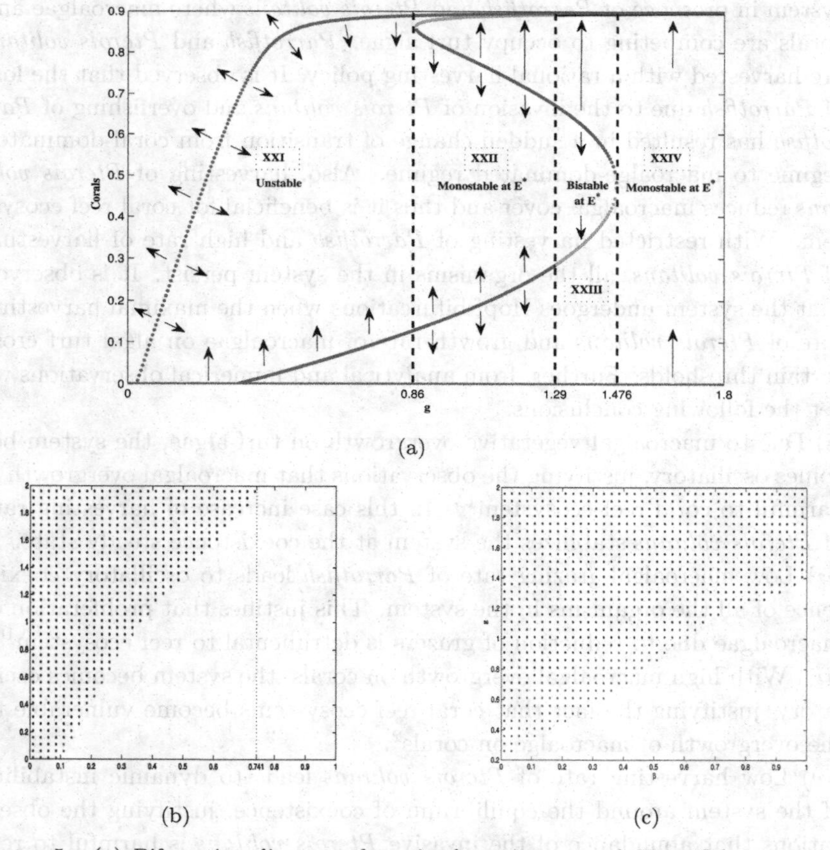

Figure 5. (a) Bifurcation diagram of maximal grazing rate of *Parrotfish* versus the equilibrium value of coral cover. (b) Coexistence region in $g - \alpha$ parameter space. (c) Coexistence region in $g - \beta$ parameter space.

Combined effects of macroalgal overgrowth on corals and grazing: From Fig. 5(b) we see that the stable coexistence is not possible for low grazing rate and high macroalgal growth rate on corals. Also, for $\alpha > 0.741$, coexistence is not possible even with high grazing rate.
Combined effects of macroalgal growth on turf algae and grazing: From Fig. 5(c) we see that the stable coexistence is not possible for low grazing rate and high macroalgal growth rate on turf algae.

7. Discussion

We have considered a model to study the dynamics of coral reef benthic system in presence of *Parrotfish* and *Pterois volitans* where macroalgae and corals are competing to occupy turf algae. *Parrotfish* and *Pterois volitans* are harvested with a rational harvesting policy. It is observed that the loss of *Parrotfish* due to the invasion of *Pterois volitans* and overfishing of *Parrotfish* has resulted in a sudden change of transition from coral-dominated regime to macroalge-dominated regime. Also, harvesting of *Pterois volitans* reduces macroalgal cover and thus it is beneficial for coral reef ecosystem. With restricted harvesting of *Parrotfish* and high rate of harvesting of *Pterois volitans*, all the organisms in the system persist. It is observed that the system undergoes Hopf bifurcations when the maximal harvesting rate of *Pterois volitans* and growth rate of macroalgae on algal turf cross certain thresholds. Further, from analytical and numerical observations we get the following conclusions:
(i) Due to macroalgal vegetative overgrowth on turf algae, the system becomes oscillatory, justifying the observations that macroalgal overgrowth is harmful to coral reef ecosystem[14]. In this case increase of harvesting rate of *Pterois volitans* stabilizes the system at the coexistence steady state.
(ii) Low macroalgal-grazing rate of *Parrotfish* leads to oscillatory coexistence of all the organisms in the system. This justifies that proliferation of macroalgae due to reduction of grazers is detrimental to reef ecosystem[16].
(iii) With high macroalgal overgrowth on corals, the system becomes oscillatory, justifying the fact that coral reef ecosystems become vulnerable to the overgrowth of macroalge on corals[2].
(iv) Low harvesting rate of *Pterois volitans* leads to dynamic instability of the system around the equilibrium of coexistence, justifying the observations that abundance of the invasive *Pterois volitans* is harmful to reef ecosystem[11].
(v) With high invasiveness of *Pterois volitans*, the system becomes oscilla-

tory. This represents the phenomenon of ecological imbalance in presence of the invasive *Pterois volitans* in coral reef ecosystem, justifying the observations that high predation rates of *Pterois volitans* is harmful to coral reef ecosystems[11]. The dynamic instability can be controlled by increasing the maximal harvesting rate of *Pterois volitans.* This, too, justifies that harvesting of these species is beneficial for coral reef ecosystem[13].
Throughout the article an attempt is made to search for a suitable way to control the growth of macroalgae, corals, *Parrotfish* and *Pterois volitans* for stable coexistence of all the species in the system. From analytical and numerical observations, it is seen that lowering the rate of harvesting of *Parrotfish* and increasing the harvesting rate of *Pterois volitans* induces a coral-dominated regime. Also, with macroalgal vegetative overgrowth, corals can recover even after extensive mortality if the harvesting of *Pterois volitans* is maximized.

Appendix

Proof of Theorem 3.1:
$\frac{dP}{dt} \leq sP\left(1 - \frac{P}{1-C}\right) \leq sP(1-P)$ implies $P(t) \leq 1$ as $t \to \infty$.
Therefore, corresponding to $\epsilon_1 > 0$, there exists $t_{\epsilon_1} > 0$ such that $P(t) \leq 1 + \epsilon_1$, for all $t \geq t_{\epsilon_1}$.
Let $\Sigma(t) = P(t) + L(t)$
Then $\frac{d}{dt}(\Sigma(t)) < s(1+\epsilon_1) - sP - d_3 L \leq s(1+\epsilon_1) - (\Sigma(t))d_0$, where $d_0 = \min\{s, d_3\}$, for all $t \geq t_{\epsilon_1}$.
Let $u(t)$ be a solution of $\frac{du}{dt} + d_0 u = s$, satisfying $u(0) = P(0) + L(0)$.
Then $u(t) = \frac{s}{d_0} + \left\{P(0) + L(0) - \frac{s}{d_0}\right\} e^{-d_0 t} \to \frac{s}{d_0}$ as $t \to \infty$.
By comparison, it follows that $P(t) + L(t) \leq \frac{s}{d_0}$ as $t \to \infty$.
Since $M(t) + C(t) \leq 1$, for all $t \geq 0$, it follows that $M(t) + C(t) + P(t) + L(t) \leq 1 + \frac{s}{d_0}$ as $t \to \infty$.
Proof of Theorem 3.2:
(*i*) Since $\lim_{t\to\infty} P(t) \leq 1$, it follows that for all $\epsilon > 0$, there exists $t_\epsilon > 0$ such that $P(t) \leq 1 + \epsilon$, for all $t \geq t_\epsilon$.
If $em \leq d_3$, then $\frac{dL}{dt} \leq -\frac{ad_3 L}{a+P} < 0$.
Thus we get, $\int_{L(t_\epsilon)}^{L(t)} \frac{d\xi}{\xi} \leq \frac{-ad_3(t-t_\epsilon)}{1+a+\epsilon}$ which gives $L(t) \leq L(t_\epsilon)e^{\frac{-ad_3(t-t_\epsilon)}{1+a+\epsilon}} \to 0$ as $t \to \infty$.
Therefore, if $em \leq d_3$, then $\lim_{t\to\infty} L(t) = 0$.
(*ii*) For $em > d_3$, we have $\int_{L(t_\epsilon)}^{L(t)} \frac{d\xi}{\xi} \leq \frac{(em-d_3)(1+\epsilon-\lambda_1)(t-t_\epsilon)}{a}$ which gives $L(t) \leq L(t_\epsilon)e^{-\frac{(em-d_3)(\lambda_1-1-\epsilon)(t-t_\epsilon)}{a_L}}$.

In this case if $\lambda_1 > 1+\epsilon$ for all $t \geq t_\epsilon$, then $L(t) \leq L(t_\epsilon)e^{-\frac{(em-d_3)(\lambda_1-1-\epsilon)(t-t_\epsilon)}{a_L}} \to 0$ as $t \to \infty$.
This implies, if $\lambda_1 > 1$ and $em > d_3$, then $L(t) \to 0$ as $t \to \infty$.

Proof of Theorem 3.3:

$\frac{dM}{dt} \geq M\left\{\alpha C - d_1 - \frac{g}{k(1-C)}\right\} \geq 0$ if $c_0 \leq C(t) \leq C_0$ and $g < \frac{k\alpha}{4}\left(1-\frac{d_1}{\alpha}\right)^2$,
where
$c_0 = \frac{\alpha+d_1-\sqrt{(\alpha-d_1)^2-\frac{4\alpha g}{k}}}{2\alpha}$ and $C_0 = \frac{\alpha+d_1+\sqrt{(\alpha-d_1)^2-\frac{4\alpha g}{k}}}{2\alpha}$.
Therefore, there exists $T_1 > 0$ and $0 < m_0 < 1$ such that $m_0 \leq M(t) \leq U$ for all $t \geq T_1$.
$\frac{dP}{dt} \geq P\left[s\left\{1-\frac{P}{k(1-C_0)}\right\} - \frac{mU}{a} - \frac{h_1}{c_1}\right]$ gives $\frac{dP}{dt}|_{P=p_0} \geq 0$, where $0 < p_0 = k(1-C_0)\left(1-\frac{mc_1U+ah_1}{asc_0}\right)$.
Therefore, if $h_1 < c_1\left(s-\frac{mU}{a}\right)$, there exists $T_2 > 0$ such that $0 < p_0 \leq P(t) \leq U$, for all $t \geq \max\{T_1, T_2\}$.
$\frac{dL}{dt} \geq L\left(\frac{emp_0}{a+p_0} - \frac{h_2}{c_2+L} - d_3\right)$ gives $\frac{dL}{dt}|_{L=l_0} \geq 0$, where $l_0 = \frac{h_2(a+p_0)}{(em-d_3)(p_0-\lambda_1)} - c_2$.
Therefore, if $h_2 > \frac{c_2(em-d_3)(p_0-\lambda_1)}{a+p_0}$, there exists $T_3 > 0$ such that $0 < l_0 \leq L(t) \leq U$, for all $t \geq \max\{T_1, T_2, T_3\}$.
Therefore, for large t, if $g < \frac{k\alpha}{4}\left(1-\frac{d_1}{\alpha}\right)^2$, $h_1 < c_1\left(s-\frac{mU}{a}\right)$ and $h_2 > \frac{c_2(em-d_3)(p_0-\lambda_1)}{a+p_0}$, there exists $m_0 > 0$ such that all the solutions of (2) enter into the set $\{(M,C,P,L) : m_0 \leq M(t) \leq U, c_0 \leq C(t) \leq C_0, p_0 \leq P(t) \leq U, l_0 \leq L(t) \leq U\}$ and will remain there forever.

Proof of Theorem 4.1:

The variational matrix at E_0 is

$$\begin{bmatrix} \beta - d_1 - 2\beta M_0 - \gamma & M_0(\alpha-\beta)-\gamma & -\frac{gM_0}{k} & 0 \\ 0 & r(1-M_0)-\alpha M_0 - d_2 & 0 & 0 \\ 0 & 0 & s-\frac{h_1}{c_1} & 0 \\ 0 & 0 & 0 & -\frac{h_2}{c_2}-d_3 \end{bmatrix}$$

At E_0, the eigenvalues of the variational matrix are $-\frac{h_2}{c_2}-d_3, s-\frac{h_1}{c_1}, r(1-M_0)-\alpha M_0-d_2$ and $\beta-d_1-2\beta M_0-\gamma$.
Since at E_0 we have $\beta - d_1 = \beta M_0 + \gamma\left(1-\frac{1}{M_0}\right)$, it follows that $\beta - d_1 - 2\beta M_0 - \gamma = -\beta M_0 - \frac{\gamma}{M_0} < 0$.
Therefore, the system is stable at E_0 if $h_1 > sc_1$ and $\alpha > \frac{r(1-M_0)-d_2}{M_0}$.

Proof of Theorem 4.2:

The variational matrix at E_1 is

$$\begin{bmatrix} \xi_1 & M_1(\alpha-\beta)-\gamma-\frac{gP_1M_1}{k} & -\frac{gM_1}{k} & 0 \\ 0 & r(1-M_1)-\alpha M_1-d_2 & 0 & 0 \\ 0 & -\frac{sP_1^2}{k} & s\left(1-\frac{2P_1}{k}\right)-\frac{h_1c_1}{(c_1+P_1)^2} & -\frac{mP_1}{a+P_1} \\ 0 & 0 & 0 & \frac{emP_1}{a+P_1}-\frac{h_2}{c_2}-d_3 \end{bmatrix}$$

where $\xi_1 = \beta - d_1 - 2\beta M_1 - \gamma - \frac{gP_1}{k}$.

At E_1, the eigenvalues of the variational matrix are $\frac{emP_1}{a+P_1} - \frac{h_2}{c_2} - d_3, \beta - d_1 - 2\beta M_1 - \gamma - \frac{gP_1}{k}, r(1-M_1) - \alpha M_1 - d_2$ and $s\left(1-\frac{2P_1}{k}\right) - \frac{h_1c_1}{(c_1+P_1)^2}$.

Since at E_1 we have $\beta - d_1 - \frac{gP_1}{k} = \beta M_1 + \gamma\left(1 - \frac{1}{M_1}\right)$, it follows that $\beta - d_1 - 2\beta M_1 - \gamma - \frac{gP_1}{k} = -\beta M_1 - \frac{\gamma}{M_1} < 0$.

Therefore, the system is stable at E_1 if $\alpha > \frac{r(1-M_1)-d_2}{M_1}, h_1 < \frac{s(c_1+P_1)^2}{k}$ and $h_2 > \frac{c_2(em-d_3)(P_1-\lambda_1)}{a+P_1}$.

Proof of Theorem 4.3:

The variational matrix at E_2 is

$$\begin{bmatrix} \xi_2 & M_2(\alpha-\beta)-\gamma-\frac{gP_2M_2}{k} & -\frac{gM_2}{k} & 0 \\ 0 & r(1-M_2)-\alpha M_2-d_2 & 0 & 0 \\ 0 & -\frac{sP_2^2}{k} & \frac{mP_2L_2}{(a+P_2)^2}+\frac{h_1P_2}{(c_1+P_2)^2}-\frac{sP_2}{k} & -\frac{mP_2}{a+P_2} \\ 0 & 0 & \frac{emaL_2}{(a+P_2)^2} & \frac{h_2L_2}{(c_2+L_2)^2} \end{bmatrix}$$

where $\xi_2 = \beta - d_1 - 2\beta M_2 - \gamma - \frac{gP_2}{k}$.

At E_2, two eigenvalues of the variational matrix are $\beta - d_1 - 2\beta M_2 - \gamma - \frac{gP_2}{k}$ and $r(1-M_2) - \alpha M_2 - d_2$.

The other two eigenvalues are given by the equation

$\lambda^2 - \lambda\left\{\frac{mP_2L_2}{(a+P_2)^2} + \frac{h_1P_2}{(c_1+P_2)^2} - \frac{sP_2}{k} + \frac{h_2L_2}{(c_2+L_2)^2}\right\} + \frac{h_2L_2}{(c_2+L_2)^2}\left\{\frac{mP_2L_2}{(a+P_2)^2} + \frac{h_1P_2}{(c_1+P_2)^2} - \frac{sP_2}{k}\right\} + \frac{em^2aP_2L_2}{(a+P_2)^3} = 0.$

Since at E_2 we have $\beta - d_1 - \frac{gP_2}{k} = \beta M_2 + \gamma\left(1 - \frac{1}{M_2}\right)$, it follows that $\beta - d_1 - 2\beta M_2 - \gamma - \frac{gP_2}{k} = -\beta M_2 - \frac{\gamma}{M_2} < 0$.

Now, $r(1-M_2) - \alpha M_2 - d_2 < 0$ if $\alpha > \frac{r(1-M_2)-d_2}{M_2}$, $\frac{h_2L_2}{(c_2+L_2)^2}\left\{\frac{mP_2L_2}{(a+P_2)^2} + \frac{h_1P_2}{(c_1+P_2)^2} - \frac{sP_2}{k}\right\} + \frac{em^2aP_2L_2}{(a+P_2)^3} > 0$ if $h_2 < \frac{em^2a(c_2+L_2)^2}{(a+P_2)^3}\left\{\frac{s}{k} - \frac{mL_2}{(a+P_2)^2} - \frac{h_1}{(c_1+P_2)^2}\right\}^{-1}$ and $\frac{mP_2L_2}{(a+P_2)^2} + \frac{h_1P_2}{(c_1+P_2)^2} - \frac{sP_2}{k} + \frac{h_2L_2}{(c_2+L_2)^2} < 0$ if $h_1 < (c_1+P_2)^2\left\{\frac{s}{k} - \frac{mL_2}{(a+P_2)^2} - \frac{h_2L_2}{P_2(c_2+L_2)^2}\right\}$.

Therefore, the system is stable at E_2 if $\alpha > \frac{r(1-M_2)-d_2}{M_2}, h_1 < (c_1 +$

$P_2)^2 \left\{ \frac{s}{k} - \frac{mL_2}{(a+P_2)^2} - \frac{h_2 L_2}{P_2(c_2+L_2)^2} \right\}$ and $h_2 < \frac{em^2 a(c_2+L_2)^2}{(a+P_2)^3} \left\{ \frac{s}{k} - \frac{mL_2}{(a+P_2)^2} - \frac{h_1}{(c_1+P_2)^2} \right\}^{-1}$.

Proof of Theorem 4.4:

At E_3, two eigenvalues of the variational matrix are $s - \frac{h_1}{c_1}$ and $-\frac{h_2}{c_2} - d_3$. The other two eigenvalues are given by the equation $\lambda^2 - \lambda\{(\alpha - r)C_3 - d_1 + \beta(1 - C_3) - 2\beta M_3 - \gamma\} + C_3[(r+\alpha)\{(\alpha-\beta)M_3 - \gamma\} - r\{\alpha C_3 - d_1 + \beta(1 - C_3) - 2\beta M_3 - \gamma\}] = 0$.

Since at E_3 we have $\alpha C_3 - d_1 + \beta(1-C_3) - 2\beta M_3 - \gamma = -\frac{\gamma}{M_3}(1-C_3) - \beta M_3$, it follows that $(\alpha - r)C_3 - d_1 + \beta(1 - C_3) - 2\beta M_3 - \gamma = -\beta M_3 - rC_3 - \frac{\gamma}{M_3}(1 - C_3) < 0$.

Now, $s - \frac{h_1}{c_1} < 0$ if $h_1 > sc_1$ and $C_3[(r+\alpha)\{(\alpha-\beta)M_3 - \gamma\} - r\{\alpha C_3 - d_1 + \beta(1 - C_3) - 2\beta M_3 - \gamma\}] > 0$ if $\beta < r + \alpha - \frac{\gamma}{M_3} + \frac{r\gamma(1-M_3-C_3)}{\alpha M_3^2}$.

Therefore, the system is stable at E_3 if $h_1 > sc_1$ and $\beta < r + \alpha - \frac{\gamma}{M_3} + \frac{r\gamma(1-M_3-C_3)}{\alpha M_3^2}$.

Proof of Theorem 4.5:

The variational matrix at E_4 is

$$\begin{bmatrix} \xi_4 & (\alpha-\beta)M_4 - \gamma - \frac{gP_4M_4}{k(1-C_4)^2} & -\frac{gM_4}{k(1-C_4)} & 0 \\ -(r+\alpha)C_4 & -rC_4 & 0 & 0 \\ 0 & -\frac{sP_4^2}{k(1-C_4)^2} & \frac{h_1P_4}{(c_1+P_4)^2} - \frac{sP_4}{k(1-C_4)} & -\frac{mP_4}{a+P_4} \\ 0 & 0 & 0 & \frac{emP_4}{a+P_4} - \frac{h_2}{c_2} - d_3 \end{bmatrix}$$

where $\xi_4 = -\beta M_4 - \frac{\gamma}{M_4}(1 - C_4)$.

At E_4, one eigenvalue of the variational matrix is $\frac{emP_4}{a+P_4} - \frac{h_2}{c_2} - d_3 < 0$ if $h_2 > \frac{c_2(em-d_3)(P_4-\lambda_1)}{a+P_4}$.

The other three eigenvalues are given by the equation $\lambda^3 + \eta_1\lambda^2 + \eta_2\lambda + \eta_3 = 0$, where

$$\eta_1 = rC_4 + \beta M_4 + \frac{\gamma}{M_4}(1 - C_4) - \frac{h_1P_4}{(c_1+P_4)^2} + \frac{sP_4}{k(1-C_4)}$$

$$\begin{aligned}\eta_2 = & rC_4 \left\{ (\alpha - \beta)M_4 - \gamma - \frac{gP_4M_4}{k(1-C_4)^2} + \beta M_4 + \frac{\gamma}{M_4}(1 - C_4) \right. \\ & \left. - \frac{h_1P_4}{(c_1+P_4)^2} + \frac{sP_4}{k(1-C_4)} \right\} - \alpha C_4 \left\{ (\alpha-\beta)M_4 - \gamma - \frac{gP_4M_4}{k(1-C_4)^2} \right\} \\ & - \left\{ \beta M_4 + \frac{\gamma}{M_4}(1 - C_4) \right\} \left\{ \frac{h_1P_4}{(c_1+P_4)^2} - \frac{sP_4}{k(1-C_4)} \right\}\end{aligned}$$

$$\eta_3 = -C_4\left\{\frac{h_1P_4}{(c_1+P_4)^2} - \frac{sP_4}{k(1-C_4)}\right\}\left[r\left\{\beta M_4 + \frac{\gamma}{M_4}(1-C_4)\right\}\right.$$
$$\left. -(r+\alpha)\left\{(\alpha-\beta)M_4 - \gamma - \frac{gP_4M_4}{k(1-C_4)^2}\right\}\right] + \frac{sgM_4C_4P_4^2(r+\alpha)}{k^2(1-C_4)^3}$$

and

$\eta_3 > 0$ if $g > \frac{k(1-C_4)^2(c_1+P_4)^2}{h_1M_4P_4(r+\alpha)}\left\{\frac{h_1}{(c_1+P_4)^2} - \frac{s}{k(1-C_4)}\right\}\left\{\frac{r\gamma}{M_4}(1-M_4-C_4) + \alpha M_4(r+\alpha-\beta) - \alpha\gamma\right\} = g_0.$

Therefore, the system is stable at E_4 if $g > g_0$, $h_1 < h_0$, $h_2 > \frac{c_2(em-d_3)(P_4-\lambda_1)}{a+P_4}$ and $\eta = \eta_1\eta_2 - \eta_3 > 0$.

Proof of Theorem 4.6:

The characteristic equation of the variational matrix at E^* is $\lambda^4 + \eta_1^*\lambda^3 + \eta_2^*\lambda^2 + \eta_3^*\lambda + \eta_4^* = 0$, where

$\eta_1^* = -(F_M^1|_{E^*} + F_C^2|_{E^*} + F_P^3|_{E^*} + F_L^4|_{E^*})$,

$\eta_2^* = (F_M^1|_{E^*} + F_C^2|_{E^*} + F_P^3|_{E^*})F_L^4|_{E^*} + F_M^1|_{E^*}(F_C^2|_{E^*} + F_P^3|_{E^*}) + F_C^2|_{E^*}F_P^3|_{E^*} - F_C^3|_{E^*}F_P^2|_{E^*} + F_L^3|_{E^*}F_P^4|_{E^*} - F_C^1|_{E^*}F_M^2|_{E^*}$,

$\eta_3^* = F_L^3|_{E^*}\{F_M^1|_{E^*}(F_C^2|_{E^*} + F_P^3|_{E^*}) + F_C^2|_{E^*}F_P^3|_{E^*} - F_C^3|_{E^*}F_P^2|_{E^*}\} + F_M^1|_{E^*}(F_C^2|_{E^*}F_P^3|_{E^*} - F_C^3|_{E^*}F_P^2|_{E^*}) - F_M^2|_{E^*}(F_C^1|_{E^*}F_P^3|_{E^*} - F_C^3|_{E^*}F_P^1|_{E^*}) - (F_M^1|_{E^*} + F_C^2|_{E^*})F_L^3|_{E^*}F_P^4|_{E^*} - F_C^1|_{E^*}F_M^2|_{E^*}F_L^4|_{E^*}$,

$\eta_4^* = (F_C^2|_{E^*}F_P^3|_{E^*} - F_C^3|_{E^*}F_P^2|_{E^*})F_M^1|_{E^*}F_L^4|_{E^*} - (F_M^1|_{E^*}F_C^2|_{E^*} - F_M^2|_{E^*}F_C^1|_{E^*})F_L^3|_{E^*}F_P^4|_{E^*} - (F_C^1|_{E^*}F_P^3|_{E^*} - F_C^3|_{E^*}F_P^1|_{E^*})F_M^2|_{E^*}F_L^4|_{E^*}$.

By the Routh-Hurwitz's criterion for stability it follows that the system is locally asymptotically stable at E^* if $\eta_1^* > 0, \eta_1^*\eta_2^* > \eta_3^*$ and $\eta_1^*\eta_2^*\eta_3^* > \eta_3^{*2} + \eta_1^{*2}\eta_4^*$.

References

1. S.J. Box and P.J. Mumby, Effect of macroalgal competition on growth and survival of juvenile Caribbean corals, *Mar. Ecol. Prog. Ser.* **342**, 139-149 (2007).
2. T.P. Hughes, N.A.J. Graham, J.B.C. Jackson, P.J. Mumby and R.S. Steneck, Rising to the challenge of sustaining coralreef resilience, *Trends Ecol. Evol.* **25(11)**, 633-642 (2010).
3. J.F. Bruno, H. Swetman, W.F. Precht and E.R. Selig, Assessing evidence of phase shifts from coral to macroalgal dominance on coral reefs, *Ecology* **90(6)**, 1478-1484 (2009).
4. A.J. Cheal, M.A. MacNeil, E. Cripps, M.J. Emslie, M. Jonker, B. Schaffelke and H. Sweatman, Coral-macroalgal phase shifts or reef resilience: links with diversity and functional roles of herbivorous fishes on the Great Barrier Reef, *Coral Reefs* **29**, 1005-1015 (2010).
5. D. Lirman, Competition between macroalgae and corals: effects of herbivore exclusion and increased algal biomass on coral survivorship and growth, *Coral Reefs* **19**, 392-399 (2001).

6. J.E. Smith, M. Shaw, D.O. Edwards and O. Pantos, Indirect effects of algae on coral: algae-mediated, microbeinduced coral mortality, *Ecol. Lett.* **9**, 835-845 (2006).
7. P.J. Mumby, N. Foster and E. Fahy, Patch dynamics of coral reef macroalgae under chronic and acute disturbance, *Coral Reefs* **24**, 681-692 (2005).
8. V.J. Harriott and S.A. Banks, Latitudinal variation in coral communities in eastern Australia: A qualitative biophysical model of factors regulating coral reefs, *Coral Reefs* **21**, 83-90 (2002).
9. P.J. Mumby and R.S. Steneck, Coralreef management and conservation in light of rapidly evolving ecological paradigms, *Trends Ecol. Evol.* **23(10)**, 555-563 (2008).
10. J.C. Blackwood, A. Hastings, P.J. Mumby, The effect of fishing on hysteresis in Caribbean coral reefs, *Theoretical Ecology* **5(1)**, 105-114 (2012).
11. J.A. Morris, J.L. Akins, A. Barse, D. Cerino, D.W. Freshwater, S.J. Green, R.C. Munoz, C. Paris and P.E. Whitefield, Biology and Ecology of Invasive Lionfishes, Pterois miles and *Pterois Volitans*, *Gulf and Caribbean Fisheries Institute* **61**, 1-6 (2009).
12. J.E. Smith, C.L. Hunter and C.M. Smith, The effects of top-down versus bottom-up control on benthic coral reef community structure, *Oecologia* **163(2)**, 497-507 (2010).
13. M.A. Albins and M.A. Hixon, Invasive Indo-Pacific lionfish (*Pterois Volitans*) reduce recruitment of Atlantic coral-reef fishes, *Marine Ecology Progress Series* **367**, 233-238 (2008).
14. L.J. McCook, J. Jompa and G. Diaz-Pulido, Competition between corals and algae on coral reefs: a review of evidence and mechanisms, *Coral Reefs* **19**, 400-417 (2001).
15. J.W. McManus and J.F. Polsenberg, Coral-algal phase shifts on coral reefs: ecological and environmental aspects, *Progress in Oceanography* **60**, 263-279 (2004).
16. D.B. Kramer, Adaptive harvesting in a multiple-species coral-reef food web, *Ecol. Soc.* **13(1)**, 17 (2007).
17. J.A. Morris Jr. and P.E. Whitfield, Biology, ecology, control and management of the invasive Indo-Pacific lionfish: an updated integrated assessment, *NOAA Technical Memorandum NOS-NCCOS* **99**, 1-65 (2009).
18. P. Lenzini and J. Rebaza, Nonconstant Predator Harvesting on Ratio-dependent Predator-prey Models, *Applied Mathematical Sciences* **4(13-16)**, 791-803 (2010).
19. S. Ruan, Persistence and coexistence in zooplankton-phytoplankton-nutrient models with instantaneous nutrient recycling, *J. Math. Biol.* **31**, 633-654 (1993).
20. R.S. Cantrell and C. Cosner, On the Dynamics of Predator-Prey Models with the Beddington-DeAngelis Functional Response, *J. Math. Anal. Appl.* **257**, 206-222 (2001).
21. J. Cui and Y. Takeuchi, Permanence, extinction and periodic solution of predator-prey system with Beddington-DeAngelis functional response, *J. Math. Anal. Appl.* **317**, 464-474 (2006).

DYNAMICS OF HEPATITIS C VIRAL LOAD WITH OPTIMAL CONTROL TREATMENT STRATEGY

RAM KEVAL
Department of Applied Sciences
Madan Mohan Malaviya University of Technology
Gorakhpur, Uttar Pradesh, India
E-mail: ramkeval@gmail.com

SANDIP BANERJEE*
Department of Mathematics
Indian Institute of Technology Roorkee
Roorkee 247667, Uttaranchal, India
E-mail: sandofma@iitr.ac.in

Mathematical models of hepatitis C with gompertzian as well as logistic proliferation are analyzed using Pontryagin's Maximum Principle with an objective functional that maximizes uninfected hepatocytes, minimizes infected hepatocytes and viral load such that an optimal control pair of efficacy of the drugs can be obtained, taking care of the side effects of the drugs: interferon and ribavirin. It is observed that the model with gompertzian proliferation shows a faster decline in the viral load, compared to the logistic proliferation.

1. Introduction

To understand the dynamics of viral infection, many mathematical models, using optimal control methods have been proposed and applied to obtain the optimal therapy for viral infection.[1,2,3,4,5,7,8,9] Kirschner *et al.*[1] used single control representing the percentage effect the chemotherapy has on viral production. Optimal control was also used by Fister *et al.*,[2] which represents the percentage effect of chemotherapy on the interaction of $CD4^+$ T cells with virus. Kwon[3] used one control parameter representing the effectiveness of the reverse transcriptase inhibitors (RTI) with mutant virus

*This study was supported by the Indo French Centre for Applied Mathematics (IFCAM) (Grant No. MA/IFCAM/15/03).

particles that minimize the wild-type virus and the mutant virus. Two controls has been considered by Joshi,[4] one featuring the boosting of the immune system and the other delaying HIV progression. Two optimal control strategy was also used by Garira *et al.*,[8] one to control the stimulating effect of RTI's and the other control simulating the effect of PI's, incorporating drugs efficacy. Hattaf[9] also incorporated two controls: efficacy of reverse trascriptase and protease inhibitors with two virus particles, namely, infectious virus and noninfectious virus.

In this paper, we have considered two mathematical models of Hepatitis C viral dynamics with different proliferation terms to study the optimal treatment strategy by using Pontryagin's Maximum Principle. Our goal is to maximize uninfected hepatocytes, minimize the infected hepatocytes and viral load such that an optimal control pair of efficacy of the drugs can be obtained, taking care of the side effects of the drugs: interferon and ribavirin. At the end, we compare which model is preferable compared to the other.

2. Model 1 and Its Control

The proposed mathematical model is the modification of the model by Banerjee *et al.*[10]:

$$
\begin{aligned}
\frac{dT}{dt} &= s + rT\left(1 - \frac{T+I}{T_{\max}}\right) - d_1 T - (1 - c\eta_1)\alpha V_I T \\
\frac{dI}{dt} &= (1 - c\eta_1)\alpha V_I T + rI \ln\left(\frac{T_{\max}}{T+I}\right) - d_2 I \\
\frac{dV_I}{dt} &= \left(1 - \frac{\eta_r + \eta_1}{2}\right)\beta I - d_3 V_I \\
\frac{dV_{NI}}{dt} &= \left(\frac{\eta_r + \eta_1}{2}\right)\beta I - d_3 V_{NI}
\end{aligned} \tag{1}
$$

The model describes the interaction of uninfected hepatocytes, infected hepatocytes, infectious virus, non-infectious virus and two drugs-namely interferon (IFN) and rebavirin. T, I, V_I, V_{NI} the uninfected hepatocytes, infected hepatocytes, infectious virus and non-infectious virus respectively. Uninfected hepatocytes are produced at rate s, and die at a rate d_1. Here, $rT\left(1 - \frac{T+I}{T_{\max}}\right)$ is the logistic proliferation term for uninfected hepatocytes and $\ln\left(\frac{T_{\max}}{T+I}\right)$ is the density dependent gompertzian proliferation term, which is used for infected hepatocytes. Both hepatocytes can proliferate at a rate r and $T_{\max}$ is the maximum carying capacity of T and I. Here η_1 denotes the effectiveness of interferon in blocking the release of new virions, $c\eta_1$ is fraction of effectiveness $(0 < c < 1)$ and $(1-c\eta_1)$ is the ineffectiveness of interferon. $(1-c\eta_1)\alpha V_I T$ denotes the interaction term, when uninfected hepatocytes interact with infectious virus V_I to produce infected hepatocytes (also seen in the second equation as a conversion term), d_2 is the natural death of infected hepatocytes. In the third equation, the first term, namely, $(1-\frac{\eta_r+\eta_1}{2})$, is ineffectiveness of combined therapy by interferon and ribavirin, which leads to the growth of the infectious virions. η_r is the effectiveness of ribavirin and β is infection rate constant. $(\frac{\eta_r+\eta_1}{2})$ is the combined effect of ribavirin and interferon, which restricts the growth of infectious virions and results in the growth of non-infectious virions, d_3 being the natural death of both the virions. Here, $0 \leq \eta_1 < 1, 0 \leq \eta_r < 1$ and $0 \leq (\frac{\eta_r+\eta_1}{2}) < 1$. The system of equations is analyzed with initial conditions $T(0) = T_0, I(0) = I_0, V_I(0) = V_{I0}, V_{NI}(0) = V_{N0}$.

We study the dynamics of the system (1) where the aim is to have a greater control on the treatment policy over a certain fixed time horizon. To achieve this, the objective function is defined as

$$J(\eta_1, \eta_r) = \int_0^{t_f} \left[T(t) - I(t) - V_I(t) - \frac{1}{2}\left(A\, \eta_1^2(t) + B\, \eta_r^2(t)\right) \right] dt \qquad (2)$$

where, the first term represents the maximization of uninfected hepatocytes, second and third term gives minimization of infected hepatocytes and infectious virus respectively. The last term $\frac{1}{2}\left(A\, \eta_1^2(t) + B\, \eta_r^2(t)\right)$ is the systemic cost of the drug treatment, in another word, the side effects of interferon and ribavirin. The positive constants A and B balance the size of terms. Since, the treatment is restricted to a limited periods of time, finite time horizon has been considered.

The aim is to maximize the number of uninfected hepatocytes and minimizing the systemic cost of the body. Our objective is to obtain an optimal control pair (η_1^*, η_r^*) such that

$$J(\eta_1^*, \eta_r^*) = \max\{J(\eta_1, \eta_r) | (\eta_1, \eta_r) \in N\},$$

where $N = \{(\eta_1, \eta_r) | \eta_1, \eta_r$ measurable, $a_1 \leq \eta_1 \leq b_1, a_2 \leq \eta_r \leq b_2, t \in [0, t_f]\}$ is the control set.

For this problem, existence of an optimal control is a sufficient condition[1,4] and its necessary condition can be obtained using Pontryagin's Maximum Principle. For this purpose, we define the Hamiltonian as follows:

$$
\begin{aligned}
H = & \left[T - I - V_I - \frac{1}{2}\left(A\,\eta_1^2 + B\,\eta_r^2\right)\right] \\
& + \xi_1\left[s + rT\left(1 - \frac{T+I}{T_{\max}}\right) - d_1 T - (1 - c\eta_1)\alpha V_I T\right] \\
& + \xi_2\left[(1 - c\eta_1)\alpha V_I T + rI\ln\left(\frac{T_{\max}}{T+I}\right) - d_2 I\right] \\
& + \xi_3\left[\left(1 - \frac{\eta_r + \eta_1}{2}\right)\beta I - d_3 V_I\right] + \xi_4\left[\left(\frac{\eta_r + \eta_1}{2}\right)\beta I - d_3 V_{NI}\right]. \quad (3)
\end{aligned}
$$

Thus, the Pontryagin's Maximum Principle gives the existence of adjoint variables satisfying:

$$
\begin{aligned}
\xi_1^{'} = & -\left[1 + \left(r - d_1 - \frac{2rT}{T_{\max}} - \frac{rI}{T_{\max}} - (1 - c\eta_1)\alpha V_I\right)\xi_1\right. \\
& \left. + \left(-\frac{rI}{T+I} + (1 - c\eta_1)\alpha V_I\right)\xi_2\right], \quad (4) \\
\xi_2^{'} = & -\left[-1 - \frac{rT}{T_{\max}}\xi_1 + \left(-d_2 - \frac{rI}{T+I} + r\ln\left(\frac{T_{\max}}{T+I}\right)\right)\xi_2\right. \\
& \left. + \left(1 - \frac{\eta_r + \eta_1}{2}\right)\beta\xi_3 + \left(\frac{\eta_r + \eta_1}{2}\right)\beta\xi_4\right], \quad (5) \\
\xi_3^{'} = & -\left[-1 - (1 - c\eta_1)\alpha T\xi_1 + (1 - c\eta_1)\alpha T\xi_2 - d_3\xi_3\right], \quad (6) \\
\xi_4^{'} = & -\left[-d_3\xi_4\right]. \quad (7)
\end{aligned}
$$

where $\xi_i(t)$ are the adjoint variables and $\xi_1^{'} = -\frac{\partial H}{\partial T}$, $\xi_2^{'} = -\frac{\partial H}{\partial I}$, $\xi_3^{'} = -\frac{\partial H}{\partial V_I}$, $\xi_4^{'} = -\frac{\partial H}{\partial V_{NI}}$. $\xi_i(t_f) = 0$ for $(i = 1, 2, 3, 4)$ are the transversality conditions. To obtain an optimality conditions, we have taken the variation with respect to controls (η_1 and η_r) and set it equal to zero,[4]

$$
\begin{aligned}
\frac{\partial H}{\partial \eta_1} &= -A\eta_1 + c\alpha T V_I \lambda_1 - c\alpha T V_I \lambda_2 - \frac{\beta I}{2}\xi_3 + \frac{\beta I}{2}\xi_4 = 0, \\
\frac{\partial H}{\partial \eta_r} &= -B\eta_r - \frac{\beta I}{2}\xi_3 + \frac{\beta I}{2}\xi_4 = 0.
\end{aligned}
$$

After solving both above equations, we get the optimal control pair as

$$
\eta_1^* = \frac{c\alpha T V_I(\lambda_1 - \lambda_2)}{A} + \frac{\beta I(\lambda_4 - \lambda_3)}{2A}, \quad (8)
$$

$$
\eta_r^* = \frac{\beta I(\lambda_4 - \lambda_3)}{2B}. \quad (9)
$$

Clearly our controls are bounded below by a_i and bounded above by $b_i, i = 1, 2$ (from the definition of N). Therefore we have

$$\eta_1^* = \begin{cases} a_1, & \frac{c\alpha TV_I(\lambda_1-\lambda_2)}{A} + \frac{\beta I(\lambda_4-\lambda_3)}{2A} \leq a_1 \\ \frac{c\alpha TV_I(\lambda_1-\lambda_2)}{A} + \frac{\beta I(\lambda_4-\lambda_3)}{2A}, & a_1 < \frac{c\alpha TV_I(\lambda_1-\lambda_2)}{A} + \frac{\beta I(\lambda_4-\lambda_3)}{2A} < b_1 \\ b_1, & \frac{c\alpha TV_I(\lambda_1-\lambda_2)}{A} + \frac{\beta I(\lambda_4-\lambda_3)}{2A} \geq b_1 \end{cases}$$

and

$$\eta_r^* = \begin{cases} a_2, & \frac{\beta I(\lambda_4-\lambda_3)}{2B} \leq a_2 \\ \frac{\beta I(\lambda_4-\lambda_3)}{2B}, & a_2 < \frac{\beta I(\lambda_4-\lambda_3)}{2B} < b_2 \\ b_2, & \frac{\beta I(\lambda_4-\lambda_3)}{2B} \geq b_2 \end{cases}$$

Thus the optimality system comprises of the state equation, co-state or adjoint equations, with their corresponding initial and final transversality condition, along with the following optimal control,

$$\eta_1^* = \min\left\{\max\left\{\frac{c\alpha TV_I(\lambda_1-\lambda_2)}{A} + \frac{\beta I(\lambda_4-\lambda_3)}{2A}, a_1\right\}, b_1\right\}, \quad (10)$$

$$\eta_r^* = \min\left\{\max\left\{\frac{\beta I(\lambda_4-\lambda_3)}{2B}, a_2\right\}, b_2\right\}. \quad (11)$$

3. Model 2 and the Control Problem

Our second model which has already been proposed by Dahari *et al.*,[6] is given as follows:

$$\begin{aligned} \frac{dT}{dt} &= s + r_T T\left(1 - \frac{T+I}{T_{\max}}\right) - d_T T - \beta V_I T \\ \frac{dI}{dt} &= \beta V_I T + r_I I\left(1 - \frac{T+I}{T_{\max}}\right) - \delta I \\ \frac{dV_I}{dt} &= (1-\rho)(1-\varepsilon)pI - cV_I \\ \frac{dV_{NI}}{dt} &= \rho(1-\varepsilon)pI - cV_{NI} \end{aligned} \quad (12)$$

Initial conditions: $T(0) = T_0', I(0) = I_0', V_I(0) = V_{I0}', V_{NI}(0) = V_{N0}'$.

Here, the uninfected hepatocytes are produced at a constant rate s, r_T and r_I are the proliferation rates of the uninfected and infected hepatocytes respectively; d_T, δ and c are the natural death rates per cell; ε and ρ are the respective efficacies of interferon and ribavirin; β is the rate at which the uninfected hepatocytes are infected with infectious virions and p is the rate at which infected cell releases new infectious virions.

There are two significant differences of this model from our proposed model (1). Firstly, the infected cell gompertzian proliferation term is replaced by the logistic proliferation growth term and the efficacies of the drug interferon and ribavirin are chosen in different functional term. Our aim is to determine optimal treatment strategy for this model and compare with our proposed model.

We define the objective functional to study the dynamics of the system (12) for greater control on the treatment policy over a certain fixed time horizon as

$$J(\rho,\varepsilon) = \int_0^{t_f} \left[T(t) - I(t) - V_I(t) - \frac{1}{2}\left(A_1\,\rho^2(t) + B_1\,\varepsilon^2(t)\right)\right] dt. \tag{13}$$

So, the Hamiltonian is

$$\begin{aligned} H = &\left[T - I - V_I - \frac{1}{2}\left(A_1\,\rho^2(t) + B_1\,\varepsilon^2(t)\right)\right] \\ &+ \xi_{11}\left[s + r_T T\left(1 - \frac{T+I}{T_{\max}}\right) - d_T T - \beta V_I T\right] \\ &+ \xi_{22}\left[\beta V_I T + r_I I\left(1 - \frac{T+I}{T_{\max}}\right) - \delta I\right] \\ &+ \xi_{33}\left[(1-\rho)(1-\varepsilon)pI - cV_I\right] \\ &+ \xi_{44}\left[\rho(1-\varepsilon)pI - cV_{NI}\right]. \end{aligned} \tag{14}$$

Thus, the Maximum Principle gives the existence of adjoint variables satisfying:

$$\begin{aligned} \xi_{11}' = &-\left[1 + \left(r_T - d_T - \frac{2r_T T}{T_{\max}} - \frac{r_T I}{T_{\max}} - \beta V_I\right)\xi_{11}\right. \\ &\left. + \left(-\frac{r_I I}{T_{\max}} + \beta V_I\right)\xi_{22}\right], \end{aligned} \tag{15}$$

$$\begin{aligned} \xi_{22}' = &-\left[-1 - \frac{r_T T}{T_{\max}}\xi_{11} + \left(r_I - \delta - \frac{2r_I I}{T_{\max}} - \frac{r_I T}{T_{\max}}\right)\xi_{22}\right. \\ &\left. + (1-\rho)(1-\varepsilon)p\xi_{33} + \rho(1-\varepsilon)p\xi_{44}\right], \end{aligned} \tag{16}$$

$$\xi_{33}' = -\left[-1 - \beta T\xi_{11} + \beta T\xi_{22} - c\xi_{33}\right], \tag{17}$$

$$\xi_{44}' = -\left[-c\xi_{44}\right]. \tag{18}$$

where, $\xi_{11}(t)$, $\xi_{22}(t)$, $\xi_{33}(t)$, $\xi_{44}(t)$ are the adjoint variables and $\xi_{11}^{'} = -\frac{\partial H}{\partial T}$, $\xi_{22}^{'} = -\frac{\partial H}{\partial I}$, $\xi_{33}^{'} = -\frac{\partial H}{\partial V_I}$, $\xi_{44}^{'} = -\frac{\partial H}{\partial V_{NI}}$. $\xi_{ii}(t_f) = 0$ $(i = 1,2,3,4)$ are the transversality conditions. After putting the variation with respect to controls (ρ and ε) equal to zero,[4] we get the optimality conditions as

$$\frac{\partial H}{\partial \rho} = -A_1\rho - (1-\varepsilon)pI\xi_{33} + (1-\varepsilon)pI\xi_{44} = 0,$$

$$\frac{\partial H}{\partial \varepsilon} = -B_1\varepsilon - (1-\rho)pI\xi_{33} - \rho pI\xi_{44} = 0.$$

Solving both above equations, we get the optimal control pair as

$$\varepsilon^* = \frac{(A_1\xi_{33} + (\xi_{33} - \xi_{44})^2\beta I)\beta I}{((\xi_{33} - \xi_{44})\beta I)^2 - A_1B_1}, \tag{19}$$

$$\varrho^* = \frac{(\xi_{33} - \xi_{44})(B_1 + \beta I\xi_{33})I\beta}{((\xi_{33} - \xi_{44})\beta I)^2 - A_1B_1}. \tag{20}$$

4. Numerical Results

We use forward-backward Runge-Kutta forth order method to solve the system of ordinary differential equations. The state equations (1) constitute a forward initial value problem with initial conditions $(T(0) = T_0, I(0) = I_0, V_I(0) = V_{I0}, V_{N0}(0) = V_{N0})$ while the costate equations (4)–(7) is a final value problem with final conditions $\xi_i(t_f) = 0$ $(i = 1,2,3,4)$, solved backward in time. The optimal control pair η_1^*, η_r^* (given by equations (10) and (11)) was updated at the end of each loop.

Table 1. Parameter values used for numerical simulations.

Parameters	Set A	Set B
s	8×10^5	3.7×10^4 cells/ml/day
r	0.45	0.73 /day
$T_{\max}$	0.7×10^7	0.6×10^7 cells /ml
α	0.6×10^{-7}	1.8×10^{-7} ml/day/virions
c	0.5	0.5
d_1	4.7×10^{-3}	2.4×10^{-3} /day
d_2	0.30	0.06 /day
β	5.4	13.9 virions/day
d_3	5.9	13.9 /day

Similarly, the state equations (12) constitute a forward initial value problem with initial conditions $(T(0) = T'_0, I(0) = I'_0, V_I(0) = V'_{I0}, V_{N0}(0) = V'_{N0})$ while the costate equations (15)–(18) is a final value problem with final conditions $\xi_{ii}(t_f) = 0$ $(i = 1, 2, 3, 4)$, solved backward in time. The optimal control pair ρ^*, ε^* (given by equations (19) and (20)) was updated at the end of each loop.

For simulation, we have used two set of parameters values from Dahari *et al.*[6] (see Table 1). The positive constants A, B, appeared in the objective function (2) and A_1, B_1, appeared in the objective function (13), maintain a balance between the order of V_I and the magnitude of the treatment costs. Here, we observe the effect of the optimal treatment regimen on the viral load during the treatment period.

Figure 1 gives the dynamics of hepatitis C virus with and without treatment for the set A of parameter values. Figures 1A and 1B shows the variation of treatments strategy of combined drug therapy interferon and ribavirin. Figure 1C shows that the viral load during the period of treatment, which exhibit a biphasic decline with aid of an optimal combination therapy of interferon and ribavirin.

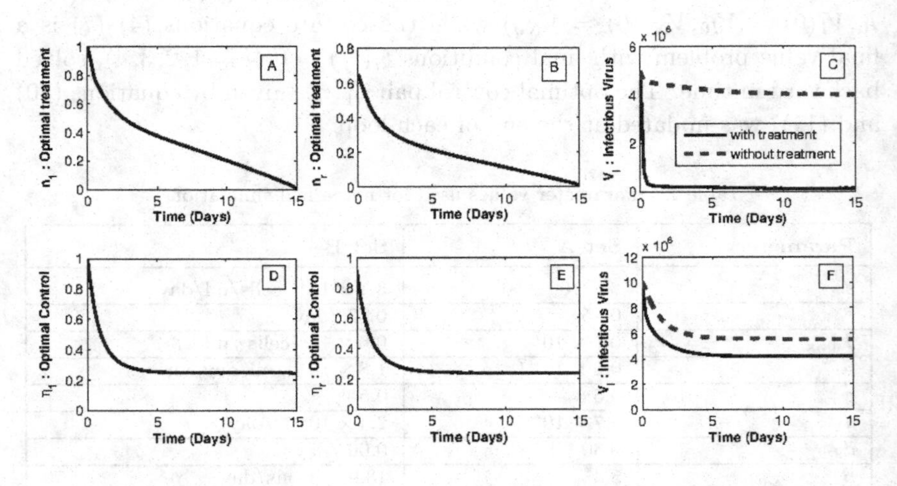

Figure 1. Figures A and B show the optimal treatment strategy of interferon and ribavirin respectively and figure C shows the dynamics of Hepatitis C virus for the set of parameter set A for the mathematical model (1) with and without treatment. The same is shown in figures D, E and F for the parameter set B.

Similar dynamics are observed when the system (1) is simulated with set B of parameter values (see Figs. 1D, 1E, 1F). Comparing both the figures (Figs. 1C and 1F), we conclude that the viral load with parameter set A shows faster decline than with parameter set B. Hence, the set A of parameter values is more effective to reduce the viral load than the set B.

A similar treatment have been done for the model (12) with the set A of parameter values and behavior of viral load is noted (Fig. 2). Comparing the dynamics of viral load for our proposed model (1) and model given by (12) (see Fig. 2D), we conclude that in our model the viral load decline as a faster rate than a model proposed in Dahari *et al.*[6]

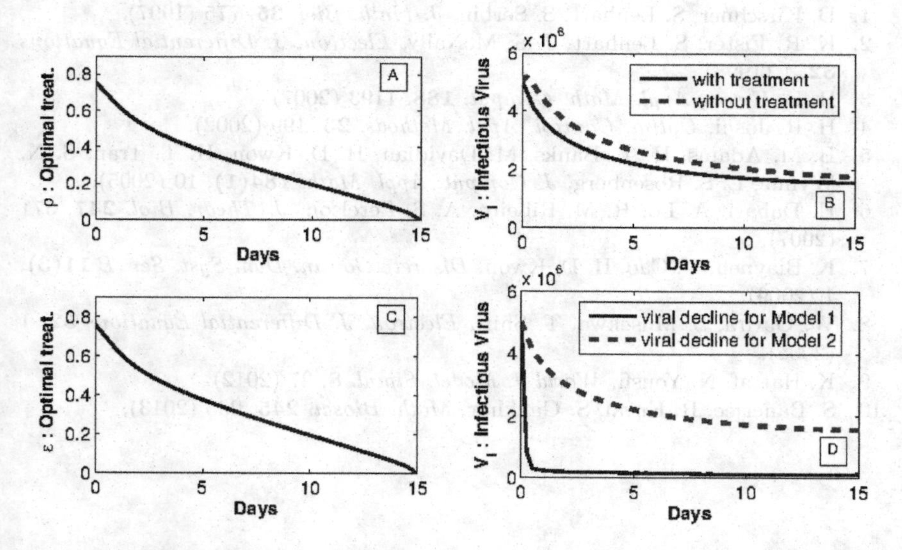

Figure 2. Dynamics of Hepatitis C virus for the parameter values of set A with and without treatment for the mathematical model (12). Figures A and B show the optimal treatment strategy of interferon and ribavirin respectively and figure C shows the viral load. Figure D shows the comparison of dynamics of hepatitis C virus between the mathematical model (1) and (12).

5. Conclusion

In this paper, we have used an optimal combined therapy paradigm to study the dynamical model of hepatitis C virus in conjunction with combination treatment of interferon and ribavirin. We have used deterministic control theory to find an optimal control pair (η_1^*, η_r^*), that is, optimal efficacies of interferon and ribavirin so that the uninfected hepatocytes are maximized and viral load along with side effects of the drugs are minimized. Numerical results indicates that the model is robust with respect to parameter set A. Comparing our model with the proposed model in Dahari *et al.*,[6] we observe that the viral load in our model declines at a faster rate than the proposed model in Dahari *et al.*[6] The optimal treatment that we have determined, causes the level of viral load to go down while at the same time keeping the side effects of drugs to be a minimal.

References

1. D. Kirschner, S. Lenhart, S. Serbin, *J. Math. Biol.* **35**, 775 (1997).
2. K. R. Fister, S. Lenhart, J. S. McNally, *Electron. J. Differential Equations* **32**, 1 (1998).
3. H. D. Kwon, *Appl. Math. Comput.* **188**, 1193 (2007).
4. H. R. Joshi, *Optim. Control. Appl. Methods.* **23**, 199 (2002).
5. B. M. Adams, H. T. Banks, M. Davidian, H. D. Kwon, H. T. Tran, S. N. Wynne, E. S. Rosenberg, *J. Comput. Appl. Math.* **184(1)**, 10 (2005).
6. H. Dahari, A. Lo, R. M. Ribeiro, A. S. Perelson, *J. Theor. Biol.* **247**, 371 (2007).
7. K. Blayneh, Y. Cao, H. D. Kwon, *Discrete Contin. Dyn. Syst. Ser. B* **11(3)**, 1 (2009).
8. W. Garira, S. Musekwa, T. Shiri, *Electron. J. Differential Equations* **52**, 1 (2005).
9. K. Hattaf, N. Yousfi, *World J. Model. Simul.* **8**, 27 (2012).
10. S. Banerjee, R. Keval, S. Gakkhar, *Math. Biosci.* **245**, 235 (2013).

SOMITOGENESIS AND TURING PATTERN

A. LEMARCHAND

Laboratoire de Physique Théorique de la Matière Condensée
Université Pierre et Marie Curie, Sorbonne Universités, CNRS UMR 7600
4 place Jussieu, case courrier 121, 75005 Paris, France
E-mail: anle@lptmc.jussieu.fr

L. SIGNON

Institut de Biologie Intégrative de la Cellule
Université Paris-Sud, CEA, CNRS UMR 9198
Avenue de la Terrasse, 91190 Gif-sur-Yvette cedex, France
E-mail: laurence.signon@i2bc.paris-saclay.fr

B. NOWAKOWSKI

Institute of Physical Chemistry, Polish Academy of Sciences
Kasprzaka 44/52, 01-224 Warsaw, Poland
Warsaw University of Life Sciences SGGW
Nowoursynowska 159, 02-776 Warsaw, Poland
E-mail: bogdan_nowakowski@sggw.pl

The adaptation of prevertebra size to embryo size is investigated in the framework of a reaction-diffusion model involving a Turing pattern. The reaction scheme is modified in order to take into account the departure from dilute conditions induced by confinement in smaller embryos. In agreement with the experimental observations of scaling in somitogenesis, our model predicts the formation of smaller prevertebrae or somites in smaller embryos. These results suggest that models based on Turing patterns cannot be automatically disregarded by invoking the question of maintaining proportions in embryonic development. In addition, our approach highlights the non trivial role that the solvent can play in biology.

1. Introduction

Scaling of pattern formation with embryo size is a universal feature observed in many organisms and the question of maintaining proportions is relevant for both vertebrates[1,2,3] and invertebrates[4,5,6,7]. However, the formation of patterns that are proportional to the size of the embryos remains a poorly understood property of development. Morphogen gradient is a widely accepted feature by which a developing tissue provides its cells with positional information[8,9]. The ability of an embryo to adapt to size variations is often related to the scaling of morphogen gradient with global embryo size[2,5,6,7]. We recently postulated that a reaction-diffusion model based on a Turing pattern could account for prevertebra or somite formation and non trivial experiments[10,11] have been reproduced[12,13,14,15,16]. In order to further investigate the validity of the model, we wish to examine if it could account for scaling of pattern formation.

Since the observation of temporal oscillations of some morphogens in the undifferentiated tissue or presomitic mesoderm[17], the clock and wavefront model[18] has been the most commonly admitted model of somitogenesis[10,19,20,21]. Nevertheless, this model has been lately challenged by recent experiments that show the formation of somites without the need of gene oscillations and clocks[22]. As an alternative to clock and wavefront type models, reaction-diffusion processes offer a minimal framework to model the formation of somites without losing the molecular scale. In this context, a Turing structure, i.e. a spatially-periodic oscillation of morphogen concentrations is supposed to develop behind a propagating chemical wave front. This prepattern is then admitted to induce a complex cascade of pathways, eventually leading to the differentiation of tissues and vertebra formation[23,24,25,26,27,28]. However, the connection between spine development and Turing instability remains a matter of debate[29]. The main criticism against Turing pattern is that it does not *a priori* account for scaling of patterning and size adaptation of the somites to the global size of an embryo. Indeed, the wavelength of a Turing structure is fixed by dynamics, the rate constants of the reactions and the diffusion coefficients of the chemical species, and not by system size[30]. Different ways to preserve proportion in Turing pattern have been achieved by introducing size-dependent dynamical parameters or additional species whose concentration depends on system size[31,32,33,34].

Furthermore, physiological media are known to suffer from confinement and we propose to address the issue of size adaptation in the general context of molecular crowding[35,36,37,38,39,40]. Morphogen signaling is known to depend on maternal factors[41,42,43,44] and it is reasonable to assume that the concentration of maternal factors is higher in smaller embryos. We start from the intuitive statement that the effect of confinement is stronger in smaller embryos. In these conditions, the usual assumptions about dilute solutions may fail and the role of water or solvent in the chemical scheme may not be ignored[45]. Consequently, the rate constants[35,36,46] and the diffusion coefficients[47] may be both modified. The effects of concentration-dependent diffusivity and enhancement of the concentration of some reactant on Turing patterns have been extensively investigated[48,49,50,51,52]. More precisely, we propose to incorporate the solvent in the chemical scheme. We recently examined the deviation to usual Fick's law of diffusion in the framework of linear irreversible thermodynamics[53,54]. We showed that the effects of the departure from ideality on diffusion have a negligible impact on the wavelength of Turing pattern[16]. In this paper, we therefore focus on the relations between the modification of the chemical scheme and the correction to the structure wavelength due to non dilute conditions. Specifically, our aim is to determine whether a strengthening of confinement may induce a decrease of structure wavelength and consequently somite size in smaller embryos. The paper is organized as follows. In section 2, we present the reaction-diffusion model and the corresponding partial differential equations in the presence of confinement. The numerical integration procedure is made precise. The results are given in section 3. Section 4 contains conclusion.

2. The Model

In order to account for the formation of somites, we recently studied a minimal reaction scheme involving two chemical species A and B and two chemostats R_1 and R_2[12,13,14,15,16], inspired from the Schnakenberg model[55] and the Gray-Scott model[56]. We make the hypothesis that the supply of species B from the chemostat R_2 simultaneously drops the solvent S into the surroundings. Conversely, the elimination of A and B through exchanges with the chemostats is supposed to require an interaction with the solvent. The reaction scheme is given by:

$$\mathrm{A} + \mathrm{S} \overset{k_1}{\rightarrow} \mathrm{R}_1 \tag{1}$$

$$2\mathrm{A} + \mathrm{B} \overset{k_2}{\rightarrow} 3\mathrm{A} \tag{2}$$

$$\mathrm{B} + \mathrm{S} \underset{k'_{-3}}{\overset{k_3}{\rightleftharpoons}} \mathrm{R}_2 \tag{3}$$

where k_i, $(i = 1, 2, 3)$ are rate constants and $k_{-3} = k'_{-3}\rho_{R_2}$ is an effective rate constant. The chemostats impose constant values on the densities ρ_{R_1} and ρ_{R_2} of the chemical species R_1 and R_2. Species A, expressed at the rostral end of the embryo, may be identified with retinoic acid (RA) and genes involved in RA signaling, whereas species B, present at the caudal end, may be related to the fibroblast growth factor (FGF) and genes involved in FGF pathway[57]. Different couples of antagonist gradients are found in the literature and may play the role of the activator A and the inhibitor B of the Turing structure, provided that B diffuses faster than species A[20,21]. The total density,

$$\rho = \rho_A(x,t) + \rho_B(x,t) + \rho_S(x,t), \tag{4}$$

is constant. The dilute solution limit,

$$\frac{\rho_A + \rho_B}{\rho} \to 0, \tag{5}$$

corresponds to large values of the density of the solvent ρ_S with respect to the densities ρ_A and ρ_B of species A and B. The reaction-diffusion equations governing the evolution of the densities of species A and B are given by:

$$\frac{\partial \rho_A}{\partial t} = -k_1\rho_A(1 - \frac{\rho_A + \rho_B}{\rho}) + k_2(\rho_A)^2\rho_B + D_A\frac{\partial^2 \rho_A}{\partial x^2} \tag{6}$$

$$\frac{\partial \rho_B}{\partial t} = -k_3\rho_B(1 - \frac{\rho_A + \rho_B}{\rho}) + k_{-3} - k_2(\rho_A)^2\rho_B + D_B\frac{\partial^2 \rho_B}{\partial x^2} \tag{7}$$

where the first term of the right-hand side of each equation becomes linear in the dilute case. The equations admit three homogeneous steady states that are denoted by $(\rho_{A,h}, \rho_{B,h})$, for $h = 1, 2, 3$, and $(\rho^0_{A,h}, \rho^0_{B,h})$ in the dilute case. Space and time are discretized. Step functions are chosen for the initial density profiles of species A and B prepared in $n(t = 0) = 150$ spatial cells as shown in the left-hand panels of Fig. 1. The 10 first spatial cells, supposed to mimic the vicinity of the head of the embryo, are prepared in the steady state

$$\rho_{A,1} = \left(k_2k_{-3} + \sqrt{(k_2k_{-3})^2 - 4(k_1)^2k_2k_3}\right)/(2k_1k_2) \tag{8}$$

$$\rho_{B,1} = (k_{-3} - k_1\rho_{A,1})/k_3 \tag{9}$$

In agreement with the preexistence of the head, zero-flux boundary conditions are chosen at the rostral end, so that the density profiles of species A and B have an extremum at this boundary, once the Turing pattern has developed. The $(n(t=0)-10)$ next spatial cells are prepared in the steady state

$$\rho_{A,3} = 0 \tag{10}$$

$$\rho_{B,3} = \frac{\rho}{2}\left(1 - \sqrt{1 - 4k_{-3}/(k_3\rho)}\right) \simeq 4.14 \tag{11}$$

which becomes $(\rho^0_{A,3} = 0, \rho^0_{B,3} = k_{-3}/k_3 = 4)$ in the dilute case. The numerical values are obtained for parameter values in the domain of stability of a Turing pattern and given in the caption of Fig. 1[14,15].

The departure from ideality or strength of confinement is defined by the ratio of the sum of the densities of species A and B, evaluated at the homogeneous steady state $(\rho_{A,3}, \rho_{B,3})$, and the total density:

$$\delta = \frac{\rho_{A,3} + \rho_{B,3}}{\rho} \tag{12}$$

According to Eqs. (10,11), it reads:

$$\delta = \frac{1}{2}\left(1 - \sqrt{1 - 4k_{-3}/(k_3\rho)}\right) \tag{13}$$

which highlights that the total density ρ may be used as a convenient parameter to control the departure from ideality, without changing the rate constants and the diffusion coefficients. In particular, Eq. (13) shows that the parameter δ and the departure from ideality decrease as the total density ρ increases. The ideal solution, where the solvent is in great excess, is associated with the limit $\rho \to \infty$ for which $\delta \to 0$.

Initially, the density of species A is higher than the density of B at the rostral end and the contrary is observed at the caudal end. The autocatalytic reaction given in Eq. (2) produces A and consumes B so that species A invades and replaces species B. Hence, a travelling front emerges as a solution of the equations as shown in the middle panels of Fig. 1. At the caudal extremity, the boundary conditions are different from those chosen in references[14,15] in which unlimited, free growth was considered. The growth of embryos of finite size is obtained as follows. We add a spatial cell at the caudal end at a constant rate, that we choose smaller than the propagation speed of the wave front, imposed by the dynamical parameters. Hence, the density of species B at a given distance from the caudal end decreases. Such conditions correctly reproduce that the presomitic mesoderm, comprised between the growing caudal end and the faster travelling antagonist gradients of A=retinoic acid and B=FGF, gradually shrinks as observed

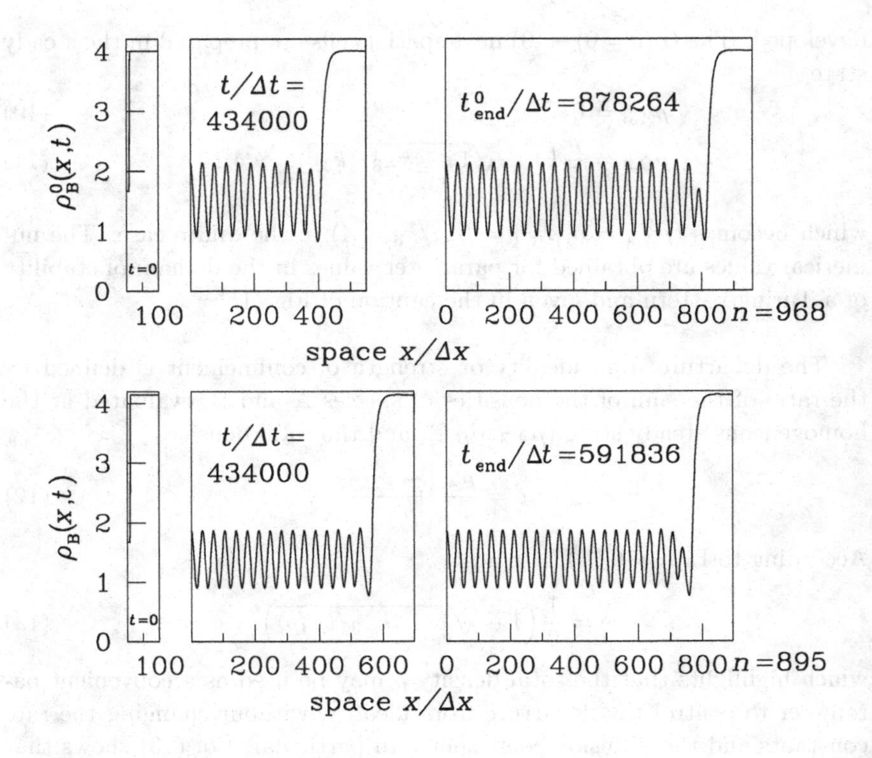

Figure 1. Spatial density profiles of species B in a dilute system (top, $\delta = 0$) and in a non dilute system (bottom, $\rho = 120$, i.e. $\delta = 0.035$) at different times, $t = 0$ (left), $t/\Delta t = 434000$ (middle), and at the end of the simulations, $t^0_{\text{end}}/\Delta t = 878264$ in the limit of a dilute solution and $t_{\text{end}}/\Delta t = 591836$ in the non dilute case. In each case, 21 somites are formed at the ending time at which the density of species B reaches the threshold $\epsilon = 3.94$ in the 100th cell before the caudal end. At the end, the system has $n = 968$ cells in the limit of a dilute solution and $n = 895$ in the non dilute case. The parameters take the following values: $k_1 = 2.57$, $k_2 = 0.88$, $k_3 = 1.93$, $k_{-3} = 7.70$, $D_A = 2.7$, $D_B = 27$, spatial cell length $\Delta x = 0.34$, time step $\Delta t = 3.45 \times 10^{-4}$.

for many vertebrates, such as zebrafish, chickens, mice, and snakes[1].

In order to stop the simulation before the travelling wave reaches the very end of the medium, we introduce a threshold ϵ, such that front propagation and somite growth are arrested when the density ρ_B of species B at a given distance from the caudal extremity falls below this threshold. Figure 2 illustrates the boundary conditions chosen to mimic the growth of an

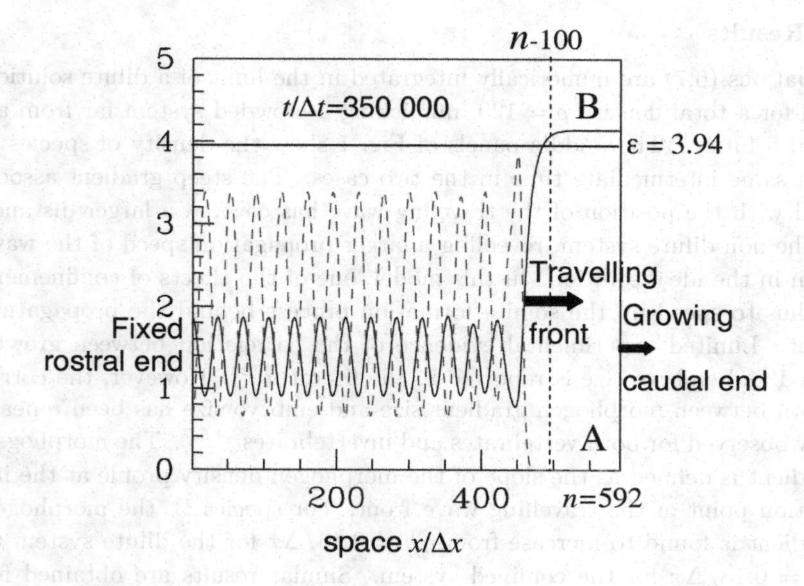

Figure 2. Snapshot of the densities $\rho_A(x,t)$ (dashed line) and $\rho_B(x,t)$ (solid line) of species A and B, solutions of Eqs. (6,7), versus space $x/\Delta x$. The parameter values are given in the caption of Fig. 1. The scheme illustrates the numerical procedure followed to solve the equations. The caudal end has reached the spatial cell $n = 592$ at time $t/\Delta t = 350000$. The travelling gradients of species A and B propagate faster than the system grows and the density of species B, $\rho_B((n-100)\Delta x, t)$, at a given distance (100 spatial cells) from the caudal end, tends to decrease. The threshold $\epsilon = 3.94$ is not yet crossed in the spatial cell $n - 100 = 492$ and growth is not stopped. Twelve somites, located between the maxima of ρ_B, are already formed.

embryo as long as the threshold ϵ has not been crossed in a system of n spatial cells. The system continues to grow as long as $\rho_B((n-100)\Delta x, t) > \epsilon$. In other words, the numerical resolution is stopped at time $t = t_{\text{end}}$ when the density of species B in the spatial cell $(n-100)$ becomes smaller than the threshold ϵ for an embryo of total size n. Using this trick, we are able to assign a length $L = n\Delta x$ to the system. Periodic spatial oscillations of the densities ρ_I of species I=A,B are formed according to Turing instability which develops behind the propagating wave front. The value of the threshold ϵ is chosen on the basis of a trial and error procedure in such a way that the same number of wavelengths, fixed at 21, are formed in the non dilute system with $\rho = 120$ and the dilute limit case when the numerical resolution stops, as shown in the right-hand panels in Fig. 1.

3. Results

Equations (6,7) are numerically integrated in the limit of a dilute solution and for a total density $\rho = 120$, mimicking a crowded system far from an ideal solution. The middle panels of Fig. 1 show the density of species B at a same intermediate time in the two cases. The steep gradient associated with the position of the traveling wave has covered a larger distance in the non dilute system, revealing a larger propagation speed of the wave than in the ideal solution. In this model, one of the effects of confinement is thus to speed up the somite formation process behind the propagating front. Limited experimental evidence of the correlation between growth speed and embryo size is reported in the literature[58]. However, the correlation between morphogen gradient size and embryo size has been repeatedly observed for both vertebrates and invertebrates[2,4,5,7]. The morphogen gradient is defined as the slope of the morphogen density profile at the inflection point of the travelling wave front. For species B, the morphogen gradient is found to increase from $g_B^0 = 0.16/\Delta x$ for the dilute system to $g_B = 0.18/\Delta x$ for the confined system. Similar results are obtained for species A. The increase of the front propagation speed is directly correlated with the increase of the travelling gradients of species A and B when the solution departs from ideality. Hence, the model predicts that an embryo of smaller size, suffering from a strengthening of confinement, is associated with a wave front of steeper gradient, in agreement with experiments on vertebrates[2] and invertebrates[4,7].

As shown in the right-hand panels of Fig. 1 and as a result of the faster wave front speed, the same final number of wavelengths are formed at a shorter time in the non dilute system than in the dilute solution limit. The total length of the non dilute system at the end time and, consequently, the wavelength of the structure are then smaller. Figure 3 gives the variation of the system length L with the departure δ from dilute conditions.

Although the approach is deterministic, the results look noisy because of the procedure used to stop the numerical solution. With a finer spatial discretization, i.e. more cells per wavelength, a better accuracy on the system length L would be obtained. As expected, the behavior of the dilute solution is recovered when solving Eqs. (6,7) for $(\rho_A + \rho_B)/\rho \to 0$. When the departure from ideality is large, for $\delta > 0.035$, i.e. $\rho < 120$, the decrease of system length is nonlinear. Interestingly, a linear relationship between system length L and the strength of confinement is obtained in the interval

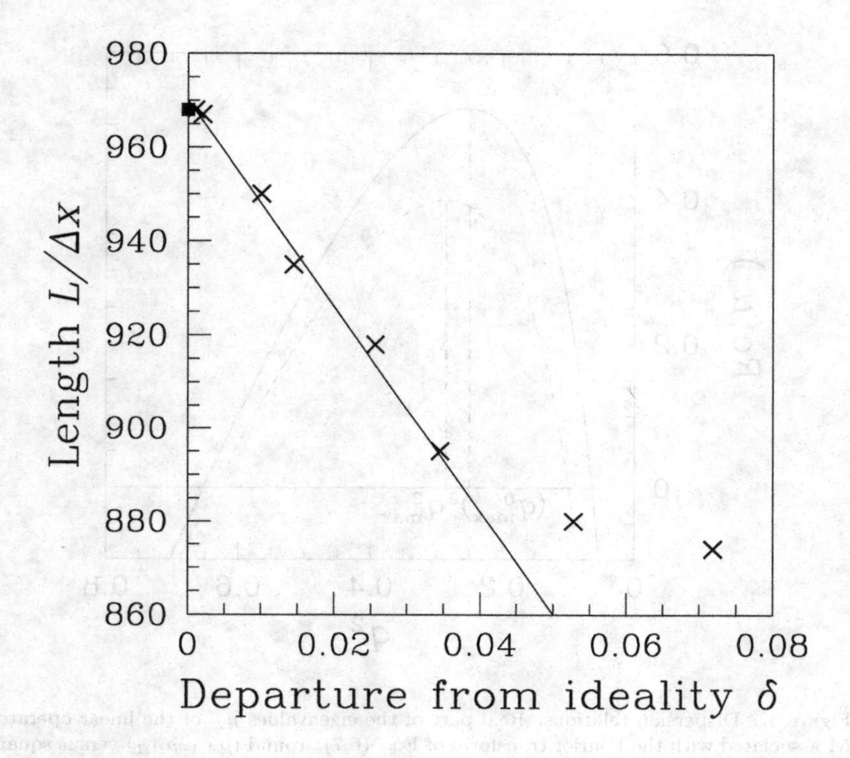

Figure 3. System length $L/\Delta x$ reached at the ending time t_{end} versus departure from dilute conditions $\delta = \frac{\rho_{A,3}+\rho_{B,3}}{\rho}$. The crosses are obtained by solving Eqs. (6,7) for a crowded system for the parameter values given in the caption of Fig. 1 and a variable total density ρ. The square is obtained in the limit of a dilute system. The line is a fit of the data in the domain $\delta \in [0, 0.035]$.

$\delta \in [0, 0.035]$, which defines the domain of validity of the approach. An embryo of acceptable size L is supposed to be associated with a departure δ from ideality inside the interval $[0, 0.035]$. We have chosen to optimize the difference between the system sizes observed in Fig. 1, and show the results obtained for the boundaries of the domain of validity, i.e. the limit $\delta \to 0$ of a dilute solution, which leads to the largest embryo, and the non ideal case associated with $\delta = 0.035$ ($\rho = 120$), which leads to a small embryo.

In order to provide an analytical evaluation of the wavelength of Turing pattern, we use the Fourier transforms, $I_q(t) = \int_{-\infty}^{\infty} \rho_I(x,t)\mathrm{e}^{-iqx}\mathrm{d}x$ for $I =$

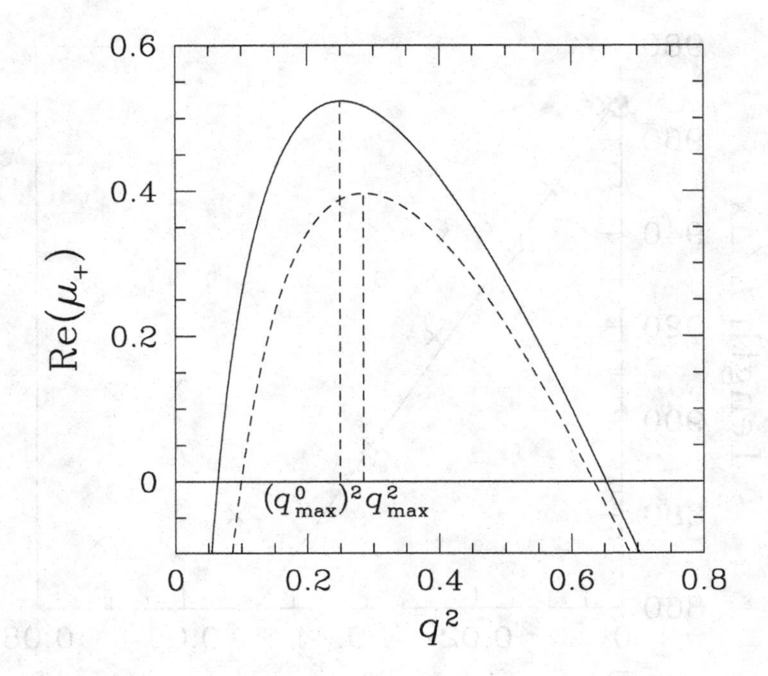

Figure 4. Dispersion relations. Real part of the eigenvalues $\mu_\pm$ of the linear operator $\mathbf{M}$ associated with the Fourier transform of Eqs. (6,7) around $(\rho_{A,1}, \rho_{B,1})$ versus square q^2 of the Fourier mode. The parameter values are given in the caption of Fig. 1. The selected wavelengths, $\lambda^0 = 2\pi/q^0_{\max}$ for the dilute system (solid line) and $\lambda = 2\pi/q_{\max}$ for the non dilute system (dashed line, $\rho = 120$) are deduced from the maximum of the corresponding curves.

A, B of the densities and determine the linear stability of the homogeneous steady state $(\rho_{A,1}, \rho_{B,1})$ with respect to inhomogeneous perturbations[59] of wave number q. The linear stability operator, $\mathbf{M}$, obeys

$$\mathbf{M} = \begin{pmatrix} M_{11} = k_1(1 - \rho_{B,1}/\rho) - D_A q^2 & M_{12} = k_1 \rho_{A,1}/\rho + k_2(\rho_{A,1})^2 \\ M_{21} = k_3 \rho_{B,1}/\rho - 2k_1(1 - \rho_{A,1}/\rho - \rho_{B,1}/\rho) & M_{22} = -k_{-3}/\rho_{B,1} + k_3\rho_{B,1}/\rho - D_B q^2 \end{pmatrix}$$

where $(\rho_{A,1}, \rho_{B,1})$ is given in Eqs. (8,9). The eigenvalues $\mu_\pm$ of $\mathbf{M}$ are:

$$\mu_\pm = \left(M_{11} + M_{22} \pm \sqrt{(M_{11} - M_{22})^2 + 4M_{12}M_{21}}\right)/2 \qquad (14)$$

The dispersion relations obtained for the non dilute system, $\delta = 0.035$, and in the dilute solution limit, $\delta = 0$, are given in Fig. 4. The selected wavelength of the structure is associated with the mode for which the dispersion relation is maximum. The mode $q_{\max}$, selected in the confined system, is larger than the mode $q^0_{\max}$ selected in the dilute system. Consequently, the selected wavelength, $\lambda = 2\pi/q_{\max} \simeq 35.3\Delta x$, is smaller than the wavelength $\lambda^0 = 2\pi/q^0_{\max} \simeq 38\Delta x$ in the dilute solution limit. The analytical values compare well with the numerical values observed in Fig. 1 when solving Eqs. (6,7).

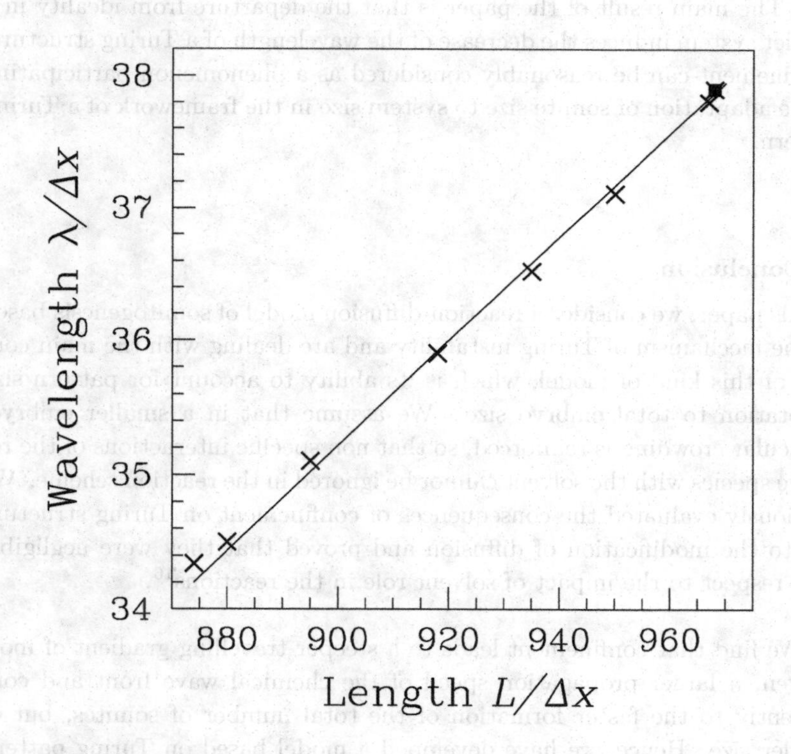

Figure 5. Wavelength $L/\Delta x$ of the Turing pattern versus final system length $L/\Delta x$. The crosses are obtained by solving Eqs. (6,7) for a crowded system for the parameter values given in the caption of Fig. 1 and a variable total density ρ corresponding to a departure from ideality $\delta \in [0, 0.035]$. The square is obtained in the limit of a dilute system. The line is a fit of the data.

In the structured region shown in Fig. 1, the maxima of ρ_B are supposed to initiate the formation of boundaries between somites so that somite size may be evaluated by the wavelength of the spatial structure. Hence, confinement in the smaller, non dilute system leads to the formation of somites of smaller size than in the unperturbed system. The connection between the departure from ideality δ and system size L has been proven in Fig. 3. Figure 5 shows the correlation between system size L and the wavelength λ of the Turing structure. The linear relationship between L and λ clearly reproduces the scaling behavior observed between embryo size and somite size. The main result of the paper is that the departure from ideality in a smaller system induces the decrease of the wavelength of a Turing structure. Confinement can be reasonably considered as a phenomenon participating in the adaptation of somite size to system size in the framework of a Turing pattern.

4. Conclusion

In this paper, we consider a reaction-diffusion model of somitogenesis based on the mechanism of Turing instability and are dealing with the main concern of this kind of model, which is its ability to account for pattern size adaptation to total embryo size. We assume that in a smaller embryo, molecular crowding is reinforced, so that non specific interactions of the reactive species with the solvent cannot be ignored in the reaction scheme. We previously evaluated the consequences of confinement on Turing structure due to the modification of diffusion and proved that they were negligible with respect to the impact of solvent role in the reactions[16].

We find that confinement leads to a steeper travelling gradient of morphogen, a larger propagation speed of the chemical wave front and consequently to the faster formation of the total number of somites, but of smaller size. Hence, we have developed a model based on Turing pattern which incorporates the effects of crowding and predicts the formation of smaller somites in smaller embryos, in agreement with the experimental observations.

Our results prove that Turing modeling of somitogenesis cannot be discarded by invoking the question of scaling in embryonic development. Moreover, the reaction-diffusion model we propose has the advantage to be based on elementary microscopic processes and to display some universal features, in so far as it may be used to model spine formation as well as segmentation of invertebrates. Our approach sheds some light on the role that the solvent can play in biological phenomena, in which it is often disregarded.

References

1. C. Gomez, E. M. Ozbudak, J. Wunderlich, D. Baumann, J. Lewis, and O. Pourquié, Nature **454**, 335 (2008).
2. D. Ben-Zvi, B. Z. Shilo, A. Fainsod, and N. Barkai, Nature **453**, 1205 (2008).
3. V. M. Lauschke, C. D. Tsiairis, P. François, and A. Aulehla, Nature **493**, 101 (2013).
4. S. J. Day and P. A. Lawrence, Development **127**, 2977 (2000).
5. T. Gregor, W. Bialek, R. R. de Ruyter van Steveninck, D. W. Tank, and E. F. Wieschaus, Proc. Natl Acad. Sci. USA **102**, 18403 (2005).
6. S. Restrepo and K. Basler, Curr. Biol. **21**, R815 (2011).
7. D. Cheung, C. Miles, M. Kreitman, and J. Ma, Development **141**, 124 (2014).
8. H. Teimouri and A. B. Kolomeisky, J. Chem. Phys. **140**, 085102 (2014).
9. B. Bozorgui, H. Teimouri, and A. B. Kolomeisky, *J. Chem. Phys.* **143**, 025102 (2015).
10. R. E. Baker, S. Schnell, and P. K. Maini, Developmental Biology **293**, 116 (2006).
11. J. Dubrulle, M. J. McGrew, and O. Pourquié, Cell **106**, 219 (2001).
12. A. Lemarchand and B. Nowakowski, EPL **94**, 48004 (2011).
13. P. Dziekan, A. Lemarchand, and B. Nowakowski, J. Chem. Phys. **137**, 074107 (2012).
14. P. Dziekan, L. Signon, B. Nowakowski, and A. Lemarchand, J. Chem. Phys. **139**, 114107 (2013).
15. P. Dziekan, L. Signon, B. Nowakowski, and A. Lemarchand, Commun. Theor. Phys. **62**, 622 (2014).
16. L. Signon, B. Nowakowski, and A. Lemarchand, *Phys. Rev. E* **93**, 042402 (2016).
17. A. J. Terry, M. Sturrock, J. K. Dale, M. Maroto, and M. A. J. Chaplain, PLoS One **6**, e16980 (2011).
18. J. Cooke and E. C. Zeeman, J. Theor. Biol. **58**, 455 (1976).
19. M.-L. Déquéant and O. Pourquié, Nature Reviews Genetics **9**, 370 (2008).
20. A. Goldbeter and O. Pourquié, J. Theor. Biol. **252**, 574 (2008).
21. Y. J. Jiang, B. L. Aerne, L. Smithers, C. Haddon, D. Ish-Horowicz, and J. Lewis, Nature **408**, 475 (2000).
22. A. S. Dias, A. S., I. de Almeida, J. M. Belmonte, J. A. Glazier, and C. D. Stern, Science **343**, 791 (2014).
23. A. M. Turing, Philos. Trans. R. Soc. London, Ser. B **237**, 37 (1952).
24. H. Meinhardt and A. Gierer, BioEssays **22**, 753 (2000).
25. Y. Schiffmann, Prog. Biophys. Mol. Biol. **84**, 61 (2004).
26. R. E. Baker, S. Schnell, and P. K. Maini, Int. J. Dev. Biol. **53**, 783 (2009).
27. S. Kondo and T. Miura, Science **329**, 1616 (2010).
28. J. B. A. Green and J. Sharpe, Development **142**, 1203 (2015).

29. A. Gonzalez and R. Kageyama, Gene Regul. Syst. Bio. **1**, 35 (2007).
30. J. D. Murray, *Mathematical Biology, I. An Introduction* (Springer, New York, 2002).
31. H. G. Othmer and E. Pate, Proc. Natl Acad. Sci. USA **77**, 4180 (1980).
32. A. Hunding and P. G. Sorensen, J. Math. Biol. **26**, 27 (1988).
33. S. Ishihara and K. Kaneko, J. Theor. Biol. **238**, 683 (2006).
34. S. Werner, T. Stückemann, M. Beiran Amigo, J. C. Rink, F. Jülicher, and B. M. Friedrich, Phys. Rev. Lett. **114**, 138101 (2015).
35. A. P. Minton, J. Biol. Chem. **276**, 10577 (2001).
36. R. J. Ellis and A. P. Minton, Nature **425**, 27 (2003).
37. J. Sun and H. Weinstein, J. Chem. Phys. **127**, 155105 (2007).
38. Z.-R. Xie, J. Chen, and Y. Wu, J. Chem. Phys. **140**, 054112 (2014).
39. P. M. Kekenes-Huskey, C. Eun, and J. A. McCammon, J. Chem. Phys. **143**, 094103 (2015).
40. P. Nalecz-Jawecki, P. Szymanska, M. Kochanczyk, J. Miekisz, and T. Lipniacki, J. Chem. Phys. **143**, 215102 (2015).
41. A. F. Schier and W. S. Talbot, Annu. Rev. Genet. **39**, 561 (2005).
42. E. Spiegler, Y. K. Kim, L. Wassef, V. Shete, and L. Quadro, Biochim. Biophys. Acta **1821**, 88 (2012).
43. M. Rhinn and P. Dollé, Development **139**, 843 (2012).
44. G. J. Hausman, D. R. Campion, and F. C. Buonomo, Growth Dev. Aging **55**, 43 (1991).
45. P. Ball, Chem. Rev. **108**, 74 (2008).
46. R. Grima and S. Schnell, Biophys. Chem. **124**, 1 (2006).
47. K. Takahashi, S. N. V. Arjunan, and M. Tomita, FEBS Letters **579**, 1783 (2005).
48. C. Varea, J. L. Aragon, and R. A. Barrio, Phys. Rev. E **56**, 1250 (1997).
49. M. G. Clerc, E. Tirapegui, and M. Trejo, Phys. Rev. Lett. **97**, 176102 (2006).
50. N. Kumar and W. Horsthemke, Phys. Rev. E **83**, 036105 (2011).
51. W.-S. Li, W.-Y. Hu, Y.-C. Pang, T.-R. Liu, W.-R. Zhong, and Y.-Z. Shao, Phys. Rev. E **85**, 066132 (2012).
52. E. P. Zemskov, K. Kassner, M. J. B. Hauser, and W. Horsthemke, Phys. Rev. E **87**, 032906 (2013).
53. S. R. de Groot and P. Mazur, *Non-Equilibrium Thermodynamics* (North Holland, Amsterdam, 1962).
54. D. Bullara, Y. De Decker, and R. Lefever, Phys. Rev. E **87**, 062923 (2013).
55. J. Schnakenberg, J. Theor. Biol. **81**, 389 (1979).
56. P. Gray and S. K. Scott, Chem. Engng Sci. **39**, 1087 (1984).
57. I. Olivera-Martinez and K. G. Storey, Development **134**, 2125 (2007).
58. P. P. L. Tam, J. Embryol. Exp. Morph. **65** (Suppl), 103 (1981).
59. G. Nicolis and I. Prigogine, *Self-Organization in Nonequilibrium Systems* (Wiley, New York, 1977).

AN EPIDEMIOLOGICAL MODEL OF VIRAL INFECTIONS IN A *VARROA*-INFESTED BEE COLONY: THE CASE OF A BEE-DEPENDENT MITE POPULATION SIZE*

SARA BERNARDI, EZIO VENTURINO
Dipartimento di Matematica "Giuseppe Peano"
via Carlo Alberto 10
Università di Torino, Italy
E-mail: s.bernardi@unito.it, ezio.venturino@unito.it

In recent years the spread of the ectoparasitic mite *Varroa destructor* has become the most serious threat to worldwide apiculture. In the model presented here we extend the bee population dynamics with mite viral epidemiology examined in an earlier paper by allowing a bee-dependent mite population size. The results of the analysis match field observations well and give a clear explanation of how *Varroa* affects the epidemiology of certain naturally occurring bee viruses, causing considerable damages to colonies. The model allows only four possible stable equilibria, using known field parameters. The first one contains only the thriving healthy bees. Here the disease is eradicated and also the mites are wiped out. Alternatively, we find the equilibrium still with no mite population, but with endemic disease among the thriving bee population. Thirdly, infected bees coexist with the mites in the *Varroa* invasion scenario; in this situation the disease invades the hive, driving the healthy bees to extinction and therefore affecting all the bees. Coexistence is also possible, with both populations of bees and mites thriving and with the disease endemically affecting both species. The analysis is in line with field observations in natural honey bee colonies. Namely, these diseases are endemic and if the mite population is present, necessarily the viral infection occurs. Further, in agreement with the fact that the presence of *Varroa* increments the viral transmission, the whole bee population may become infected when the disease vector is present in the beehive. Also, a low horizontal transmission rate of the virus among the honey bees will help in protecting the bee colonies from *Varroa* infestation and viral epidemics.

*This work was supported by the Gruppo Nazionale per il Calcolo Scientifico (GNCS-INdAM) and by the European Union Seventh Framework Programme (FP7/2007-2013) under grant agreement nr. 600841.

1. Introduction

Current evidence shows that Earth loses between one and ten percent of biodiversity per decade, a major event of extinction of biological diversity. The causes are mainly habitat loss, pest invasions, pollution, overharvesting and diseases[11], thereby endangering natural ecosystem services that are vital for humanity.

All products of agriculture depend on pollination, that is performed by wild, free-living organisms such as bees, butterflies, moths and flies. To this end, there is also the availability of commercially-managed bee species. But bees represent the most important group of pollinators in most geographical regions, also for economical reasons[11]. An estimation of the United Nations Food and Agriculture Organisation (FAO) sets the bee-pollinated crops at 71 out of the 100 species providing 90% of food worldwide. In Europe, there are 264 crop species and 84% of them are animal-pollinated; further, 4000 vegetable varieties thrive due to the pollination activities of bees[11].

As the bee group is the most important pollinator worldwide, this paper focuses on the instability of bee populations and, in particular, on the the most serious threat to apiculture globally: the external parasitic mite *Varroa destructor*, discovered in Southeast Asia in 1904. This mite is of the size of a pinhead; it feeds on the bees circulatory fluid. It is an invasive species, spreading from one hive to another one. Today it is present nearly in the whole world[11]. The serious damage to the bee colonies does not derive from the parasitic action of the mite but, above all, from its action as vector of many viral diseases. It increases the transmission rate of diseases such as acute paralysis (ABPV) and deformed wing viruses (DWV), that are considered among the main causes of Colony Collapse Disorder (CCD). This action has been reinforced since about thirty years ago when the mite has shifted hosts from *Apis cerana* to *Apis mellifera*. With no control, the infestations are bound to cause the untimely death of bee colonies within three years.

Mathematical modelling represents a powerful tool for investigating the triangular relationship between honey bees, *Varroa* and viral disease. It allows the exploration of the beehive system without the need of unfeasible or costly field studies. In particular, mathematical models of the epidemiology of viral diseases on *Varroa*-infested colonies would allow us to explore the host population responses to such stressful situations.

In the next Section, the model is presented and some basic properties are investigated. In Section 3, equilibria are analysed for feasibility and stability. A sensitivity analysis on the model parameters is performed in Section 4.4. A final discussion concludes the study.

2. The Model

Let B denote healthy bees, I the infected ones, M the healthy mites and N the infected ones and let the population vector be $X = (B, I, M, N)^T$.

In the model proposed here we essentially extend the epidemiology of a beehive infested by *Varroa* mites examined earlier in[2], to account for a Leslie Gower term in the mite populations, to model the *Varroa* population size as a bee population-dependent function, a step also undertaken in the recent paper[9]. In fact, the mites essentially carry out their life on the hosts body. These changes are therefore reflected in the last two equations, for the mites evolution. The resulting model reads as follows:

$$X' = f(X) = (f_B, f_I, f_M, f_N)^T, \quad f : \mathcal{D}^0 \to \mathbb{R}^4_+, \tag{1}$$

where

$$\frac{dB}{dt} = f_B = b\frac{B}{B+I} - \lambda BN - \gamma BI - mB$$

$$\frac{dI}{dt} = f_I = b\frac{I}{B+I} + \lambda BN + \gamma BI - (m+\mu)I$$

$$\frac{dM}{dt} = f_M = r(M+N) - nM - \frac{p}{h(B+I)}M(M+N) - M(\beta I - \delta N - eB)$$

$$\frac{dN}{dt} = f_N = -nN - \frac{p}{h(B+I)}N(N+M) + \beta MI + \delta MN - eNB.$$

To prevent in the right-hand side of (1) the vanishing of some terms in the denominator, we define $\mathcal{D}^0$ as the domain of f, explicitly

$$\mathcal{D}^0 = \left\{X = (B, I, M, N) \in \mathbb{R}^4_+ : B + I \neq 0\right\}. \tag{2}$$

The first two equations describe respectively the evolution of healthy and infected bee populations. They are born healthy or infected in proportion to the fraction of healthy or infected bees in the colony, with constant b. The infection process for the larvae indeed occurs mainly through their meals of contaminated royal jelly. Healthy bees can contract the virus by infected mites at rate λ, second terms in the equations. The next terms model horizontal transmission of the virus among adult bees, occuring at rate γ via small wounds of the exoskeleton, e.g. as a result of hair loss, or ingestion of faeces. In the last terms we find the bees' natural mortality m

and the disease-related mortality μ, taken into account for the infected bee population only.

The last two equations contain the *Varroa* dynamics, partitioned among susceptible and infected states, that vector the viral disease and once infected remain so for their lifetime. The mites are always born healthy at rate r, as no viruses are passed vertically from the parents to offsprings. In the extended model the first three terms describe *Varroa* growth, natural mortality and intraspecific competition. Note that instead in the previously introduced system[2], the mites grow logistically. Here if the mites reproduction rate r is smaller than their natural mortality n, i.e.

$$r < n, \tag{3}$$

it is easy to see that the total mite population becomes extinct. The effect of intraspecific competition among mites is described by the Leslie Gower term at rate p. It accounts for the mite population dependence on the total number of adult bees, the "resource for which they compete" in the colony[8]. The parameter h expresses the average number of mites per bee. Note that because the virus does not cause any harm to the infected *Varroa* population, competition among healthy and infected mites occurs at the same rate, in other words we find the terms $M(M+N)$ and $N(M+N)$ with the same "weight" in both equations. If infected were weakened by the disease, we would rather have in the bracket $M+cN$, with some $c<1$. Healthy mites can become infected by infected bees at rate β, the fourth terms in the last two equations, but they can also acquire the virus at rate δ also horizontally from other infected mites. The last term in the last two equations models the grooming behavior of healthy bees at rate e. We assume that the infected bees do not groom because they are weaker due to the disease effects. Note that for healthy and infected mites the damage due to grooming is the same, e, as this activity depends only on the bees.

2.1. *Well-posedness and boundedness*

We now address the issue of well-posedness, following basically the path of[10]. The solutions trajectories are shown to be always at a finite distance from the set $\Theta = \{(B,I,M,N) \in \mathbb{R}^4_+ : B+I=0\}$, where the total bee population vanishes.

Theorem 2.1. *Well-posedness and boundedness.*

Let $X_0 \in \mathcal{D}^0$*. A solution of* (1) *defined on* $[0,+\infty)$ *with* $X(0)=X_0$ *exists uniquely. Also, for an arbitrary* $t>0$*, it follows* $X(t) \in \mathcal{D}^0$*, and,* *indicating by L a positive constant,*

$$\frac{b}{\widetilde{m}} \leq \liminf_{t\to+\infty}(B(t)+I(t)) \leq \limsup_{t\to+\infty}(B(t)+I(t)) \leq \frac{b}{m}, \quad \widetilde{m} = m+\mu; \quad (4)$$

$$M(t)+N(t) \leq L, \quad \forall t \geq 0. \quad (5)$$

Proof. We follow the arguments of[10,2] for the first part of the proof. Global Lipschitz continuity of the right-hand side of the system holds in $\mathcal{D}^0$, thereby implying existence and uniqueness of the solution of system (1) for every trajectory at a finite distance of this boundary. Now (4) and (5) hold for all the trajectories starting at any point with $B+I \neq 0$. The boundedness of the variables entail consequently that all trajectories exist at all times in the future and are bounded away from the set Θ.

Adding the two equations for the bees in (1), at any point in $\mathcal{D}^0$ we have

$$B'+I' = b - mB - (m+\mu)I \geq b - \widetilde{m}(B+I), \quad \widetilde{m} = m+\mu.$$

Integration of this differential inequality between $X(0) = X_0$ and $X(t)$, points belonging to a trajectory such that $X(\tau) \in \mathcal{D}^0$ for all $\tau \in [0,t]$, we obtain the following lower bound, with positive right hand side at any $t > 0$,

$$B(t)+I(t) \geq \frac{b}{\widetilde{m}}(1-e^{-\widetilde{m}t}) + (B(0)+I(0))e^{-\widetilde{m}t}. \quad (6)$$

Similarly, but bounding from above, we find $B'+I' \leq b - m(B+I)$, and therefore

$$B(t)+I(t) \leq \frac{b}{m}(1-e^{-mt}) + (B(0)+I(0))e^{-mt}. \quad (7)$$

The inequalities in (4) for any portion of the trajectory belonging to $\mathcal{D}^0$ follow from (6) and (7).

We now turn to the *Varroa* populations M and N, and their the whole mite population $V = M+N$, summing their corresponding equations of (1) and using (4), for any point within $\mathcal{D}^0$ we have the upper bound

$$V' + \alpha V = (r-n+\alpha)V - \frac{p}{h(B+I)}V^2 - eBV \quad (8)$$

$$\leq \left[r-n+\alpha - \frac{p}{h(B+I)}V\right]V \quad \leq \left[r-n+\alpha - \frac{pm}{hb}V\right]V = \phi(V) \leq \phi(\overline{V}),$$

where $\phi(\overline{V})$ represents the maximum value of the parabola $\phi(V)$, namely $\overline{V} = hb(r-n+\alpha)(2pm)^{-1}$.

A consequence of (8) is the differential inequality $V' \le \phi(\overline{V}) - \alpha V$ and thus integrating it we obtain the following inequality, for all $t \ge 0$, proving (5),

$$V(t) \le \frac{\phi(\overline{V})}{\alpha}\left(1 - e^{-\alpha t}\right) + V(0)e^{-\alpha t} \le \max\left\{V(0), \frac{\phi(\overline{V})}{\alpha}\right\}. \qquad (9)$$

From (6), (7) and (9) the boundedness of all populations is thus ensured[10]. Thus all trajectories originating in $\mathcal{D}^0$ remain in $\mathcal{D}^0$ for all $t > 0$. □

Let $\mathcal{D}^1$ be the largest subset of $\mathcal{D}^0$ satisfying the inequalities of Theorem (2.1),

$$\mathcal{D}^1 \doteq \left\{(B, I, M, N) \in \mathbb{R}^4_+ : \frac{b}{\widetilde{m}} \le B + I \le \frac{b}{m}, 0 \le M + N \le L\right\}.$$

As a consequence of Theorem 2.1, it is a compact set, positively invariant for all the system's trajectories. All the system's equilibria belong to $\mathcal{D}^1$, as it is shown later. Thus the equilibria analysis is sufficient to explain the whole system's behavior.

3. Equilibria

The equilibria $E_k = (B_k, I_k, M_k, N_k)$ of (1) are

$$E_1 = \left(\frac{b}{m}, 0, 0, 0\right), \quad E_2 = \left(0, \frac{b}{m + \mu}, 0, 0\right),$$

which are always feasible; the points

$$E_3 = \left(\frac{b}{m}, 0, \left(r - \frac{eb}{m}\right)\frac{bh}{rm}, 0\right), \quad E_4 = \left(\frac{\mu^2 - b\gamma + m\mu}{\mu\gamma}, \frac{b\gamma - m\mu}{\mu\gamma}, 0, 0\right),$$

the former feasible for

$$r \ge \frac{eb}{m}, \qquad (10)$$

the latter feasible whenever the following condition is satisfied

$$0 < b\gamma - m\mu < \mu^2. \qquad (11)$$

Then, there is the coexistence equilibrium $E_* = (B_*, I_*, M_*, N_*)$. However, it is not analytically tractable and will therefore be investigated numerically, with the help of simulations.

Finally, we find the equilibrium $E_5 = (0, I_5, M_5, N_5)$ with no healthy bees. It is obtained as intersection of two conic sections that lie in the fist quadrant of the M - N phase plane. We discuss it in the next subsection.

3.1. *The healthy-bees-free equilibrium*

From the second equation we find $I = b(m+\mu)^{-1}$. Substituting the value of I, the last two equations can be rewritten as

$$\psi : (r-n)M + rN - \frac{p(m+\mu)M(M+N)}{hb} - \frac{\beta bM}{m+\mu} - \delta MN = 0, \quad (12)$$

and

$$\eta : -nN - \frac{p(m+\mu)N(N+M)}{hb} + \frac{\beta bM}{m+\mu} + \delta MN = 0. \quad (13)$$

Thus, the equilibrium follows from determining the intersection of ψ and η in the fist quadrant of the $M-N$ phase plane.

We begin by analyzing the curve ψ.

Solving (12) for the variable N, we find

$$\psi : N = M\left[\frac{p(m+\mu)^2M + \beta hb^2 + (n-r)hb(m+\mu)}{(m+\mu)(hbr - p(m+\mu)M - \delta hbM)}\right] = M\tilde{\psi}(M) \quad (14)$$

The conic ψ crosses the vertical axis at the origin and the horizontal one at the absissa

$$M_0 = \frac{hb[(r-n)(m+\mu) - \beta b]}{p(m+\mu)^2} \quad (15)$$

Further, there is the vertical asymptote

$$M_\infty = \frac{hbr}{p(m+\mu) + \delta hb} > 0. \quad (16)$$

From (1), it is easy to assess the signs of the numerator and the denominator, respectively $\mathcal{N} > 0$ for $M > M_0$ and $\mathcal{D} > 0$ for $M < M_\infty$. Depending on the sign of $\mathcal{N}$ and $\mathcal{D}$, ψ_1 thus shows three different shapes, Figure 1.

We now turn to the second conic section, η.

Rearranging (13) for M, we get the explicit form

$$M = N\frac{[hbn + p(m+\mu)N](m+\mu)}{\beta b^2h + [\delta hb(m+\mu) - p(m+\mu)^2]N} = N\tilde{\eta}(N)$$

Proceeding as before, we determine the intersections of η with the axes. The conic η goes through the origin and crosses the N axis at

$$N_0 = -\frac{hbn}{p(m+\mu)} < 0.$$

Again, we get one vertical asymptote

$$N_\infty = \frac{\beta b^2 h}{(m+\mu)[p(m+\mu) - \delta hb]}.$$

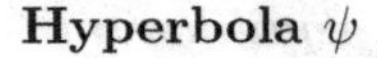

Hyperbola ψ

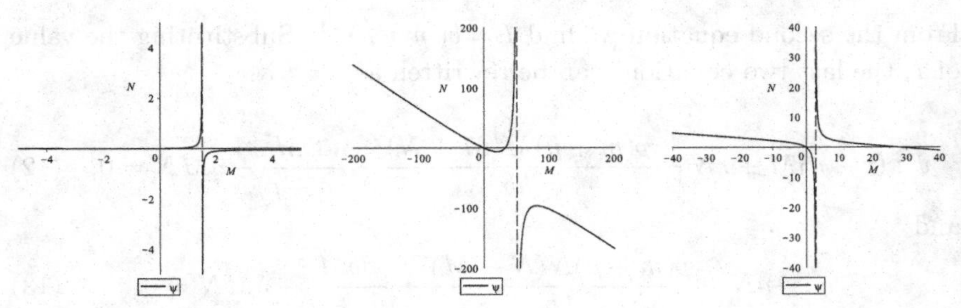

Figure 1. Left: Case 1. $M_0 < 0 < M_\infty$, for the parameters values $r = 0.06$, $e = 0.001$, $\lambda = 0.03$, $b = 2150$, $\beta = 0.0002$, $m = 0.023$, $\mu = 8$, $\delta = 0.04$, $p = 0.05$, $h = 0.9$, $n = 0.007$. Center: Case 2. $0 < M_0 < M_\infty$, for the parameters values $r = 0.2$, $e = 0.001$, $\lambda = 0.03$, $b = 150$, $\beta = 0.0002$, $m = 0.023$, $\mu = 8$, $\delta = 0.001$, $p = 0.05$, $h = 0.9$, $n = 0.1$. Right: Case 3. $0 < M_\infty < M_0$, for the parameters values $r = 0.06$, $e = 0.001$, $\lambda = 0.03$, $b = 250$, $\beta = 0.0002$, $m = 0.04$, $\mu = 8$, $\delta = 0.02$.

Now from (13), the sign of the numerator is positive, $\mathcal{N} > 0$, for $N > N_0$, while for the denominator $\mathcal{D}$ two cases arise. Namely, if $\delta hb < p(m + \mu)$, i.e. $N_\infty > 0$, we get $\mathcal{D} > 0$ for $N < N_\infty$. Otherwise, if $\delta hb > p(m + \mu)$, i.e. $N_\infty < 0$, to have $\mathcal{D} > 0$ the opposite condition $N > N_\infty$ must hold.

Figure 2 sums up the three possible shapes for η.

Hyperbola η

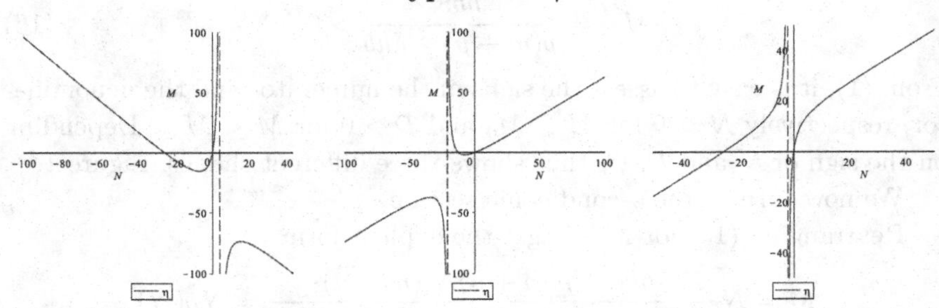

Figure 2. Left: Case A. $N_0 < 0 < N_\infty$, for the parameters values $r = 0.2$, $e = 0.001$, $\lambda = 0.03$, $b = 100$, $\beta = 0.001$, $m = 0.023$, $\mu = 8$, $\delta = 0.001$, $p = 0.05$, $h = 1$, $n = 0.1$. Center: Case B. $N_\infty < N_0 < 0$, for the parameters values $r = 0.2$, $e = 0.001$, $\lambda = 0.03$, $b = 100$, $\beta = 0.01$, $m = 0.023$, $\mu = 8$, $\delta = 0.01$, $p = 0.05$, $h = 1$, $n = 0.05$. Right: Case C. $N_0 < N_\infty < 0$, for the parameters values $r = 0.2$, $e = 0.001$, $\lambda = 0.03$, $b = 100$, $\beta = 0.001$, $m = 0.023$, $\mu = 8$, $\delta = 0.01$, $p = 0.05$, $h = 1$, $n = 0.1$.

Finally, by graphically plotting both the hyperbolae in the M-N plane, they meet at the origin and they further intersect at another point located in the first quadrant. This immediately provides the unconditional existence and feasibility of the equilibrium point E_5.

3.2. *Stability*

The Jacobian J of (1) is the following matrix

$$\begin{pmatrix} J_{11} & -\frac{bB}{(B+I)^2} - \gamma B & 0 & -\lambda B \\ J_{21} & \frac{b}{B+I} - \frac{bI}{(B+I)^2} + \gamma B - (m+\mu) & 0 & \lambda B \\ J_{31} & \frac{pM(M+N)}{h(B+I)^2} - \beta M & J_{33} & r - \frac{pM}{h(B+I)} - \delta M \\ J_{41} & \frac{pN(M+N)}{h(B+I)^2} + \beta M & J_{43} & -n - \frac{p(M+2N)}{h(B+I)} + \delta M - eB \end{pmatrix}$$

with

$$J_{11} = \frac{b}{B+I} - \frac{bB}{(B+I)^2} - \lambda N - \gamma I - m, \quad J_{21} = -\frac{bI}{(B+I)^2} + \lambda N + \gamma I,$$

$$J_{31} = \frac{pM(M+N)}{h(B+I)^2} - eM, \quad J_{33} = r - n - \frac{p(2M+N)}{h(B+I)} - \beta I - \delta N - eB,$$

$$J_{41} = \frac{pN(M+N)}{h(B+I)^2} - eN, \quad J_{43} = -\frac{pN}{h(B+I)} + \beta I + \delta N.$$

At E_1 two eigenvalues of the the Jacobian evaluated at this point, $J(E_1)$, are negative, $-m$ and $-n-ebm^{-1}$. The remaining two provide the stability conditions

$$\gamma b < \mu m, \quad mr < eb + nm. \tag{17}$$

At equilibrium E_2 the Jacobian $J(E_2)$ has two explicit eigenvalues, $-(m+\mu) < 0$ and $\mu - \gamma b(m+\mu)^{-1}$ while the Routh-Hurwitz conditions on the remaining minor show that if (3) is not satisfied, the equilibrium is unstable, because in such case $\det(J(E_2)) < 0$,

$$-\mathrm{tr}(J(E_2)) = 2n + \frac{b\beta}{m+\mu} - r, \quad \det(J(E_2)) = (n-r)\left(n + \frac{b\beta}{m+\mu}\right).$$

The Jacobian $J(E_3)$ once again gives two explicit eigenvalues, $-m < 0$ and another one providing the first stability condition $r < n + 2pM_3(hB_3)^{-1} + eB_3$, i.e. explicitly

$$2eb < m(r-n), \tag{18}$$

and the Routh-Hurwitz criterion on the remaining minor $\widetilde{J}_3$ gives the further conditions

$$-\mathrm{tr}(\widetilde{J}_3) = -\frac{b}{B_3} - \gamma B_3 + m + \mu + n + \frac{pM_3}{hB_3} - \delta M_3 + eB_3 > 0, \quad (19)$$

$$\det(\widetilde{J}_3) = \left(\frac{b}{B_3} + \gamma B_3 - m - \mu\right)\left(\delta M_3 - eB_3 - n - \frac{pM_3}{hB_3}\right) - \beta\lambda B_3 M_3 > 0. \quad (20)$$

Note that if (3) holds, E_3 is unstable, because in this case (18) cannot be satisfied.

At the point E_4 the characteristic equation factorizes into the product of two quadratic equations, for which the Routh-Hurwitz conditions provide the following pairs of inequalities, to be satisfied for stability

$$2m + \mu + \gamma I_4 > \gamma B_4 + \frac{b}{B_4 + I_4}, \quad (21)$$

$$B_4 I_4 \left[\gamma^2 - \frac{b^2}{(B_4 + I_4)^4}\right] + \left(\frac{b}{B_4 + I_4} - m\right)^2 + \left(\frac{b}{B_4 + I_4} - m\right)\left[\gamma B_4 - \frac{bI_4}{(B_4 + I_4)^2} - \mu - \frac{bB_4}{(B_4 + I_4)^2}\right] > \left(\gamma I_4 + \frac{bB_4}{(B_4 + I_4)^2}\right)\left[\gamma B_4 - \mu - \frac{bI_4}{(B_4 + I_4)^2}\right], \quad (22)$$

and

$$2n + 2eB_4 + \beta I_4 > r, \quad (n + \beta I_4 + eB_4)(n + eB_4) > r[(n + eB_4) + \beta I_4]. \quad (23)$$

Two eigenvalues are also explicitly found at E_5, $-m - \mu < 0$ and the other one giving the first stability condition

$$\frac{b}{I_5} < \gamma I_5 + m + \lambda N_5. \quad (24)$$

The Routh-Hurwitz conditions for stability on the remaining quadratic become then

$$2n + \frac{3p(M_5 + N_5)}{hI_5} + \beta I_5 + \delta N_5 > r + \delta M_5, \quad (25)$$

$$\left(r - n - \frac{p(2M_5 + N_5)}{hI_5} - \beta I_5 - \delta N_5\right)\left(\delta M_5 - n - \frac{p(M_5 + 2N_5)}{hI_5}\right) + \left(\beta I_5 - \frac{pN_5}{hI_5} + \delta N_5\right)\left(\frac{pM_5}{hI_5} + \delta M_5 - r\right) > 0.$$

4. Results

After describing the set of parameter values used, the chosen initial conditions and the field data available, we perform the sensitivity analysis. The simulations have been performed using the Matlab built-in ordinary differential equations solver ode45.

4.1. *Model parameters from the literature*

The model parameters that are known from the literature[6,4,1] and[3], §8.2.3.5, are fixed in the simulations, while the remaining ones are changed over a suitable range. In the sensitivity analysis however we will vary also the known ones, to simulate possible environmental variations, due perhaps to climatic changes or other external disruptions.

We set the time unit to be the day. The worker honey bees birth rate is $b = 1500$, their natural mortality rate instead is $m = 0.023$, which implies a life expectancy of 43.5 days in the adult stage[6]. There are no precise values for the grooming behavior in the literature. A possible range of e is presumed to be in the interval $[10^{-6}, 10^{-5}]$[4].

The *Varroa* population reproduces exponentially fast, doubling every month during the spring and summer. We then take $r \approx 30^{-1} \ln 2$, i.e. $r = 0.02$[3], p. 225. In the same season, the mite natural mortality rate in the phoretic phase, i.e. when attached to an adult bee, is presumed to have the value $n = 0.007$[1].

In Table 1 we list all the reference values for the numerical experiments. The other model parameters are freely chosen, with hypothetical values.

Table 1. For these model parameters the values are obtained from the literature. Bee and mite populations are measured in pure numbers.

Parameter	Interpretation	Value	Unit	Source
b	Bee daily birth rate	1500	day^{-1}	6
e	Healthy bee grooming rate	$10^{-6} - 10^{-5}$	day^{-1}	4
m	Bee natural mortality rate	0.023	day^{-1}	6
r	*Varroa* growth rate	0.02	day^{-1}	3
n	*Varroa* natural mortality rate in the phoretic phase	0.007	day^{-1}	1

4.2. *Setting of initial conditions and free parameters*

The colony conditions at the beginning of the spring come from field data: all the bees are healthy. Indeed the infected ones do not survive the winter, because they have a lower life expectancy. In addition, the colony treatments with acaricide are usually performed in the late autumn, to allow the mite eradication. We can safely assume then the mite population is around 10 units at the start of the spring.

4.3. *Use of field data*

Before proceeding with the sensitivity analysis, we check that the feasibility and stability conditions of the equilibrium points are satisfied by the known parameter values, see Table 2.

In particular, we remark that the disease-free equilibrium point E_3 never occurs. Namely, the feasibility condition (10) does not hold in field conditions.

This result highlights the close connection between *Varroa* and viruses. Further, it matches well with beekeeper observations: the bees viral infection is bound to occur whenever the mite population is present. The same conclusion has been obtained in our previous study[2].

Table 2. Summary of the equilibria: feasibility and stability conditions.

Equilibrium	**Feasibility**	**Stability**	**In field conditions**
E_1	always	(17)	allowed
E_2	always	unstable	unstable
E_3	(10)	(18), (19), (20)	infeasible
E_4	(11)	(21), (22), (23)	allowed
E_5	always	(24), (25)	allowed
E_*	numerical simulations	numerical simulations	allowed

4.4. *Sensitivity analysis*

In this section we investigate the behavior of the system responses when the parameters change their values within an appropriate range. We compute the various populations equilibrium values as function of a pair of parameters at a time, thus obtaining surfaces in all the possible pairs of parameter spaces. In order to do this, for each pair of parameters we combine the respective ranges to build a equispaced grid of values and then we plot the four surfaces resulting from the values assumed by the populations at each point of the grid. Since our model contains 12 parameters, there are 66 possible cases. Of these, we present the most interesting results starting from the best situation for the hive, i.e. the cases in which the mite-and-disease-free equilibrium E_1 is stably attained. They arise for the following parameter pairs: $h - \gamma$, $b - \mu$ and $m - \mu$.

In Figure 3 (left) the system shows different transcritical bifurcations, as the parameter γ increases. Specifically, for smaller values of the horizontal trasmission rate γ the system settles to E_1, the healthy beehive scenario, then the infected bee population appears in the system, i.e. we find the epidemics among the bees, E_4, and finally E_5, where the healthy bees population disappears and the mites invade the hive. An increase in h instead positively affects the mite populations, as expected.

From Figure 3 (right), increasing the bees disease-related mortality rate μ drives the system into a safer scenario, all bees are healthy and the colony becomes mite-free. A higher bees birth rate seems to favor the infected mites and bees populations, when the former appear in the ecosystem and also the healthy bees, but this is true only for low values of b. Instead the susceptible mites decrease for an increasing b. Here transcritical bifurcations relate all the equilibrium points. The results confirm what already remarked in[2]. Namely, the equilibrium E_1 is reached for very small values of the transmission rate γ combined with a high enough bees disease-related mortality μ. The behavior of the sensitivity surface in the $m - \mu$ case is very similar.

In the parameter spaces $b - \lambda$ and $b - \gamma$, not shown, a low γ or λ favors the healthy bees and both healthy and infected mites, while b fosters all the system's populations. For $b - \beta$ and $b - \delta$ the behavior is similar in terms of b, but decreasing the other parameter depresses instead the infected mites.

In Figure 4 (left) as the *Varroa* growth rate r increases, the healthy bee population decreases while the mite populations reasonably increase. The dynamics of the infected bee population, instead, is almost insensitive to the parameter r, but it grows linearly with the daily bee birth rate b. Furthermore, this kind of behavior is observed whenever the parameter b is considered, regardless of the other parameter being taken into consideration. In fact, a larger b means a greater number of bees in the colony and

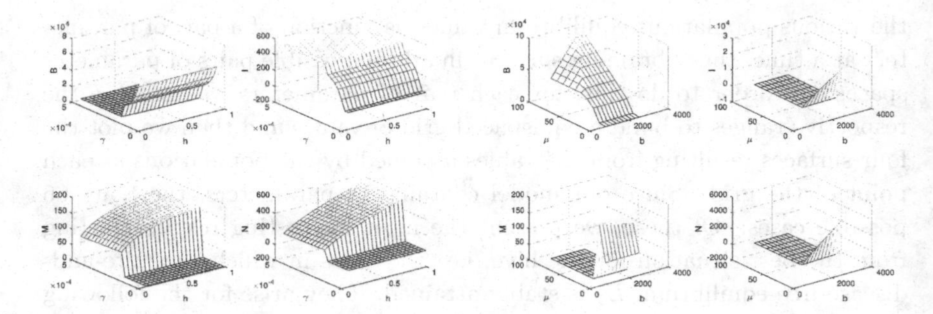

Figure 3. Sensitivity surfaces, in terms of the pair of parameters $h-\gamma$, (left), and $b-\mu$ (right). The other parameter values are choosen as $r = 0.02$, $b = 1500$, $m = 0.023$, $n = 0.007$, $e = 0.000001$, $\mu = 3$, $\gamma = 0.001$, $\lambda = 0.004$, $\delta = 0.00008$, $\beta = 0.00005$, $h = 0.25$, $p = 0.013$. Initial conditions $B = 15000$, $I = 0$, $M = 6$, $N = 4$.

thus proportionally also more infected. As in the previous Figure, infected mites benefit from a higher b, while the healthy ones are instead depressed.

Another transcritical bifurcation is shown in Figure 4 (right) connecting the mite-free equilibrium E_4 and coexistence. The bigger the *Varroa* growth rate, the greater the grooming rate needed to wipe out the mites from the beehive. If this resistance mechanism is not high enough, the mites invade the colony. In the $r-n$ parameter space, all the populations behave like in this $r-e$ case.

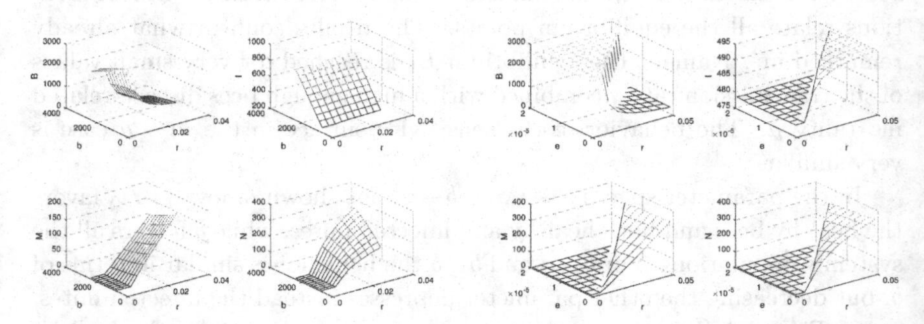

Figure 4. Sensitivity surfaces, in terms of the pair of parameters $r-b$, (left), and $r-e$ (right). The other parameter values are choosen as $r = 0.02$, $b = 1500$, $m = 0.023$, $n = 0.007$, $e = 0.000001$, $\mu = 3$, $\gamma = 0.001$, $\lambda = 0.004$, $\delta = 0.00008$, $\beta = 0.00005$, $h = 0.25$, $p = 0.013$. Initial conditions $B = 15000$, $I = 0$, $M = 6$, $N = 4$.

In Figure 5 (left), the combined effect of the mites growth and of the horizontal transmission of the virus among bees are reported. Only for re-

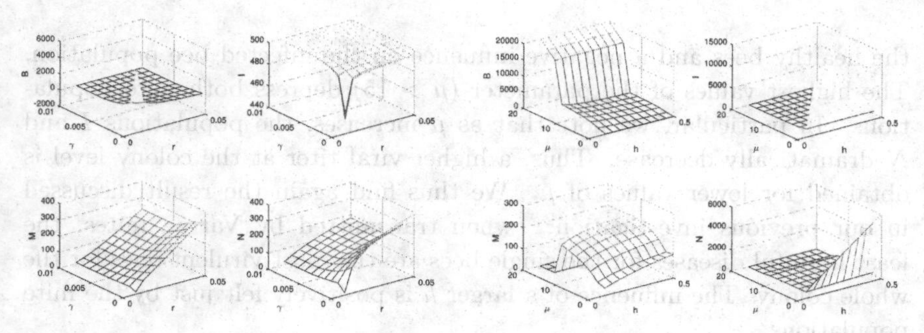

Figure 5. Sensitivity surfaces, in terms of the pair of parameters $r-\gamma$, (left), and $h-\mu$ (right). The other parameter values are choosen as $r = 0.02$, $b = 1500$, $m = 0.023$, $n = 0.007$, $e = 0.000001$, $\mu = 3$, $\gamma = 0.001$, $\lambda = 0.004$, $\delta = 0.00008$, $\beta = 0.00005$, $h = 0.25$, $p = 0.013$. Initial conditions $B = 15000$, $I = 0$, $M = 6$, $N = 4$.

ally small values of both r and γ the healthy bee population can survive. An increase of both γ and r has a negative influence on the healthy bee population and a positive impact on the infected bee population. In this region of the parameter domain, the system approaches the mite-free attractor E_4. For slightly larger values of anyone of these parameters, the mites establish themselves in the system. As a result, we find coexistence followed, for even larger values of both such parameters, by the healthy-bees-free equilibrium E_5. This last transition occurs in particular when the bifurcation parameter γ crosses the critical value $\gamma^{\dagger} \approx 0.005$. Note that as r increases, the mites populations resonably increase too, while they are less sensitive to changes in γ. The bees surfaces look also alike in the cases $\lambda - \gamma$ and $\delta - \beta$, not shown, but the mites have a peak at the origin and are instead depressed by an increase in either one of the parameters.

In the parameter spaces $r-\beta$, $r-\delta$ and $r-\lambda$ the bee populations behave similarly, not shown. A similar picture is also found in the $h-\delta$ parameter space. A low value of r independently of the other parameter leads to the healthy-bee-only point. Then we find coexistence and ultimately tends toward the susceptible-bee-free equilibrium. Larger values of the parameters β and δ increase both mites populations but above all the infected; instead for the latter the opposite occurs for λ.

Figure 5 (right) better shows the influence of the disease-related mortality μ on the system dynamics. Starting from low values of the bifurcation parameter μ, we first find the healthy-bees-free equilibrium E_5, then coexistence followed by the healthy-bees-only equilibrium E_1. Evidently, a higher value of the bees disease-related mortality has a positive impact on the healthy bees and a negative influence on the infected bee population. The highest values of the parameter ($\mu > 15$) depress both mite populations. In particular, we note that as μ increases, the populations I and N dramatically decrease. Thus, a higher viral titer at the colony level is obtained for lower values of μ. We thus find again the result discussed in our previous investigation[2]: when transmitted by Varroa mites, the least harmful diseases for the single bees are the most virulent ones for the whole colony. The influence of a larger h is positively felt just by the mite populations.

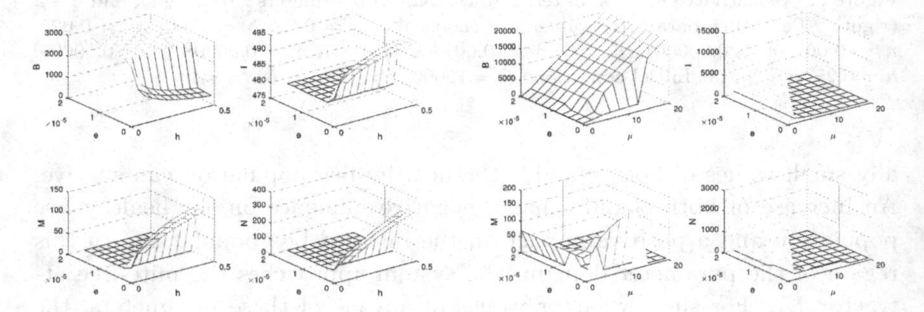

Figure 6. Sensitivity surfaces, in terms of the pair of parameters $h - e$, (left), and $e - \mu$ (right). The other parameter values are choosen as $r = 0.02$, $b = 1500$, $m = 0.023$, $n = 0.007$, $e = 0.000001$, $\mu = 3$, $\gamma = 0.001$, $\lambda = 0.004$, $\delta = 0.00008$, $\beta = 0.00005$, $h = 0.25$, $p = 0.013$. Initial conditions $B = 15000$, $I = 0$, $M = 6$, $N = 4$.

In Figure 6 (left) we explore the effectiveness of the grooming behavior e as a resistance mechanism against to the *Varroa* infestation. The behavior of sensitivity surfaces is encouraging: regardless of the average number of mites per bee, if the grooming behavior is performed strongly enough mites are eventually wiped out of the system. Indeed, a transcritical bifurcation between E_4 and the coexistence equilibrium occurs when the grooming rate crosses the bifurcation threshold $e^{\dagger} \approx 0.000007$. The higher the average number of mites per bee h, the larger both infected populations become, as well as the susceptible mites, while healthy bees are depressed. A similar behavior occurs for the case $h - n$. Also for $m - n$ and $m - e$ we find the equilibrium surfaces to look as for the $h - e$ case, with the only difference that in the last two cases the infected bees benefit from a low m and at least in the range explored, they do not vanish.

The bees in cases $\mu - n$ and $\mu - e$ behave similarly, see Figure 7 right. Mites and infected bees tend to disappear for larger values of μ.

For $\mu - \lambda$, $\mu - \gamma$, $\mu - \delta$ and $\mu - \beta$ the infection in both populations is hindered by larger values of the mortality, while healthy bees benefit from it. The remaining parameter plays a role only on susceptible mites, with low values fostering their growth.

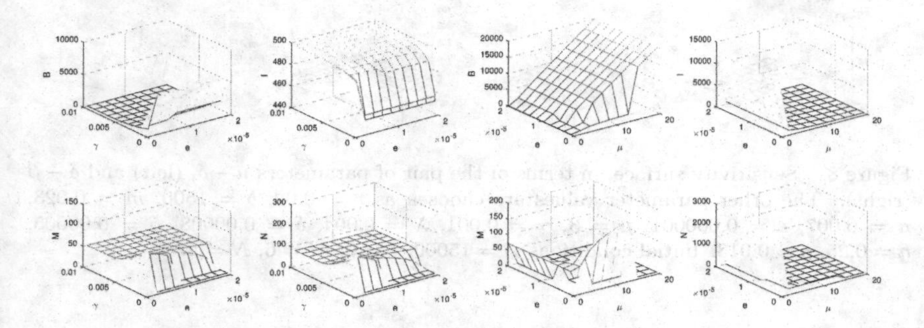

Figure 7. Sensitivity surface, in terms of the pair of parameters $e - \gamma$, (left) and $\mu - e$ (right). The other parameter values are choosen as $r = 0.02$, $b = 1500$, $m = 0.023$, $n = 0.007$, $e = 0.000001$, $\mu = 3$, $\gamma = 0.001$, $\lambda = 0.004$, $\delta = 0.00008$, $\beta = 0.00005$, $h = 0.25$, $p = 0.013$. Initial conditions $B = 15000$, $I = 0$, $M = 6$, $N = 4$.

The surfaces in the parameter spaces $n - \lambda$, $n - \beta$, $e - \lambda$, $e - \delta$, $e - \beta$ and $n - \delta$ again look similar, see Figure 8 left. The mite-free equilibrium is attained for relatively large values of n (or e), low values of the remaining parameter help only the healthy mites.

Cases $\lambda - \delta$, $\lambda - \beta$ are again similar, with larger λ favoring the infection in the bees and low values of the other parameter helping the healthy mites.

Again in the $\beta - \gamma$ and $\delta - \gamma$ cases the most influencial parameter is γ, large values leading to the healthy-bee-free point. The other parameter helps the susceptible populations if it is small.

Cases $p - h$, $p - r$ and $p - m$ are similar, a large p depresses all the populations but the healthy bees, low values of the other parameter helping them too. Also in the parameter spaces $p - \delta$ and $p - \beta$ the surfaces have similar shapes. For $p - \lambda$ and $p - \gamma$ we find a similar behavior in terms of p, but here the infected bees are much less affected by its growth. For $p - b$ similar considerations hold, but again we find that a decrease in b reduces the infected bees, while p on them has scant effect.

In Figure 8 right, we can observe the influence of the transmission rates δ and β. As their values increase, both the healthy bees and mites populations decrease while the infected ones increase. As a result, the level of infection in the hive dramatically grows and almost all the *Varroa* mites become

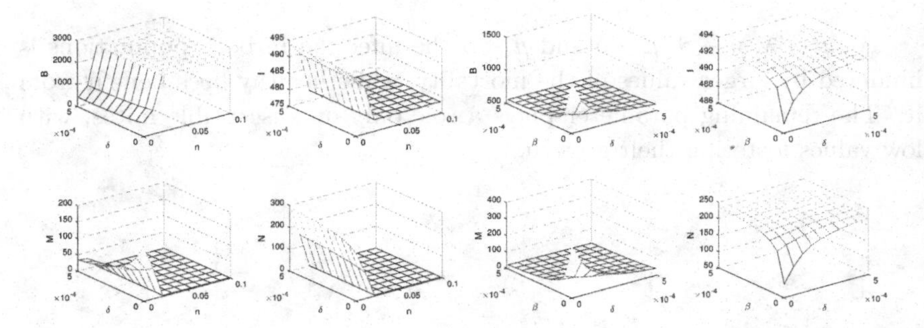

Figure 8. Sensitivity surface, in terms of the pair of parameters $n - \delta$, (left) and $\delta - \beta$ (right). The other parameter values are choosen as $r = 0.02$, $b = 1500$, $m = 0.023$, $n = 0.007$, $e = 0.000001$, $\mu = 3$, $\gamma = 0.001$, $\lambda = 0.004$, $\delta = 0.00008$, $\beta = 0.00005$, $h = 0.25$, $p = 0.013$. Initial conditions $B = 15000$, $I = 0$, $M = 6$, $N = 4$.

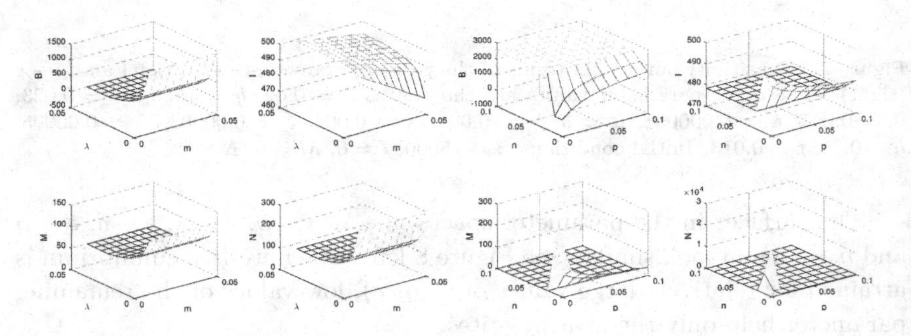

Figure 9. Sensitivity surface, in terms of the pair of parameters $m - \lambda$, (left) and $p - n$ (right). The other parameter values are choosen as $r = 0.02$, $b = 1500$, $m = 0.023$, $n = 0.007$, $e = 0.000001$, $\mu = 3$, $\gamma = 0.001$, $\lambda = 0.004$, $\delta = 0.00008$, $\beta = 0.00005$, $h = 0.25$, $p = 0.013$. Initial conditions $B = 15000$, $I = 0$, $M = 6$, $N = 4$.

virus carriers.

In Figure 9 left, another transcritical bifurcation occurs. An increase of the parameter λ has its greatest influence on the healthy bee population and makes it completely vanish from the system. Here the transition between the coexistence and the no-healthy-bees equilibrium point is observed. Conversely, the transmission rate between infected *Varroa* and healthy bees has a positive effect on the infected bees population. Furthermore, also the mites populations decrease but evidently this is due to the reduction of the host population.

Finally, from Figure 9 right, we note the combined effect of two parameters that hinder the infestation spreading. If the *Varroa* natural mortality is higher than the threshold value $n^\dagger \approx 0.023$ the mites disappear from the hive and we find equilibrium E_4. Instead, the influence of *Varroa* intraspecific competition is shown for smaller values of n: as p increases, both the mites populations and the infected bees decrease while the healthy bees increase. This behavior occurs also in the $p - e$ parameter space.

Although a sufficient rate of the grooming behavior can really help the colony to control the *Varroa* infestation, Figure 6 (right) suggests that the disease-related mortality μ has a more significant influence on the shape of the surfaces and thus on the system dynamics than the grooming rate e. Indeed the shape of the sensitivity surfaces is mostly determined by the changing parameter μ. Note here the following transcritical bifurcations: starting for low values of μ we find E_5, the healthy-bee-free point, then, as μ increases, the coexistence equilibrium and finally the mite-free equilibrium E_4. This chain of transitions reemphasizes once again the effect of the parameter μ.

Furthermore, from Figure 7 left, the effectiveness of the grooming behavior can be relevant only in the presence of a really low value of the parameter γ, i.e. in this region where the transcritical bifurcation mentioned above between coexistence and E_4 is observed, with e as bifurcation parameter. Conversely, for higher values of the horizontal transmission rate among bees, the colony is driven from E_4, the mite-free situation, through coexistence, to E_5 with the extinction of healthy bees.

The knowledge of these most relevant parameters could suggest the ones which would maybe be affected by human external measures in order to drive the system to settle in a possibly safer position. To sum up, the parameters most affecting the system are r, μ, γ and e. Our findings elucidate their influence on the system. These are however all ecosystem-related parameters, that perhaps can hardly be influenced by man-undertaken measures, although they might depend on other external factors such as for instance climatic changes. Theoretically, the sensitivity surfaces show that the population of healthy bees would highly benefit from a reduction of the *Varroa* growth rate, as well as a reduction of the horizontal transmission rate among bees. Also a higher bees disease-related mortality and a larger grooming rate would help in protecting the colonies.

5. Conclusion

We have introduced a model for bees and mites, modifying our previous approach[2], allowing a link between the carrying capacity of the mites and the bees population, while in the former investigation these were not bound together. Our results agree with empirical evidence, in addition to describing the fundamental role played by the mite in this process. The ecosystem described here can settle only to the following possible outcomes: the disease-and-mite-free environment, the ideal situation; the mite-free situation, in which however part of the bees are endemically infected; the coexistence equilibrium in which both bees and mites are present, all affected by the disease; the *Varroa* invasion leading to extinction of the healthy bees, while the remaining ones are all infected.

Among our findings, we observe that the endemic disease cannot affect all the bees in a *Varroa*-free colony. Indeed, the infected-bees-only situation, described by equilibrium E_2 in our analysis, turns out to be unstable, if the mite reproduction rate exceeds their mortality rate, i.e. the opposite of condition (3) holds. Conversely, the *Varroa* invasion scenario can become possible where the bees are all infected and mites invade the hive, the point E_5 above. In agreement with the fact that the presence of *V. destructor* increments the viral transmission, this result indicates that the whole bee population may become infected when the disease vector is present in the beehive.

Also, the two healthy populations cannot survive together, in the absence of infection. This result agrees with field observations, in which the bee colony is considered infected when *V. destructor* is present in the beehive. Thus, the discovery of *Varroa* in the hive necessarily implies that at least part of the bees are virus-affected.

The findings of this study also indicate that a low horizontal virus transmission rate among honey bees in beehives will help in protecting the bee colonies from the *Varroa* infestation and the viral epidemics. In fact, from the first condition in (17), a low γ is necessary for the disease- and mite-free equilibrium E_1 to be stable.

The sensitivity analysis allows us to identify the parameters most affecting the system: r, μ, γ and e. Specifically, for small changes of these parameters the system experiences transcritical bifurcations between all the equilibria. These results emphasize the importance of keeping the *Varroa* growth under control. In order to do this, the sensitivity surfaces suggest that a decrease of γ combined with an increase of e will benefit the colonies. Furthermore, the analysis substantiates the empirical remark that the most harmful diseases at the colony level are those that least affect the single bees. In fact, we can observe that for a very low bee mortality μ, the equilibrium toward which the system always settles is the healthy-bee-free point, in which only infected bees and mites thrive. The counterintuitive effect of the parameter μ on the system's dynamics agrees with the experimental findings of[7]. By artificially infecting the larval colonies of *Apis mellifera* and *Apis cerana*, the researchers have compared the *Varroa* evolution in both colonies. They found that in the european bees it does not change much in comparison with the colonies that are not infected, while the oriental bees were hindered, with a high larval mortality. They conclude that the higher vulnerability of the latter to the mite could imply a higher resistance at the colony level, because the infected larvae are more easily spotted and killed by the worker bees. This also agrees with the general consensus that the *Varroa* represents the most important factor affecting the survival of *Apis mellifera* colonies, while the oriental bees thrive relatively easily in its presence.

In a similar way, a higher bees mortality induced by the diseases vectored by *Varroa* maintains the total viral load in the colonies at a relatively moderate level. This thus enhances their survival chances, because the individual infected bees have less time available for horizontally transmitting the virus thereby infecting other individuals. Both[7] and this investigation show that vulnerable individuals may help the superorganism, against the common assumption that it is just the presence of "strong" individuals to ensure the colony survival.

Comparing the results of this investigation with the outcomes of the model[2], it appears that the present model, more mathematically elaborated in that it contains a term of Leslie-Gower type, namely the mite bee-dependent carrying capacity, is not really fundamentally necessary, as its qualitative analysis is captured well by the former system[2].

The bees are the most important pollinators worldwide. In recent years managed honey beehives have been subject to decline mainly due to infections caused by the invading parasite *Varroa destructor*, leading to widspread colony collapse[5]. In turn, this has caused alarm among the scientists for apiculture and even more for agriculture. This model represents a further step beyond the basic model presented in[2] for the understanding of this problem. But there is the need of the model validation using real data. At present, we are currently working on this aspect of the research, gathering the field data. From the collected field information we plan to obtain an estimate of the model parameters of most interest for the beekeepers: first of all the *Varroa* growth rate, a parameter most needed to assess the amount of acaricides to be used and the timings for their most effective application. In order to measure real field data we have started field experiments, in collaboration with beekeepers of the Cooperative "Dalla Stessa Parte", Turin, and from "Aspromiele", the Association of Piedmont Honey Producers, Turin. These experiments represent probably the first collaboration between beekeepers and mathematicians on this topic in our country and they are really precious for the research, since real field data are not yet presently available.

References

1. M. A. Benavente, R. R. Deza, M. Eguaras, Assessment of Strategies for the Control of the *Varroa destructor* mite in *Apis mellifera* colonies, through a simple model, MACI, II Congreso de matemática aplicada, computacional e industrial, Editors E. M. Mancinelli, E. A. Santillán-Marcus, D. A. Tarzia, Rosario, Argentina, December 14th-16th 2009, 5-8.
2. S. Bernardi, E. Venturino *Viral epidemiology of the adult Apis Mellifera infested by the Varroa destructor mite*, HELIYON (2016).
3. E. Carpana, M. Lodesani, Editors. Patologia e avversità dell'alveare (Pathologies and adversity of beehives), springer, 2014.
4. J. S. Figueiró, F. C. Coelho, The role of resistance behaviors in the population dynamics of Honey Bees infested by *Varroa Destructor*, Abstract Collection, Models in Population Dynamics end Ecology, Ezio Venturino (Editor), International Conference, Università di Torino, Italy, August 25th-29th 2014, 23.
5. E. Genersch, M. Aubert, Emerging and re-emerging viruses of the honey bee (*Apis mellifera L.*), Veterinary Research 41, (2010), 54.
6. M. Milito, Biologia della api, IZSLT http://www.izslt.it/apicoltura/wp-content/uploads/2012/07/BIOLOGIA-DELLE-API.pdf.
7. P. Page, Z. Lin, N. Buawangpong, H. Zheng, F. Hu, P. Neumann, P. Chantawannakul, V. Dietemann, Social apoptosis in honey bee superorganisms, Scientific Reports 6, Article number: 27210 (2016) doi:10.1038/srep27210.
8. V. Ratti, P. G. Kevan, H. J. Eberl, A mathematical model for population dynamics in honeybee colonies infested with *Varroa destructor* and the *Acute Bee Paralysis Virus*, Can. Appl. Math. Q. 21 (2013), no. 1, 63-93. DOI 10.1007/s11538-015-0093-5.
9. V. Ratti, P. G. Kevan, H. J. Eberl, A Mathematical Model of the Honeybee—*Varroa destructor*—Acute Bee Paralysis Virus System with Seasonal Effects, Bull Math Biol DOI 10.1007/s11538-015-0093-5.
10. J. F. Santos, F. C. Coelho, P. J. Bliman, Behavioral modulation of the coexistence between *Apis mellifera* and *Varroa destructor*: A defense against colony collapse? PeerJ PrePrints 3:e1739, (2015). doi.org/10.7287/peerj.preprints.1396v1.
11. United Nations Environment Programme, Global honey bee colony disorders and other threats to insect pollinators, UNEP Emerging Issues, (UNEP, Nairobi) (2010).

DOES SENSITIVITY ANALYSIS VALIDATE BIOLOGICAL RELEVANCE OF PARAMETERS IN MODEL DEVELOPMENT? REVISITING TWO BASIC MALARIA MODELS

SANDIP MANDAL

Public Health Foundation of India
Gurgaon, Delhi NCR 122002, India
E-mail: sandip.mandal@phfi.org

SOMDATTA SINHA

Department of Biological Sciences
Indian Institute of Science Education and Research (IISER) Mohali
Punjab 140306, India
E-mail: ssinha@iisermohali.ac.in

Several increasingly complex epidemiological models have been developed to understand the complicated transmission dynamics of malaria in human population over the past hundred years. The primary aim of these models is to offer a realistic account for the prevalence of malaria, by including more features of the host-pathogen interactions through inclusion of new parameters and interaction functions, and predict suitable control strategies. For this it is essential to identify the rank of model parameters representing these processes, according to their influence on model variables, so that those having stronger influence in the prevalence of the disease can be determined. It is, however, not clear how inclusion of new biologically realistic processes that change the model structure minimally in terms of dynamics and parameters, can affect the ranking of the common parameters, and thereby influence control policies.

Towards this, two basic and closely related models of malaria (Ross and Macdonald models) have been studied, which differ by only one parameter, and both exhibit stable temporal dynamics with variation only in final prevalence level. Comparative sensitivity analysis indicates the change in ranking of important parameters, thereby proving the importance of specific processes that influence disease transmission, and predict different control strategies for the two models. Thus we propose that sensitivity analysis can be critical in delineating the influence of parameters in closely related models showing similar dynamics, and may be used as an effective approach to understand and frame disease control strategies from more complex models.

1. Introduction

Malaria, a vector-borne, macroparasitic infectious disease, continues to cause millions of deaths worldwide every year (WHO) [1]. It is caused by the protozoan parasites of genus Plasmodium, and spread by mosquito vectors of the Anopheles genus. Plasmodium parasite spends two stages of its life cycle in its human host and in the mosquito gut. Infected mosquitoes can spread the disease to an uninfected human by injecting sporozoites of the parasite while taking blood meals. The disease can transmit from an infected human to an uninfected mosquito by taking the gametocyte of the parasite from the blood of infected humans. After infection, there is about a 10-day latent period during which the infected mosquito is not infectious. In human body this latent period for parasite development is nearly 21 days [2]. There are several factors, like host immunity, host heterogeneity, socio-economic structure, ecological and environmental factors, and climate and weather change that makes this 'two host — one parasite' interaction quite complicated [3-11].

Mathematical modeling in epidemiology helps to understand the underlying mechanisms that influence the spread and control of the disease. To analyse the transmission dynamics of malaria, Sir Ronald Ross proposed a mathematical model in 1911 (Ross model) with infected human hosts and vectors [12, 13]. The main conclusions from this model was that there is a causal relationship between the mosquito density and the number of infected humans, and that it was not necessary to kill every mosquito to end transmission [14]. Nearly after four decades of Ross's model, recognizing the epidemiological importance of the latent period of parasite development in the mosquito and mosquito longevity, George Macdonald, who was then Director of the Ross Institute and a member of the World Health Organization expert panel on malaria, modified this model (Macdonald model) [15]. The number of parameters in the two models differs only by one (latency of parasite development) with other six parameters being the same, and both exhibit stable temporal prevalence dynamics with variation only in final prevalence level. Based on his model analysis, Macdonald coordinated a massive World Health Organization (WHO)-managed campaign that focused on killing mosquitoes for the elimination of malaria among 500 million people in Africa using the insecticide dichlorodiphenyltrichloroethane (DDT) [15, 16]. This was a highly successful strategy with tremendous short- term success. Due to the absence of any effective vaccine, severe negative environmental effects of long term usage of DDT, and the evolution of resistance towards drugs (Chloroquine, Artemisinin), the primary method of controlling malaria has still remained focused on vector elimination. Several control strategies for vector control (killing mosquitoes, reducing larval development sites, using insecticide treated net, drug treatment, etc.) have been made for eradication of this deadly disease. Thus, modelling has played a major role in developing control strategies of malaria [17, 18].

Though mathematical models provide a framework to study disease transmission in a population and are useful tools to aid in decision-making in different epidemiological processes, yet there remain several uncertainties associated with this method of analysis. Not all the parameters contribute equally to the model behaviour — changes in some parameters can induce larger variations in the model variables, and for some parameters the model may be quite robust, both in terms of changes in dynamics and changes in the actual value. Parameter sensitivity of the model can be addressed either numerically introducing small random changes in the parameter values locally near a given point and assessing their effect on the variable; or methodically, over a small region in the physically-reasonable parameter space, using sensitivity analysis, which determines the parameters that are the key drivers of a model behaviour. The first has been used in different climate models [19, 20] and in complex models of malaria [21-23] to arrive at some understanding of parameter sensitivity. In the latter approach, the derivatives of the model output with respect to the parameters (called sensitivity functions) at a specific location in the parameter space are considered [24, 25]. Such sensitivity analysis has proven to be quite useful in extracting important information in models of malaria [26] and of other infectious diseases [27, 28].

In this paper we have chosen the two basic models of malaria by Ross and Macdonald, and studied their parameter dependence using the local sensitivity analysis technique [25, 29-31]. Our aim is to use simple models to show the usefulness of this analysis, which makes the idea of applying them for more complex models as a strategy for identifying important parameters for intervention.

In the next section, the two basic models of malaria (Ross and Macdonald) and the methods of linear sensitivity analysis are described. Numerical simulation of these models and the effects of interventions by changing important parameter values are described in the Results section. This section also deals with the linear sensitivity analysis of these models by forming sensitivity functions in matrix form. The conclusion of this analysis is given in the last section.

2. Models and Methods

2.1. *Epidemiological compartment models*

In mathematical epidemiology, depending on the infection status of the individuals, a population is divided into epidemiological compartments, such as — Susceptible (S), Exposed (E), Infected (I) and Recovered (R). Individuals who are susceptible to the disease lie in the S class and those infected, but not infectious, lie in the E class. Similarly, individuals who are infected as well as infectious belong to the I class, and those who have recovered from the disease, with permanent or temporary immunity, go to the R class. The transition of proportion of population between these compartments provides an epidemiological transmission model.

2.2. *Ross model (RM)*

Sir Ronald Ross introduced the first differential equation based model of malaria by dividing the human and mosquito populations into susceptible (S_h and S_m) and infected (I_h and I_m) compartments [12, 13]. In his simple model, infected human, on recovery, returns back to the susceptible class again, which lead to the SIS structure for human population. Mosquitoes do not recover from infection due to their short life span, and hence, the mosquito population in the Ross model follows the SI structure. Mosquito death rate is considered to be constant in this simple model, although recent studies have shown mortality rate of mosquitoes to be age-dependent [32, 33]. Both human and mosquito population sizes are considered to be constant ($S + I = 1$). In this model, the time evolution of the infected classes in human host (I_h) and infected mosquitoes (I_m) are given as:

$$\frac{dI_h}{dt} = abmI_m(1 - I_h) - \gamma I_h$$

$$\frac{dI_m}{dt} = acI_h(1 - I_m) - \mu I_m$$

In this model the six parameters are — a: the biting rate of female mosquito, m: mosquito to human ratio, b and c: probability of getting infection to human and mosquito, respectively, upon an infectious bite, γ: the human recovery rate, and μ: the natural death rate of adult mosquito.

2.3. *Macdonald model (MM)*

Malaria parasites entering the mosquito body through a blood meal from an infected human take some time (latency or delay) to develop to the infectious stage. The Ross model (RM) did not consider the latency period of the parasite in mosquitoes and their survival during that period. This resulted in the model predicting a rapid progress of the disease in human population, and a higher equilibrium prevalence of infectious mosquitoes (see Fig. 1). Including the delay τ in disease transmission, caused by the latency in mosquito and introducing an additional Exposed (E_m) class in the mosquitoes, Macdonald remodeled malaria transmission dynamics. In his model, mosquitoes were divided into three compartments (SEI) keeping the human to (SIS). The set of equations representing Macdonald model (MM) is given by:

$$
\begin{aligned}
\frac{dI_h}{dt} &= abmI_m(1-I_h) - \gamma I_h \\
\frac{dE_m}{dt} &= acI_h(1 - E_m - I_m) \\
&\quad - acI_h(t-\tau)[1 - E_m(t-\tau) - I_m(t-\tau)]e^{-\mu\tau} - \mu E_m \\
\frac{dI_m}{dt} &= acI_h(1-\tau)[(1 - E_m(1-\tau) - I_m(1-\tau))]e^{-\mu\tau} - \mu I_m
\end{aligned}
$$

This model with an additional equation depends on 7 parameters of which 6 are the same as of the Ross model, and 3 variables with a new class E_m. In comparison to RM, R_0 in this model also depends on the extra parameter, τ (latency in mosquito).

The model dynamics for RM and MM are stable, as shown in Fig. 1, for parameter values obtained from literature [2, 8, 9]. Even though both models show lower equilibrium prevalence for infected mosquitoes (I_m) compared to that of human (I_h), MM shows about 4% reduced equilibrium prevalence for I_h and nearly 40% reduced value for equilibrium prevalence for I_m when compared to RM.

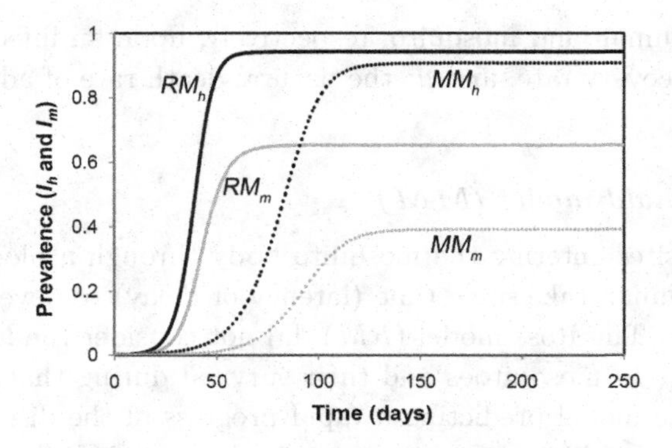

Figure 1. Prevalence pattern from Ross model (RM) and Macdonald model (MM). Solid line, and dotted line, are for Ross model (RM_h, RM_m) and Macdonald model (MM_h, MM_m) respectively, for infected human (black line) and infected mosquito (grey line) population. Parameters are: $a = 0.2$, $b = 0.5$, $c = 0.5$, $\mu = 0.05$, $\gamma = 0.02$, $m = 5$, $\tau = 10$.

2.4. *Local sensitivity analysis*

Sensitivity analysis of a model aims to understand the response of the variables to changes in parameters [26]. For a set of equations describing the time evolution of variables y_i in a model is given by

$$\frac{dy_i(t)}{dt} = f(y_i(t), \phi_j)$$

where, $i = 1, 2, ..., n$ are the variables, and $\phi_j = 1, 2, ..., m$ are the parameters. Local sensitivity analysis defines a sensitivity function, which is measured as the rate of change of the dependent variable (y_i) with respect to the selected parameter (ϕ_j). The sensitivity function (S'_{ij}) is defined by the $(n \times m)$ matrix as,

$$S'_{ij} = \frac{dy_i(t)}{d\phi_j}$$

Parameters (ϕ_j) may differ by orders of magnitude, and may have different dimensions. Therefore, to compare sensitivity function for different parameters one needs to make it dimensionless. This is done by multiplying with a weighing/scaling factor. This function, then becomes

$$S_{ij} = \frac{dy_i(t)}{d\phi_j} . \frac{w_{\phi j}}{w_{yi}}.$$

Here, w_{yi} is the weighting factor for variable y_i (usually equal to its value) and $w_{\phi j}$ is the scaling of parameter ϕ_j (usually equal to the parameter value) [34].

Additional information that can be obtained from other important functions of the sensitivity matrix S_{ij} for each variable. These are:

$$Mean(S_{ij}) = \frac{1}{T}\sum_{t=1}^{T} S_{ij},$$

$$L_1 = \frac{1}{T}\sum_{t=1}^{T} |S_{ij}|,$$

$$L_2 = \frac{1}{T}\sqrt{\sum_{t=1}^{T} S_{ij}^2}.$$

The norms of the columns of S_{ij} i.e., L_1 (vector-1 norm) and L_2 (vector-2 norm), give a measure of the importance of individual parameters. For a particular variable, the correlation coefficient between the sensitivity functions of different parameters can give us an idea how pairwise these parameters are related to that particular variable. Sensitivity analyses of the models (RM and MM) have been performed using the FME package in R [34].

3. Results and Discussion

In the following sections, we first compare the transmission dynamics of infected human (I_h) and infected mosquitoes (I_m) in both models for similar changes in the six common parameters, one at a time, and then present the results of different aspects of sensitivity analyses for variation of all parameters. The constant parameter values as used in these analyses are given in Table 1.

Table 1. List of parameters and their values as used in these analyses.

Parameter	a	b	c	μ	γ	m	τ
Values	0.2	0.5	0.5	0.05	0.02	5	10

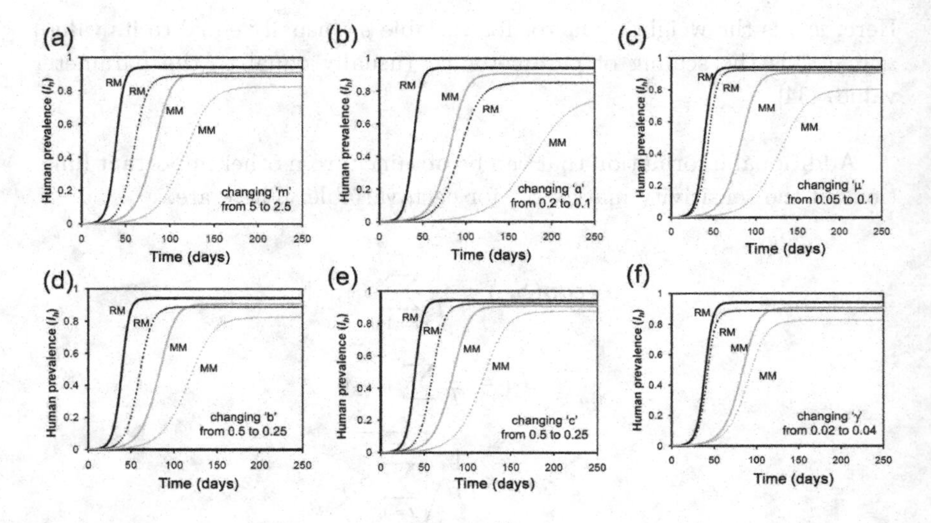

Figure 2. Effect of interventions on human prevalence in Ross Model (black) and Macdonald Model (Grey). On changing, (a) mosquito density by half, (b) biting rate by half, (c) adult mosquito mortality by double, (d) probability of human infection by half, (e) probability of mosquito infection by half, and (f) human recovery rate by double. Basic parameter values are as in Fig. 1. Solid lines are for the basic and dotted lines for the changed parameter values.

3.1. *Model dynamics on parameter variation*

Figs. 2(a-f) show the effect of changing the six common parameters on infection prevalence in humans in RM (Ross model) and MM (Macdonald model). In this study, specific parameter values are simply doubled or reduced by half, and time variation of I_h shown. Reducing the mosquito density (m) by half results in a decrease in the final prevalence of infection in both models, as shown in the 'dotted' curve of Fig. 2a, but the RM shows comparatively lower (5.66%) reduction of human prevalence (I_h) compared to the reduction in MM (9.04%). Similarly, if the mosquito biting rate (a) is decreased to half, the percentage of reduction in I_h is 9.59% for RM and 15.46% for MM (Fig. 2b). On the other hand, increasing adult mosquito mortality rate (μ) by double reduces I_h by a mere 2.04% for RM but by 11.39% for MM (Fig. 2c). Reducing b and c, the probability of getting infection in human and mosquito respectively, by half results in a decrease in the final prevalence of I_h in both the models (Fig. 2d and 2e). The extent of reduction for reducing b is the same as observed for m, but for c

it is 2% in RM and 3.4% in MM. Fig. 2f shows the effect of increase of human recovery rate γ. Doubling recovery rate (γ), shown in Fig. 2f, has similar effect on I_h as observed by reducing the parameters m or b by half, although the transient dynamics of I_h is quite different.

The above results indicate that the relative effects of different interventions (i.e., changing one or another parameters) by a similar amount have different effects on the malaria prevalence in human hosts. Generally, the MM shows higher reduction in I_h for all parameter changes when compared to the RM. For RM, reducing 'biting rate' has maximum effect in reducing infection in humans than 'mosquito density' or 'mosquito mortality rate'. MM also shows high reduction in infection in humans for reducing the biting rate, but in comparison to the RM, the maximum change (11.39/2.04 = 5.58) occurs for the parameter 'adult mosquito mortality rate', whereas the minimum change (9.04/5.66 = 1.59) occurs for the parameter 'mosquito density'. Thus, even though both these models show similar kind of infection pattern, similar amount of parameter variations did not reduce the prevalence of I_h equally. In other words, the dependence of these model variables on the same parameters is different due to the inclusion of another epidemiological compartment in mosquito (E_m) and the latency period (τ).

3.2. *Sensitivity analysis*

The study in the earlier section does not offer a general understanding of how the models behave on variations of different parameters for larger range of values, and also when all parameters are varied together. Here the sensitivity functions of the two basic malaria models (RM and MM) are studied for all parameters and their dependence on the output variables are described. The correlations between pairs of parameters are also studied from the sensitivity matrices.

3.3. *Sensitivity functions for different parameters*

Time variations of the sensitivity functions for I_h and I_m in RM (Fig. 3a) and MM (Fig. 3b) for variation in different parameters clearly show that the same parameters contribute differentially to the output variables of the two models. Positive (negative) values of sensitivity function indicate that the increase (decrease) in parameter values and the resultant output

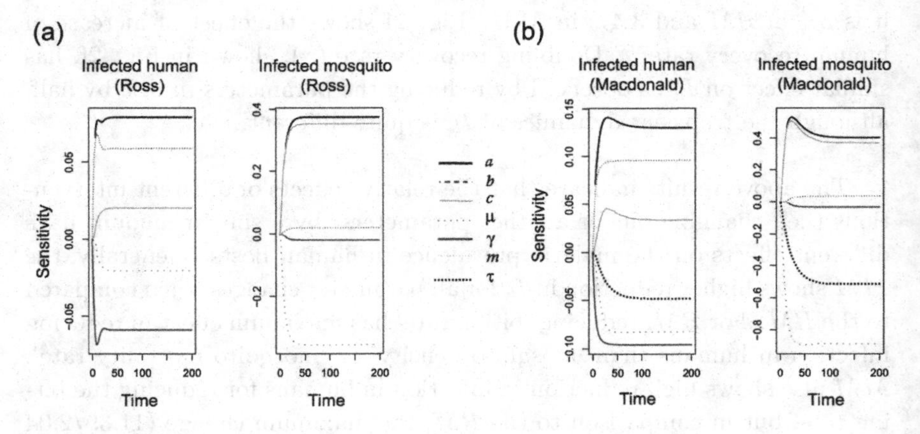

Figure 3. Time variation of sensitivity functions. For infected humans and infected mosquitoes in (a) Ross model and (b) Macdonald model. Basic parameter values are as in Fig. 1.

variables are positively (negatively) correlated. Table 2 lists the saturation level values of the sensitivity functions for I_h and I_m corresponding to the parameters a, b, c, μ, γ and m in RM and MM (including τ). In the Ross model, the parameters a, b, c and m have positive effect, and, μ, and γ have negative effect, on I_h and I_m. Thus, these intuitively clear phenomena are also validated by sensitivity analysis.

Table 2. Sensitivity functions for different parameters at equilibrium level.

Models		a	b	c	μ	γ	m	τ
Ross	I_h	0.079	0.058	0.020	−0.020	−0.058	0.058	–
	I_m	0.374	0.020	0.354	−0.354	−0.020	0.020	–
Macdonald	I_h	0.130	0.096	0.034	−0.082	−0.096	0.096	−0.048
	I_m	0.401	0.034	0.367	−0.884	−0.034	0.034	−0.517

In quantitative terms, from Table 2 one can see that infected mosquito (I_m) is more sensitive to the parameters a (4.7 times), c (17.7 times) and μ (17.7 times) in comparison with infected human (I_h). For the parameters b,

γ and m, sensitivities of I_h are only 2.9 times more than I_m. Interestingly, these results indicate that human infection is affected more by the change in mosquito density (or, mosquito to human ratio, m) than the infection in mosquito. MM shows similar correlations (positive or negative) between all common parameters for I_h and I_m as observed for the RM, though the sensitivity values (in Table 2) are much higher in this model. Sensitivity for the additional parameter τ in MM is higher for I_m compared to I_h, and therefore, the equilibrium prevalence of infected mosquitoes is much lower than that of infected humans as shown in Fig. 1. Introduction of τ also influences the model sensitivity to other parameters. Table 2 also shows that I_h and I_m in these models are equally sensitive for the parameters b, m and γ, though the latter shows negative correlation.

3.4. *Other sensitivity parameters*

Three other functions (Mean, L_1 and L_2 — see Models and Methods) that summarize the overall model sensitivity (i.e., considering both I_h and I_m variables together) with respect to different parameters are calculated for both the models and plotted in Fig. 4. Value of L_1 and L_2 can be used to rank the importance of parameters in influencing the output variable. Large differences between L_1 and L_2 indicate a high variability or outliers in the elements of sensitivity matrix for that parameter. 'Mean' of sensitivity functions for a particular variable may be positive or negative, as it represents the average of sensitivity values over time. But the norm L_1 is the modulus of sensitivity values, and therefore, is always positive. A comparison of L_1 and 'Mean' shows whether all the elements of sensitivity matrix have the same sign, and 'Mean' gives information on the sign of the averaged effect a change in parameter has on the model output.

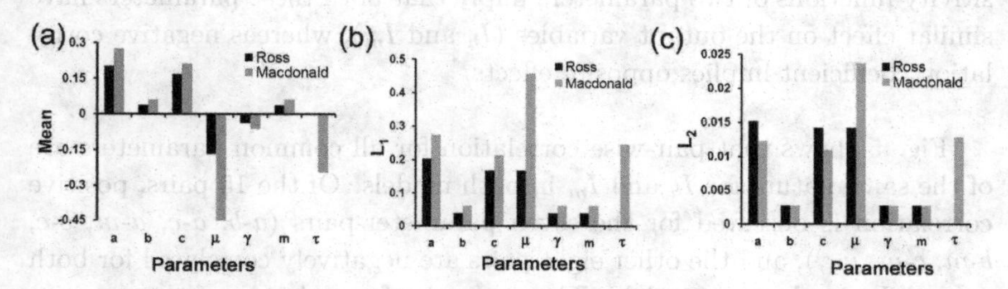

Figure 4. Plot of sensitivity parameters (Mean, L_1, L_2) for Ross model and Macdonald model. Basic parameter values are as in Fig. 1.

'Mean' of sensitivity functions for Ross and Macdonald models (Fig. 4a) shows that μ, γ, and τ have negative effect on the model variables for both the models, whereas other parameters have positive effect. From Figs. 4b and c, for the Ross model, the rank of parameters for model sensitivity is: $a > (c, \mu) > (b, \gamma, m)$, but for the Macdonald model it is: $\mu > a > \tau > c > (b, \gamma, m)$, where the parameters in bracket have the same L_1 and L_2 values. The maximum variability in sensitivity (L_1-L_2) is maximum for the parameter mosquito biting rate, a (0.187), in Ross model, whereas for the Macdonald model it is maximum for the parameter death rate of mosquitoes, μ (0.433). Therefore, though both the models are least sensitive to the same three parameters (b, γ, m), but their rank of parameter sensitivity are quite different. In Ross model changes in the parameters a and c have more effect on model output compared to μ, whereas Macdonald model is more sensitive to 'mosquito mortality rate' (μ) than 'mosquito biting rate' (a). So, from sensitivity analysis one can conclude that increasing mosquito mortality rate will be more effective in Macdonald model than reducing mosquito-biting rate, which is the more important parameter in Ross model. It may also be noted in Fig. 4 that the L_1 values for Macdonald model for most parameters are higher and more variable compared to the Ross model indicating stronger sensitivity of the variables to parameter variations.

3.5. *Pair-wise sensitivity functions*

The sensitivity matrices can also give pair-wise relationship between any two parameters for infected humans (Fig. 5a) and infected mosquitoes (Fig. 5b) for the RM and MM. Positive correlation coefficients between the sensitivity functions of two parameters imply that both these parameters have similar effect on the output variables (I_h and I_m), whereas negative correlation coefficient implies opposite effects.

Fig. 5 shows that pair-wise correlation for all common parameters are of the same nature for I_h and I_m in both models. Of the 15 pairs, positive correlation is observed for the seven parameter-pairs (a-b, a-c, a-m, b-c, b-m, c-m, μ-γ), and the other eight pairs are negatively correlated for both I_h and I_m in the two models. The strength of correlation of the sensitivity functions between parameters b-γ, b-m and γ-m is 1 in both models. This implies that, for the positively (negatively) correlated pairs, effect of

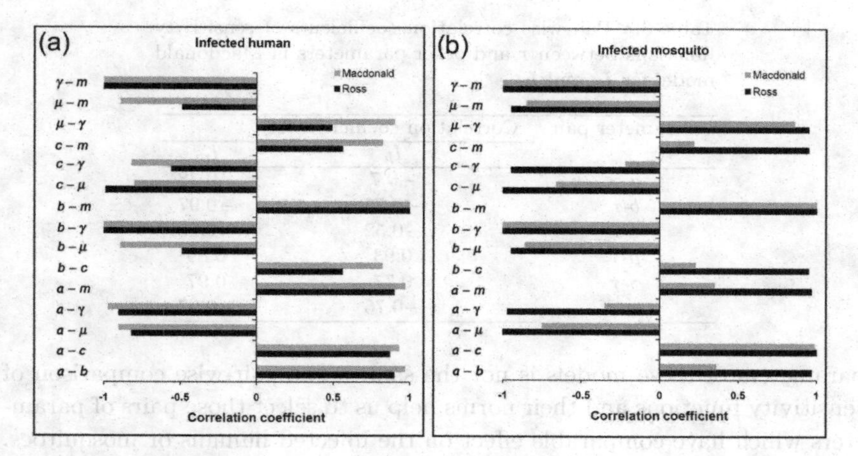

Figure 5. Correlation coefficients of sensitivity functions between pairs of parameters in Ross model and Macdonald model. (a) infected humans and (b) infected mosquitoes.

increasing one parameter of the pair on the output variable can be completely negated by decreasing (increasing) the other by the same amount. Pair-wise correlations are much higher in MM for I_h, whereas RM shows strong correlations for all pairs of parameter for I_m. For I_h in RM, only the sensitivity functions for 'c-μ' show higher correlation than the MM. For I_m, the correlation coefficients are reduced by more than 60% between parameter pairs (c-m, c-γ, b-c, a-m, a-γ and a-b) in MM. These results show that the pair-wise parameter sensitivity is also quite different for the two models, even though their dynamics are similar and robust to parameter variations.

The correlation coefficients of sensitivity functions between the parameter τ, and other six-parameters in MM is given in Table 3. The parameters μ and γ have positive correlation with τ for both the variables I_h and I_m, and the other parameters have negative correlation with τ. In comparison to the others, the correlation coefficient of the sensitivity functions between c and τ is lower for both I_h and I_m. Though pair-wise correlations for all other parameters are much higher in MM for I_h, with τ (latent period) it is higher for I_m except for a and c.

Though the dynamics of the model outputs are robust (stable) on addition of a new parameter, our results show that the model sensitivity of

Table 3. Pair-wise correlation coefficients of sensitivity functions between τ and other parameters in Macdonald model for I_h and I_m.

Parameter pair	Correlation coefficient for	
	I_h	I_m
a-τ	−0.7	−0.38
b-τ	−0.76	−0.97
c-τ	−0.52	−0.25
μ-τ	0.93	0.89
γ-τ	0.77	0.97
m-τ	−0.76	−0.97

parameters of these models is not the same. The pair-wise comparison of sensitivity functions and their norms help us to select those pairs of parameters which have comparable effect on the infected humans or mosquitoes. Thus, this approach clearly can distinguish those parameters (processes) in the models, which are important for intervention and control in the spread of malaria. These models are simple enough to arrive at the above-mentioned results intuitively or by simple comparison, even though ranking of the parameters is not straightforward.

4. Conclusions

Mathematical models provide a framework to study disease transmission in a population, and are useful tools to help in decision-making in different epidemiological processes. Specifically, modelling has played a major role in developing control strategies for malaria worldwide [17, 18]. It is true that "Mathematics cannot replace mental, qualitative intuition, but can expand into more formal and quantitative realms in which components and their interconnections are made more specific, assumptions and objectives more precise" [35], and mathematical and statistical methods, like Sensitivity Analysis, can be useful in assessing the contribution of different parameters in nonlinearly interacting processes in mathematical models.

Models are described by variables and parameters representing relevant biological processes. It is not clear how the inclusion of new biologically realistic processes, that change a model structure minimally in terms of parameters and dynamics, can affect the sensitivity of the output variables to variations in common parameters. Using the approach of sensitivity analysis, we have tested this with two basic models of malaria (the Ross model and the Macdonald model), which differ by only one parameter (delay in parasite development in the host), and both exhibit stable temporal prevalence dynamics with alteration only in the long term disease prevalence level.

The higher the sensitivity function value, the more important the parameter is, and thus the magnitudes of sensitivity summary values are used to rank the importance of parameters on disease transmission. Through rank analysis it is observed that the parameter, which was most important in Ross model, does not occupy the same rank in Macdonald model. Ross model shows the 'biting rate' as the most important parameter whereas according to Macdonald model 'active mosquito killing' is found to be the most important one. Thus, Ross model sensitivity analysis indicates that reducing mosquito bites should be the most effective measure in controlling the disease prevalence, which can be achieved by using bed-nets, mosquito-repellents, etc. A similar analysis on a more complex model, with no latency included, also shows 'mosquito biting rate' to be the most important parameter [26]. The Macdonald model's sensitivity analysis ranks 'mosquito mortality rate' to have the maximum sensitivity. Thus, the inclusion of time delay (as latency in parasite development) in Macdonald model induces the change in the sensitivity rank of this important parameter. This result theoretically justified WHO's successful Global Malaria Eradication Programme (GMEP, 1955-1969) for killing adult mosquitoes to control malaria. To add a note of caution from a different aspect, this theoretically-justified change lead to a critical change in control/eradication policies with far-reaching global consequences (e.g., massive use of DDT inducing evolution of resistance in mosquitoes), which were outside the consideration of the models.

Our results also delineate the relative importance of the other common parameters in the two models, and clearly group the ones (b, m, γ), which have least effect on model variables. This analysis gives a clear idea of the parameters and their sensitivities for both infected humans and mosquitoes. Thus, such an approach can give well-defined prescriptions about control strategies to be opted specifically for the human host or vector. The pair-wise plot of sensitivity functions gives an overview about the correlation between these parameters. This helps in identifying processes that may interfere with each other in disease prevalence. The results of these analyses with the simple models do not differ from what can be understood intuitively or through simple comparisons as has been obtained in earlier studies. We use them to justify that sensitivity analysis technique not only validates the model results, it is also the most comprehensive method as it simultaneously gives the rank of parameters, along with their pairwise comparisons. Therefore, this method justifies the importance of parameter selection, which, when performed on any model, can help to determine suitable control strategies in any disease control programme.

Acknowledgments

SM thanks IISER Mohali for Research Associateship where part of the study was done. SS acknowledges support from the Department of Science and Technology, India as J C Bose Fellowship.

References

1. World Health Organization (WHO) [http://www.who.int/topics/malaria/en/].
2. R. M. Anderson and R. M. May, Infectious diseases of humans: dynamics and control *(London: Oxford University Press)* (1991).
3. M. Amexo, R. Tolhurst, G. Barnish, I. Bates, *Lancet.* **364**, 1896-1898 (2004).
4. P. W. Gething, D. L. Smith, A. P. Patil, A. J. Tatem, R. W. Snow, S. I. Hay, *Nature* **465**, 342-346 (2010).
5. S. Gupta, A. V. S. Hill, *Proc. Roy. Soc. Lond. B* **261**, 271-277 (1995).
6. S. Gupta, J. Swinton, R. M. Anderson, *Proc. Roy. Soc. Lond. B* **256**, 231-238 (1994).
7. T. M. Lunde, M. N. Bayoh, B. Lindtjrn, *Parasite. Vectors.* **6**, 20 (2013).
8. S. Mandal, R. R. Sarkar, S. Sinha, *Malar. J.* **10**, 202 (2011).
9. S. Mandal, S. Sinha, R. R. Sarkar, *Bull. Math. Biol.* **75**, 2499-2528 (2013).
10. K. P. Paaijmans, A. F. Read, M. B. Thomas, *Proc. Nat. Acad. Sci.* **106**, 13844-13849 (2009).
11. J. Sachs, P. Malaney, *Nature* **415**, 680-685 (2002).
12. R. Ross, The prevention of malaria (London: John Murray) 651-686 (1911).
13. R. Ross, *Nature* **87**, 466-467 (1911).
14. A. Lotka, *Am. J. Hyg.* **3** (Suppl 1), 38-54 (1923).
15. G. Macdonald, *Bull. World Health Org.* **15**, 613-626 (1956).
16. E. J. Pampana, A textbook of malaria eradication (London: Oxford University Press) (1963).
17. J. L. Aron, R. M. May, The population dynamics of malaria; in Population Dynamics of Infectious Disease (RM Anderson Ed.) *(London: Chapman and Hall)* 139-179 (1982).
18. K. Dietz, In Principles and Practice of Malariology. Edited by: Wernsdorfer, W., McGregor, Y. (Edinburgh: Churchill Livingston) 1091-1133 (1988).
19. J. E. Campbell, G. R. Carmichael, T. Chai, M. Mena-Carrasco, Y. Tang, D. R. Blake, N. J. Blake, S. A. Vay, G. J. Collatz, I. Baker, J. A. Berry, S. A.

Montzka, C. Sweeney, J. L. Schnoor, C. O. Stanier, *Science* **322**, 1085-1088 (2008).
20. J. Murphy, D. Sexton, D. Barnett, G. Jones, M. Webb, M. Collins, D. Stainforth, *Nature* **430**, 768-772 (2004).
21. O. J. T. Briët, D. Hardy, T. A. Smith, *Malar. J.* **11**, 20 (2012).
22. J. A. N. Filipe, E. M. Riley, C. J. Darkeley, C. J. Sutherland, A. C. Ghani, *PLoS Comp. Biol.* **3**, 2569-2579 (2007).
23. T. M. Lunde, D. Korecha, E. Loha, A. Sorteberg, B. Lindtjrn, *Malar. J.* **12**, 28 (2013).
24. N. T. J. Bailey, J. Duppenthaler, *J. Math. Bio.* **10**, 113-131 (1980).
25. R. Brun, P. Reichert, *Water Resources Res.* **37(4)**, 1015-1030 (2001).
26. N. Chitnis, J. M. Hyman, J. M. Cushing, *Bull. Math. Bio.* **70**, 1272-1296 (2008).
27. D. M. Bortz, P. W. Nelson, *Bull. Math. Biol.* **66(5)**, 1009-1026 (2004).
28. C. Okas, S. Roche, M. L. Kürzinger, B. Riche, H. Bricout, T. Derrough, F. Simondon, R. Ecochard, *Vaccine* **28(51)**, 8132-8140 (2010).
29. D. G. Cacuci, Sensitivity and Uncertainty Analysis: Theory, Volume 1 (Chapman and Hall) (2003).
30. A. Grievank, A. Walther, Evaluating derivatives: principles and techniques of algorithmic differentiation, 2nd Edition (SIAM publisher) (2008).
31. D. M. Hamby, *Environ. Monit. Assess.* **32(2)**, 135-154 (1994).
32. S. E. Bellan, *PLoS ONE* **5(4)**, e10165 (2010).
33. V. N. Novoseltsev, A. I. Michalski, J. A. Novoseltseva, A. I. Yashin, J. R. Carey JR, A. M. Ellis, *PLoS ONE* **7(6)** e39479 (2012).
34. K. Soetaert, T. Petzoldt, *J. Stat. Software* **33(3)**, 1-28 (2010).
35. F. E. McKenzie, *Parasitol. Today* **16(12)**, 511-516 (2000).

SUBCORTICAL HOMEOSTATIC CIRCUITRY MODULATES BRAIN WAVES AND BEHAVIORAL ADAPTATION: RELEVANCE FOR THE EMERGING MULTIDISCIPLINE OF SOCIAL NEUROSCIENCE*

LIMEI ZHANG DE BARRIO†, VITO S. HERNÁNDEZ

Department of Physiology, Faculty of Medicine
National Autonomous University of Mexico (UNAM)
Av. Universidad 3000, Mexico City, 04510, Mexico
E-mail: limei@unam.mx, vitohdez@unam.mx

LEE E. EIDEN

Section on Molecular Neuroscience
Intramural Research Program (SMN-IRP)
National Institute of Mental Health
National Institutes of Health (NIMH, NIH)
9000 Rockville Pike, Bethesda, MD, United States
E-mail: eidenl@mail.nih.gov

*Partially supported by: DGAPA-UNAM-PAPIIT-IN216214, CONACYT-CB-176919 & CB-238744 to LZ and NIMH-IRP-1ZIAMH002386 to LEE.

†LZ is a fulbright visiting scholar, also supported by PASPA-DGAPA-UNAM fellowships for her sabbatical research stay hosted by LEE of SMN-NIMH-NIH-USA (limei.zhangdebarrio@nih.gov).

This chapter is an overview of our recent research concerning the effects of centrally-acting neuropeptides and neurosteroids, and the circuits that synthesize and release them. These systems simultaneously control basic physiological functions, such as energy and hydro-electrolyte balance, fight-flight reaction and reproduction, and also the motivational and behavioral pathways that allow mammals to prioritize seeking of food, water, safety, and sex. This work is broadly situated in the branch of current neuroscience research that is modern neuroendocrinology. In particular, we study the re-organizational effects of early life stress (ELS) on the homeostatic circuitry which controls water-electrolyte balance, i.e. the hypothalamic vasopressinergic magnocellular neurosecretory neurons (VPMNNs). We have found that these neurons also project to the limbic regions of the brain, and their activation modifies the range of cortical oscillations, as well as their coherence between distinct cortical regions. ELS upregulates this VPMNN system, producing aberrant extensive innervation patterns in limbic regions which could contribute to altered emotional processing, and even emotional instability, in adulthood. Our own perspective, a medical and neuroscientific one, is that modern neuroendocrinology has an increasingly important role to play in understanding how brain information processing, with dramatic changes in the *millisecond time-scale*–for instance, the action potential firing probability, frequency patterns and population recruitment–can be potently modulated by mid- and long-term neuropeptide and neurosteroid expression patterns. These, in turn, are modulated by genetic, developmental and environmental factors. Intriguing new questions and challenges emerge from recent studies on neuropeptide and neurosteroid effects in the brain, at several levels, i.e. 1) how neuroendocrine signaling affects neuronal firing probabilities; 2) how changed postsynaptic neuron firing probability determines downstream physiological network recruitment; 3) what the consequences of local network activation on inter-regional oscillation coherence and information flow are; and 4) how these changes in inter-regional information flow ultimately affect behaviors, especially those involving inter-individual interactions. These questions will probably best be addressed proactively through *concerted multidisciplinary studies*. Hence we will put forward the argument, at the conclusion of this chapter, that a formal multidisciplinary social neuroscience should be created. Colleagues with expertise in biology, medicine, mathematics, physics and computational sciences should propose theoretical frameworks and perform large scale computational analysis based on physical principles. As this is an urgent social priority, it must obviously be approached with honest and genuinely committed trans-disciplinary participation, and a deep understanding of, and appreciation for, the detailed experimental facts driving extrapolation to human society and behavior.

1. Brain Waves

The brain is perpetually active, even in the absence of environmental stimuli. The main cellular components of the brain are the neurons (critically supported by an equally numerous cell type-the glia), which communicate with one another via a generally polarized cellular architecture. At the receiving end of the neuron are the dendrites, often studded with so-called spines; these communicate with the cell soma, which integrates the electrical inputs from the dendrites at the axon hillock. The neuron sends this integrated information via action potentials propagated down the axon, which terminates in axon terminals which release neurotransmitter to continue signal propagation, via the synapse, to the next neuron or neurons in the network.

Both neurons and glial cells are bounded by an *excitable* membrane. Neuronal (and glial) excitability implies that the cell contains energy consuming, *electrical potential-generator pumps*, and protein-based *channels*. The latter can be either *voltage-gated*, or *ligand-gated* (temperature- and mechanically-gated channels also make minor but potentially critical contributions). Ion channel gating allows non-linear changes in both the general and specific properties of these channels, under specific physical-chemical conditions.

The transmembrane currents generated by gated ion channels, and electro-chemical gradients between both sides of any excitable membrane, whether it is from a spine, dendrite, soma, axon or axon terminal, contribute to the extracellular field potential. The *field potential* is the superposition of all ionic processes, from fast action potentials to the slowest fluctuations in glial cells. All electrical phenomena in the brain superimpose at any given point in space to yield an electric potential at that location. All living tissues have electrical potential at any point in space, but the electrical potentials generated in the brain are particularly dynamic: they serve as the basis for brain function, and therefore their measurement provides intimate insight into how the brain translates sensory input, and previous experience, into executive action (behavior).

The electric potential can be recorded from the surface of the scalp, via electroencephalography (EEG, one of the oldest and most widely used methods for the investigation of the electrical activity of the brain); or from the surface of the cerebral cortex, i.e. electrocorticography (ECoG), or from a deeper site within brain tissue, by inserting a metal or glass electrode, or silicon probes, to record the local field potential (LFP, also known as "micro-EEG"). Another popular method, though expensive, is magnetoencephalography (MEG), which uses superconducting quantum interface devices, from outside the skull, to measure tiny magnetic field changes in precise locations, (typically in the 10-1,000 fT range) from currents generated by the neurons. Because MEG is non-invasive and has a relatively high spatio-temporal resolution (approximately 1 ms, and 1-2 mm in principle), it has become a popular method for monitoring neuronal activity in the human brain. An advantage of MEG is that magnetic signals are much less dependent on the conductivity of the extracellular space than EEG. The scaling properties, that is the frequency versus power relationship, of EEG and MEG often show differences, typically in higher frequency bands. These differences may be partially explained by the capacitive properties of the extracellular medium (such as skin and scalp muscles) that distort the EEG signal but not the MEG signal.

The electric potentials recorded at any place, whether by the EEG, ECoG, or LFP, *oscillate* in the *time* domain, producing *brain waves.* These oscillations, although changing from brain state to state, are generated from within, and perturbations of those default patterns by external input at any given time often cause only minor departures from its robust, internally controlled program[1]. Yet, these perturbations are absolutely essential for adapting the brain's internal operations to perform useful computation and to execute precise movements that serve individual survival, and species propagation. In other words, oscillations, or brain waves, are generated by excitable membrane activities, mainly from neurons, with some contribution from glial cells, and, in turn, these oscillations influence the firing patterns of the neurons that, collectively, generate them.

2. Brain States

The history of brain physiology in the last 100 years is actually the history of a pendulum swinging back and forth, from the concept of localized function, to the concept of distributed function, through physically distributed interacting and dynamic systems.

Hal Weinberg

Living in a complex and changing environment, animals (including humans) constantly change their *brain state*, and transitioning between functional brain states allows quick and efficient adaptation to the environment. These dynamic brain states are essential for normal brain function, and their disturbance may be associated with severe neurological and psychiatric disorders[2]. Common examples include changes between the states of sleep and wakefulness, attentiveness and distraction, and sociability and aggression. These state changes are accompanied by changes in the global patterns of neural activities in many brain areas, most clearly seen in cerebral cortex, that are measurable using electrophysiological techniques[3]. The neurons of the cerebral cortex generate temporal patterns of activity that can enhance, or limit, interactions with other neural centres and with the environment. Rhythmic and synchronized cortical activities correlate with the behaviour and performance of the body[4].

In neural dynamics, the concept of an *attractor* (or *attracting set*) within a closed subset of states, a *basin*, toward which activity fluctuations converge, is used to explain the temporal mechanisms of both cyclic and chaotic changes in brain states. Recent studies have shown that the landscape of attractors available to shape the dynamics of the neural network is not fixed, but changes depending on the experiential history and the context of current events. It also depends on *network parameters* (e.g. inhibition vs. excitation, neural accommodation due to leak ion channels and neuropeptide and neurosteroid modulatory effects on synaptic dynamics etc.). Attention results from inhibitory competition, bidirectional interactive processing and multiple-constraints satisfaction. In most cases the basin of attraction is relatively large, allowing associations and movement from one basin of attraction to another more easily, facilitating exploration of the activation space. However, without neuronal accommodation (a phenomenon which mainly involves participation of voltage-dependent K^+ channels, network attractor basins are deep and narrow, hence it is difficult to move between them. Therefore, the associations are weak. In other words, if availability of neurotransmitters is too *fixed* (i.e. non-adaptable), or the attractors are too *deep*, neurodynamics may be slowed down, and entrapment within a given brain state ensues, making attention shifts difficult (leading to autism-like behaviors). In contrast, if attractors are too shallow, for instance, in conditions that the temporal pattern of transmitter release (e.g. dopamine) is too variable[5], the system may not maintain a given brain state in a stable or protracted fashion, and attentional focus may be impaired (ADHD-like behavior)[6].

Until recently, individual differences in brain function were described primarily in terms of deviations from the average variability in imaging measurements[7]. However, it is now possible to attempt an understanding of how brain systems change as a result of experience, and how these systems, and their changes, are responsible for individual responses to similar stimuli (i.e. "different individuals" being *different individuals*). MEG currently may have the best potential to measure individual brain differences, because of its time resolution, and its ability to directly measure the function of neuronal systems in real time, and without the use of high frequency or chemical impositions that could themselves affect function[8]. However, technological advances in all methods of brain imaging will soon make individual differences increasingly recognizable and documentable (for instance, see studies[9,10], allowing more powerful predictions about how those differences may impact an individuals responses and future behaviors. Enhanced behavioral predictability for the treatment of disease and behavioral disorders, and for management and control of abnormal behavioral responses, could change our current concepts of individuality and diversity in society.

It is commonly understood that the organization of appropriate behaviour for each activity is based on anatomical circuits. However, we wish to point out the useful definition of *physiological circuits*, i.e. the functional connections and downstream circuit recruitment, being notably determined by the neurotransmitter co-release and neurotransmitter switching phenomena (which will be discussed later). These physiological circuits define more broadly what we call the striking state-dependent changes in neural activity within sets of neurons, and are as versatile as alterations in neuronal firing rates. This suggests that different brain states are associated with different functions, in which case differences in current brain state could account for marked differences in behaviours between individuals, even in similar circumstances[11].

3. Neurotransmitter Co-Release and Neurotransmitter Switching: The Dynamic Synapse

Many of the dogmas concerning our current understanding of brain information processing are built on the basic notion of cellular integration of EPSPs and IPSPs (*excitatory and inhibitory post-synaptic potentials*) at the axon hillock level for a given action potential generation of a given *glutamatergic* (excitatory) or *GABAergic* (inhibitory) neuron, which then in turn leads to depolarizing or hyperpolarizing inputs to the postsynaptic neuron. Through spatio-temporal summation and frequency changes, neuronal populations and their connections (circuitry) encode and transmit the information being processed, to transform sensory input into muscle contractions and gland secretions to promote animal survival and reproduction.

In addition to the parameters outlined above, evidence accumulating since the beginning of the last decade has revealed another striking form of integration, which occurs at the single-synapse level, and consists of transmitter co-release and transmitter switching. This phenomenon encompasses *dynamic* transmitter co-release and switching, including transmitter addition, loss, or replacement of one transmitter with another[12,13,14]. These changes in transmitter identity can be driven by naturally-occurring environmental stimuli, and provoke matching changes in postsynaptic transmitter receptors. Amazingly, they often convert the synapse entirely from excitatory to frankly inhibitory, or vice versa, providing a basis for changes in behavior in those cases in which it has been examined[12,14,15,16].

As mentioned previously (*vide supra*), brain information processing is ultimately based on electrochemical events which produce dramatic electrical potential changes in millisecond time-scale, i.e. the action potential, and its firing frequency patterns and population recruitment. However, the contents released at the axons nerve terminal(s) ultimately determines *physiological* circuitry recruitment and *information flow directionality.* Neurotransmitter content, and the post-synaptic effects of transmitter release at the synapse, can be potently modulated by *neuropeptide* and *neurosteroid expression patterns* within neuron circuits, both acutely and chronically, which are controlled by genetic, developmental stage and environmental factors. Among the latter, the *stress response* is a particularly important one. For this reason, we again wish to highlight the critical distinction between physiological circuits, which are highly dynamic, and *anatomical circuits*, which are physical connections stably persisting regardless of activation states. Powerful modern ultrastructural tools and transgenic technology combined with optogenetic/pharmacogenetic activation of given cell types to decipher the anatomical connectivity are providing new insights into neuronal circuits, revealing a wealth of anatomically-defined synaptic connections. These wiring diagrams are incomplete, however, because each ultrastructural connectivity map encodes multiple *physiological* circuits, some of which are active and some of which are latent at any given time. Thus, the functional connectivity is *actively shaped* by many factors, for instance, the neuromodulators' presence, that modify neuronal dynamics, excitability, and synaptic function. Much progress has been made in identifying the factors that induce transmitter switching and in understanding the molecular mechanisms by which it is achieved. However, fundamental questions remain to be carefully addressed[17]. In this regard, it is essential to note that natural stimulation of other sensory modalities and pharmacological agents may cause changes in transmitter expression that regulate different behaviors.

4. Brain States Are Susceptible to Emotional Reactivity: Role of Neuropeptide — Containing Subcortical Homeostatic Circuitry

It seems obvious that strong sensations, feelings and emotions may motivate or promote selective activation or inhibition of physiological circuits. Aggression against con-specifics in fights for a sexual partner, to protect infants, to compete for food or water, or to defend territory is one striking example of behavioral activation (fighting) motivated by strong emotions and feelings (sexual or maternal drive, hunger and thirst, self-protection)[18]. More extremely, strong emotions can lead to non-selective aggression, accompanied apparently by an over-riding of the "rational brain", as in cases of crimes of passion, psychopathic violence, and suicide, which may be seen as an extreme aggression against oneself in a state of despair. It is critical to know how these changes in brain states are regulated. What are the biological determinants of the remarkable gender/age/emotional state differences of these functions? There is now ample evidence that circuit element recruitment for extreme behaviors is complex. How the brain prioritizes activation and inhibition of competing circuits is selectively facilitated by the activity of subsets of neurons located in the subcortical regions, e.g. hypothalamus, midbrain, brain stem, with long axonal projections that release so-called *neuromodulators*, i.e., molecules able to modify the electrical properties of neurons and their synaptic connections. Most of these molecules are peptidergic, such as vasopressin (VP), oxytocin (OT), corticotrophin releasing hormone (CRH), pituitary adenylate cyclase-activating polypeptide (PACAP), enkephalins, and several others. The study of dynamic changes in neuropeptide expression, leading to behavioral changes, has afforded an important opportunity to reveal the linkages between physiological circuitry, brain waves, brain states, and behavior. For example, our own group (unpublished) has recently shown that the up-regulation of the hypothalamic magnocellular vasopressinergic pathways increases the theta oscillation coherence between the cortical regions such as septal-temporal hippocampus, and temporal hippocampus vs hypothalamic paraventricular nucleus. *These data directly demonstrate the close interaction between subcortical brain regions central to the processing of primary stress signals and the higher cortical brain regions, which mediate the behavioural outcomes of these stress experiences.*

4.1. *Hypothalamic VPMNNs are vulnerable to early life stress*

Instinctual behaviours, such as water and food intake, fight-flight stress response, sexual behaviour, determine straightforwardly animal's survival, both at individual and species levels. These essential behaviors are directly controlled by hypothalamic homeostatic circuits, which involve importantly the participation of neuropeptides and neurosteroids. Evolutionary conservation of the hypothalamus attests to its critical role in the control of fundamental behaviors[19,20,21]. Several recent studies have found that the hypothalamus is particularly susceptible to early stress induced either by endocrine imbalances, or psychological stressors such as neonatal maternal separation (MS). We have previously reported that in response to either maternal hyperthyroidism[22,23] or neonatal maternal separation (MS)[24,25], the rat hypothalamic vasopressin system becomes permanently up-regulated, showing enlarged volume of the hypothalamic paraventricular and supraoptic vasopressin nuclei and increased cell number, with an increased sensitivity to acute stressors or anxiogenic conditions in adulthood. Another recent study by our group[26] showed that the life-long consequence of neonatal maternal separation may be imprinted in changes in cell density in several hypothalamic regions, through the modification of the activities of pro- and anti-apoptotic factors during development. Moreover, the VP innervation to amygdala, a brain lmbic region, which exerts regulatory functions on food intake, sexual behaviour, aggression and fear processing, is remarkably increased in neonatal maternal separated rats[27]. These observations clearly demonstrate that the stress of maternal separation in early life has a re-organizing effect on this subcortical structure.

The neuropeptides vasopressin and oxytocin, are neuro-hormones synthesized primarily in the paraventricular and supraoptic nuclei that (in addition to their important peripheral actions when secreted by the posterior pituitary) are released in the brain and act centrally to modulate emotional behaviour (such as motivation, anxiety and depression) and many aspects of social behaviour. Currently it is considered that vasopressin plays a role in promoting anti-social behaviour, such as aggression[28,29,30]. Conversely oxytocin facilitates pro-social actions such as affiliation, attachment and social support, and maternal behaviour[31,32,33]. Hence vasopressin is considered to tend to have anxiogenic effects while oxytocin is anxiolytic. However, these notions are apparently reductionist and lacking of scientific rigor. For instance, central actions of vasopressin and oxytocin on affiliative behaviour can be sex-specific[34]. Moreover, cases associated with chronic social defeat, which is associated with *up*-regulation of oxytocin signalling pathways and *down*-regulation of vasopressin pathways[35], and/or the extreme group attachment ("feed-forward" affiliative behaviors) can lead to behaviours of later destructive aggression.

4.2. *Hippocampal theta rhythms and subfield oscillatory coupling: Could VPMNNs play a regulatory role?*

Theta rhythm (3-12 Hz) in rodents has been postulated to be involved in arousal, behavioural functions, sensor-motor integration and emotion[36]. Although this band of oscillation has been observed in several brain regions, the hippocampal formation is considered to be the main structure involved in the generation of theta band oscillations. Two types of theta have been documented: type I the movement related theta rhythm ranging from 7-12 Hz, non-cholinergic but sensitive to anaesthetics (such as urethane or pentobarbital) and type II, the non-movement related theta rhythms (4-7 Hz)[36,37]. The physiological occurrence of type II theta rhythm has been associated with the activation of a number of structures from the reticular formation to basal part of forebrain forming the ascending brainstem hippocampal synchronizing pathway[38]. The caudal diecephalic region, primary posterior hypothalamic nuclei have reported to involved in the control of theta oscillation.

The hippocampus' dorsal part (dHi, also called septal hippocampus) and ventral part (vHi, also called temporal hippocampus) have been considered having different roles, such as for "learning and memory" vs. "stress coping and emotional control"[39]. Both segments show clear oscillations in theta bands, however, there is a low *phase-locking* relationship between the two segment under rest condition[40]. So far, little is known concerning the oscillations coherence under different brain states between these two sub-regions.

In recent years, our research group has devoted efforts to decipher the connectomic aspects (*anatomical and functional connectivity*) of magnocellular vasopressin systems. The results of our group have contributed to the current understanding of neuromodulation by this system. Using *in vivo* juxtacellular labelling together with immunohistochemical and anatomical methods, we have shown that axon collaterals co-expressesing the vesicular glutamate transporter type 2 (VGLUT2), a marker for glutamatergic pathways[41], innervate both dHi and vHi and salt loading potently increase the theta oscillation power (Fig. 1).

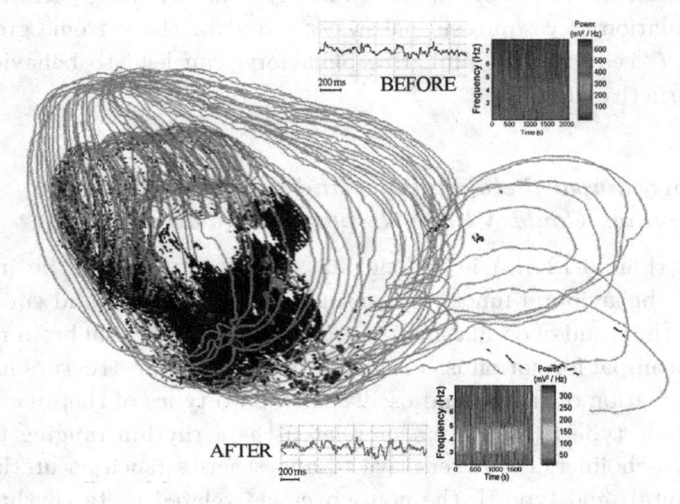

Figure 1. Vasopressinergic innervation (green lines, reconstructed from 78 serial sections using Neurolucida technique) in the subcortical regions and in the hippocampus (red lines circumscribed region. The section boundaries were outlined with blue lines and the soma of vasopressinergic cells were denoted by black dots. Upper and bottom insets show the spectrograms. Note the increase of power after salt challenge.

On the other hand, electrophysiological studies have showed that nanomolar concentrations of [Arg8]-vasopressin (AVP) induced a prolonged increase in amplitude and slope of the evoked population response in the hippocampus[42].

We have observed, with retro-and antero-grade tracers that the vasopressin neurons in PVN and SON are important sources of innervation to cortical and subcortical structures such as the epithalamic nucleus, lateral habenula, thalamic nuclei, locus coeruleus, amygdaline nuclei and the hippocampal formation. To confirm these findings, we used the technique of juxtacellular labelling to label single neurons in the PVN (Fig. 2). We observed that most of the magnocellular vasopressinergic neurons possessed multiple axons or axon collaterals that project toward different limbic structures and hippocampus (Fig. 2)[41,43] and after analysis of electrophysiological recordings we determined that this PVN region presents theta frequency electrical oscillations (Fig. 2), and these oscillations intensify with an osmotic stimulus.

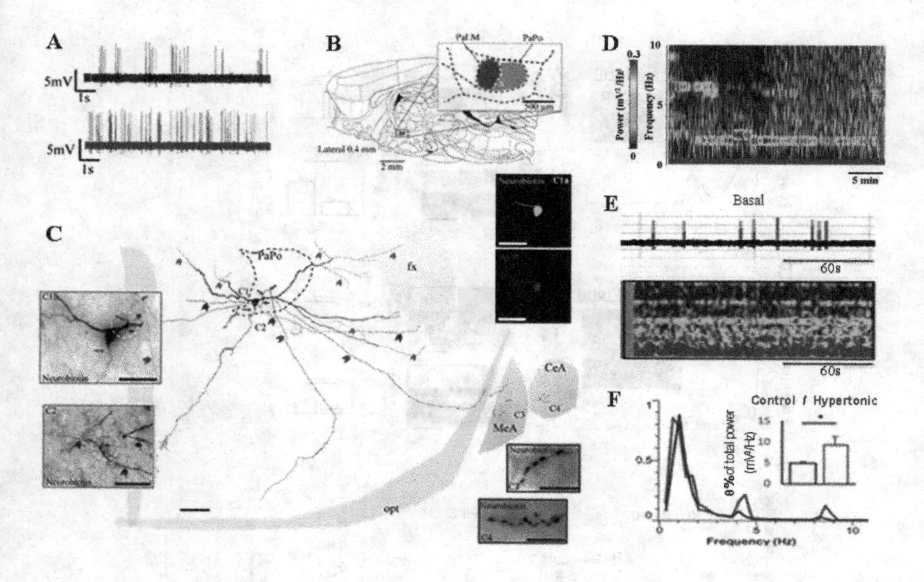

Figure 2. Extracellular recording/juxtacellular labeling revealed theta rhythms and "REM" like transition in PVN and projections into the internal capsule and fimbria-fornix conducting systems connecting with hippocampus.

In a recent study from our group (unpublished), using in vivo dual electrode extracellular recording and juxtacellular labelling in CA2 of the dorsal and ventral hippocampus, and PVN combined with signal analysis, we observed that AVP magnocellular neurons exhibit spontaneous REM-like transitions and REM episodes (Fig. 2). An osmotic stressor activates the magnocellular neurons of the PVN and increases the low-theta power and numbers of REM like episodes (Fig. 2). The theta oscillation coherence between ventral versus dorsal hippocampus and ventral hippocampus versus PVN was also increased (Fig. 2). These findings are reflecting subcortical AVP containing circuits influence on hippocampal function and suggest that a potent activation of a subcortical circuitry mainly devoted to water-electrolyte balance can influence cortical functions, such as hippocampal coordinated learning and memory (Fig. 3).

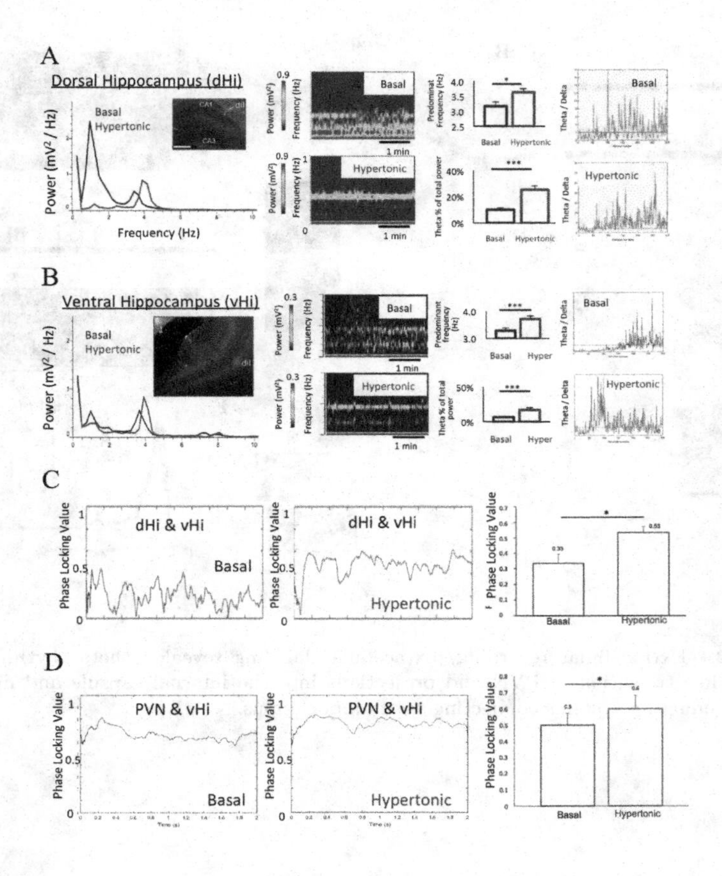

Figure 3. The dorsal (panel A) and ventral (panel b) hippocampus showed an increase in the theta power and in the predominant theta frequency after an osmotic (Hypertonic) stressor. The theta oscillation synchronization between thePVN and vHi (panel C) and between dHI and vHi (panel D) increased after the hypertonic stimulation.

4.3. *Behavioral consequences of changing VPMNN tone: Two examples*

Maternal separation (MS) has been demonstrated to up-regulate the hypothalamic vasopressin (VP) system. Intracerebrally released VP has been demonstrated to affect several types of animal behavior, such as active/passive avoidance, social recognition, and learning and memory. We investigated the effects of an osmotic challenge and a V1b receptor-specific (V1bR) antagonist, SSR149415, on spatial learning of maternally separated and animal facility reared adult male Wistar rats. Rats raised to adulthood in the animal facility with normal maternal rearing (AFR), and rats that underwent neonatal maternal separation for 3 hours during the first 15 days after birth (MS3h) were used in these studies. The latter group was shown to have a potentiated vasopressinergic system[24,25]. Morris water maze task (MWM) is a widely used behavioral test for rodent to assess the learning and memory function in which hippocampus plays a central role. In a previous study we have demonstrated that the VPMNNs send axon collaterals to innervate selectively this brain region[24,25]. The test used a black circular pool (diameter 156 cm, height 80 cm) filled with 30 cm of water (25 $\pm$ 1° C) with distant visual cues. A circular black escape platform (diameter 12 cm) was submerged 1cm below the water surface. Rats were habituated to this swimming task (without the presence of the platform) a week before the MWM. On the day of the test, rats were allowed up to 60 s to locate the escape platform. If the allowed time ended and the experimental subjects have not found the platform, they are guided to it. Once on the platform, rats were permitted to stay for 10 s and allowed to observe their location. Each rat underwent 8 sequential trials on the same day, with an inter-trial interval of approximately 5 min. The time required to locate the hidden platform in each trial was recorded. Generally, the rats learn and consolidate the location of the platform around the 3rd trial which can be seen in panel A of the figure 4, under basal condition, both experimental groups had similar learning performance. In panel B, the rats received a 900mM saline intraperitoneal injection of 2% (ml/g of bodyweight) which is known to potently upregulate the VPMNNs metabolic activity. The MS3h, which possesses a potentiated vasopressinergic system showed higher vulnerability which suggested an abnormal hippocampal functions correlated with the VPMNNs system upregulation in an upregulated system. In panel C, the rats received an injection of a selective V1b receptor antagonist and in this case the AFR group showed an enhanced vulnerability at the given doses comparing with the MS3h, which possesses a potentiated vasopressinergic system. In panel D, both groups received both hypertonic saline and the antagonist intraperitoneal injections 30 min before the behavioral test, the differences in learning observed in the previous tests were abolished. The panels E and F are curves grouped by experimental groups showing that for AFR group, with a normal VPMNN system, the condition which impairs their learning process is the pharmacological treatment (application of a selective V1b antagonist) and the condition for the MS3h group, with a potentiated VPMNN system, is the osmotic challenge, during which the

VPMNN system is further upregulated (metabolic activation due to salt loading).

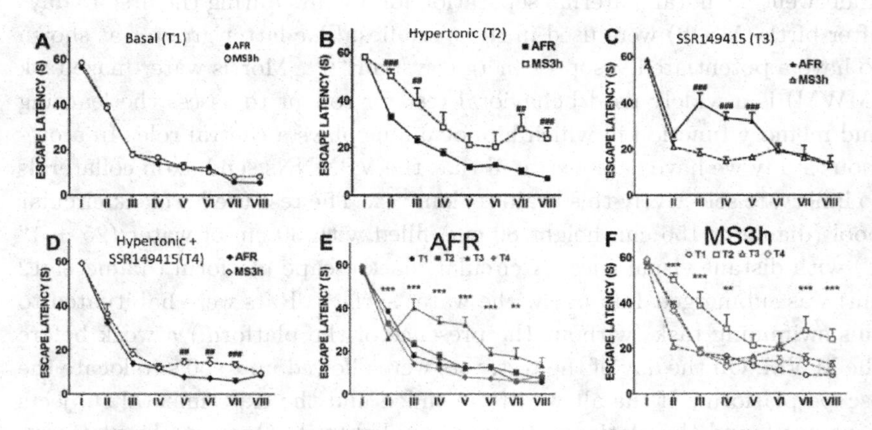

Figure 4. Physiological upregulation and pharmacological downregulation of hypothalamic vasopressinergic magnocellular neurosecretory neurons induce differential behavioral modification in Morris water maze (MWM) performance of adult animal facility reared (AFR) and maternally separated (MS3h). Under basal conditions (A, T1) both MS and AFR rats have a similar performance during the MWM test. Salt loading (900mM, 2% b.w. and i.p.; B, T2) impairs only the MWM performance of the MS3H group and the administration of SSR149415 (5 mg/kg, i.p.; C, T3) impairs only the AFR group MWM performance. When both MS3h and AFR received both hypertonic saline injection, and also the V1bR antagonist SSR149415 the differences cannot be observed (D, T4). The panels E and F are curves grouped by experimental groups showing that for AFR group, with a normal VPMNN system, the condition which impairs their learning process is the pharmacological treatment and the condition for the MS3h group, with a potentiated VPMNN system, is the osmotic challenge, during which the VPMNN system is further upregulated (metabolic activation due to salt loading). (Modified from:[24]).

It is widely accepted that the lateral habenula encodes negative motivational value and that its over-activation fosters psychomotor deficiency, a cardinal symptom of depressive behaviour[44,45,46,47]. The presence of VP-containing axons and terminals in lateral habenula (LHb) has long been observed. In a recent study we found, using anatomical and immunohistochemical methods, that LHb medial subdivision (LHbM) was a major target for AVP axon terminals, and that the latter synapsed upon a unique population of GABAergic interneurons in the habenular complex. At the electron microscopical level, AVP+ terminals were found to establish Gray type I (excitatory) synapses with AVP-containing vesicles docked on the presynaptic membrane suggesting a role for AVP and glutamate co-release during synaptic transmission at this site. With *in vivo* juxtacellular labelling and reconstruction methods, we observed that GABA containing neurons located in the LHbM possessed axons that branched extensively in the LHbM parvocellular subnucleus (LHbMPc), and that their dendrites received AVP+ fibre contacts. Fluorogold retrograde tracing from LHb and juxtacellular labelling from PVN showed that VPMNNs of PVN served as one of the main sources of AVP habenular innervation. We therefore hypothesized that modulation of thirst might significantly influence innate fear and behavioural despair processing.

It is in this tenor that we present another example (Fig. 5) whereby upregulating the metabolic activity of the VPMNNs by 24h water-deprivation (WD24), the rats reduce freezing/immobility behaviors during live predator (cat) exposure and forced swimming test, which correlated with reduced Fos expression in the whole LHb and suggested a down-regulation of LHb's functional output in response to WD24[48].

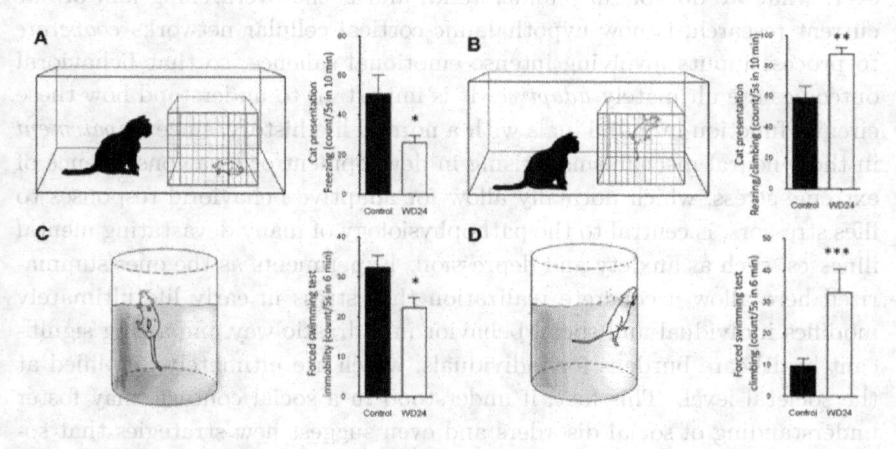

Figure 5. Active stress coping is promoted during 24 hours of water deprivation (WD24) with innate fear processing and behavioral despair. The metabolic activities of AVP containing magnocellular neurosecretory neurons in SON and PVN are increased by the physiological stimulus of WD24. The innate fear-related passive and active behaviors (A, B) are expressed by rats upon cat exposure. A reduction of freezing counts (A) and an increase of climbing and rearing behaviors (B) are then showed by rats. During a forced swimming test (FST), for behavioral despair, similar observations are obtained (C, D)

These examples illustrate the close interaction between the processing of primary stress signals within sub-cortical brain regions and their subsequent output to associational cortex, which integrates multiple inputs to mediate the behavioral outcomes of these stress experiences.

5. What Are We Learning from Our Experiments? — The Urgency of An Emergent Multidisciplinary Neuroscience in Response to a Growing Societal Demand

The profound effects of VPMNN, the hypothalamic center for physiological regulation of salt and blood volume, on the hippocampus, the cortical center for memory, or learning from experience, have much to teach us about the human brain and human behavior. We have so far established (vide supra) that sub-cortical processing of stressful experience, whether during development or in adulthood, can profoundly affect behavior. However, what we do not fully understand, and is the overarching aim of our current research, is how hypothalamic-cortical cellular networks *cooperate* to process inputs involving intense emotional salience, so that behavioral outcomes are ultimately *adaptive*. It is important to understand how these circuits function in individuals with a normal life history, since *impairment* in these neural mechanisms, arising in development or as a consequence of extreme stress, which normally allow for adaptive behavioral responses to lifes stressors, is central to the pathophysiology of many devastating mental illnesses, such as anxiety and depression. Experiments as the ones summarized here allow a concrete realization that stress in early life ultimately modifies individual and social behavior in a drastic way, imparting significant healthcare burdens for individuals, which are ultimately amplified at the societal level. This fact, if understood in a social context, may foster understanding of social disorders and even suggest new strategies that societies, through their social institutions, could use to enhance cohesion and prosocial behaviours.

Perhaps the most important conclusion to be drawn from the present work is that it is time for a multidisciplinary neuroscience that merges with social sciences and institutions. It should be one that brings our understanding of experience-dependent changes in brain function to bear on human social problems commonplace in our modern society as rapidly as possible. An alliance between multidisciplinary neuroscience, social sciences, and social institutions may be the only fruitful way to accomplish this.

Acknowledgments

This work was partially supported by: DGAPA-UNAM-PAPIIT-IN216214, CONACYT-CB-176919 & CB-238744 to LZ and NIMH-IRP-1ZIAMH002386 to LEE. LZ is a Fulbright visiting scholar, also supported by PASPA-DGAPA-UNAM fellowships for her sabbatical research stay hosted by LEE of SMN-NIMH-NIH-USA. We thank Mariana Marques and Hernn Barrio for technical assistance in some experimental works we used in this chapter and to Rafael A. Barrio and James D. Murray for critical reading and comments on an early form of this manuscript.

References

1. Llinas, R. R.; Pare, D. *Neuroscience* **1991**, 44, (3), 521-35.
2. Llinas, R. R.; Ribary, U.; Jeanmonod, D.; Kronberg, E.; Mitra, P. P. *Proc. Natl. Acad. Sci. USA* **1999**, 96, (26), 15222-7.
3. Gervasoni, D.; Lin, S. C.; Ribeiro, S.; Soares, E. S.; Pantoja, J.; Nicolelis, M. A. *J Neurosci.* **2004**, 24, (49), 11137-47.
4. Somogyi, P.; Katona, L.; Klausberger, T.; Lasztoczi, B.; Viney, T. J. *Philosophical Transactions of the Royal Society of London. Series B, Biological sciences* **2014**, 369, (1635), 20120518.
5. Badgaiyan, R. D.; Sinha, S.; Sajjad, M.; Wack, D. S. *PLoS One* **2015**, 10, (9), e0137326.
6. Duch, W.; Dobosz, K.; Jovanovic, A.; Klonowski, W., Neurodynamic Insight into Functional Connectivity, Cognition, and Consciousness In *Exploring the Landscape of Brain States, NeuroMath COST Action*, Neurosci., F., Ed. 2010.
7. Mohr, P. N.; Nagel, I. E. *The Journal of Neuroscience: The Official Journal of the Society for Neuroscience* **2010**, 30, (23), 7755-7.
8. Weinberg, H. In *The beginning of a new look in the understanding of brain function: Systems of the brain related to the processing and utilization of information may different for different individuals*, Transient Dynamic Brain States, From basic research to clinical applications, 2015.
9. Sullivan, E. V.; Lane, B.; Kwon, D.; Meloy, M. J.; Tapert, S. F.; Brown, S. A.; Colrain, I. M.; Baker, F. C.; De Bellis, M. D.; Clark, D. B.; Nagel, B. J.; Pohl, K. M.; Pfefferbaum, A. *Brain Imaging and Behavior* **2016**.
10. Pfefferbaum, A.; Rohlfing, T.; Pohl, K. M.; Lane, B.; Chu, W.; Kwon, D.; Nolan Nichols, B.; Brown, S. A.; Tapert, S. F.; Cummins, K.; Thompson, W. K.; Brumback, T.; Meloy, M. J.; Jernigan, T. L.; Dale, A.; Colrain, I. M.; Baker, F. C.; Prouty, D.; De Bellis, M. D.; Voyvodic, J. T.; Clark, D. B.; Luna, B.; Chung, T.; Nagel, B. J.; Sullivan, E. V. *Cerebral Cortex* **2016**, 26, (10), 4101-21.
11. Lee, S. H.; Dan, Y. *Neuron* **2012**, 76, (1), 209-22.
12. Dulcis, D.; Jamshidi, P.; Leutgeb, S.; Spitzer, N. C. *Science* **2013**, 340, (6131), 449-53.
13. Gutierrez, R.; Romo-Parra, H.; Maqueda, J.; Vivar, C.; Ramirez, M.;

Morales, M. A.; Lamas, M. *The Journal of Neuroscience: The Official Journal of the Society for Neuroscience* **2003**, 23, (13), 5594-8.
14. Shabel, S. J.; Proulx, C. D.; Piriz, J.; Malinow, R. *Science* **2014**, 345, (6203), 1494-8.
15. Gutierrez, R. *Journal of Neurophysiology* **2000**, 84, (6), 3088-90.
16. Zhang, S.; Qi, J.; Li, X.; Wang, H. L.; Britt, J. P.; Hoffman, A. F.; Bonci, A.; Lupica, C. R.; Morales, M. *Nature Neuroscience* **2015**, 18, (3), 386-92.
17. Spitzer, N. C. *Neuron* **2015**, 86, (5), 1131-44.
18. Russell, J. A. *Journal of Neuroendocrinology* **2002**, 14, (1), 1-3.
19. Swanson, L. W., *Brain Architecture: Understanding the Basic Plan*, 2nd ed.; Oxford University Press: New York; Oxford, 2012.
20. Elmquist, J. K.; Coppari, R.; Balthasar, N.; Ichinose, M.; Lowell, B. B. *The Journal of Comparative Neurology* **2005**, 493, (1), 63-71.
21. Saper, C. B.; Lowell, B. B. *Current Biology: CB* **2014**, 24, (23), R1111-6.
22. Zhang, L.; Hernandez, V. S.; Medina-Pizarro, M.; Valle-Leija, P.; Vega-Gonzalez, A.; Morales, T. *J Neurosci. Res.* **2008**, 86, (6), 1306-15.
23. Zhang, L.; Medina, M. P.; Hernandez, V. S.; Estrada, F. S.; Vega-Gonzalez, A. *Neuroscience* **2010**, 168, (2), 416-28.
24. Hernandez, V. S.; Ruiz-Velazco, S.; Zhang, L. *Neurosci. Lett.* **2012**, 528, (2), 143-7.
25. Zhang, L.; Hernandez, V. S.; Liu, B.; Medina, M. P.; Nava-Kopp, A. T.; Irles, C.; Morales, M. *Neuroscience* **2012**, 215, 135-48.
26. Irles, C.; Nava-Kopp, A. T.; Moran, J.; Zhang, L. *Stress* **2014**, 17, (3), 275-84.
27. Hernandez, V.; Hernndez, O.; Gomora, M.; Perez De La Mora, M.; Fuxe, K.; Eiden, L.; Zhang, L. *Frontiers in Neural Circuits* **2016**, 10, (92).
28. Ferris, C. F. *Novartis Foundation Symposium* **2005**, 268, 190-8; discussion 198-200, 242-53.
29. Veenema, A. H.; Beiderbeck, D. I.; Lukas, M.; Neumann, I. D. *Hormones and Behavior* **2010**, 58, (2), 273-81.
30. Wersinger, S. R.; Ginns, E. I.; O'Carroll, A. M.; Lolait, S. J.; Young, W. S., 3rd. *Molecular Psychiatry* **2002**, 7, (9), 975-84.
31. Hu, Y.; Scheele, D.; Becker, B.; Voos, G.; David, B.; Hurlemann, R.; Weber, B. *Scientific Reports* **2016**, 6, 20236.
32. Lukas, M.; Toth, I.; Reber, S. O.; Slattery, D. A.; Veenema, A. H.; Neumann, I. D. *Neuropsychopharmacology: Official Publication of the American College of Neuropsychopharmacology* **2011**, 36, (11), 2159-68.
33. Marsh, N.; Scheele, D.; Gerhardt, H.; Strang, S.; Enax, L.; Weber, B.; Maier, W.; Hurlemann, R. *The Journal of Neuroscience: The Official Journal of the Society for Neuroscience* **2015**, 35, (47), 15696-701.
34. Young, L. J. *Am. J. Med. Genet.* **2001**, 105, (1), 53-4.
35. Steinman, M. Q.; Duque-Wilckens, N.; Greenberg, G. D.; Hao, R.; Campi, K. L.; Laredo, S. A.; Laman-Maharg, A.; Manning, C. E.; Doig, I. E.; Lopez, E. M.; Walch, K.; Bales, K. L.; Trainor, B. C. *Biological Psychiatry* **2016**, 80, (5), 406-14.
36. Lever, C.; Kaplan, R.; Burgess, N., The Function of Oscillations in the Hip-

pocampal Formation. In *Space, Time and Memory in the Hippocampal Formation*, Derdikman, D.; Knierim, J. J., Eds. Springer Vienna: Vienna, 2014; pp 303-350.
37. Kramis, R.; Vanderwolf, C. H.; Bland, B. H. *Experimental Neurology* **1975**, 49, (1 Pt 1), 58-85.
38. Bland, B. H.; Oddie, S. D. *Neuroscience and Biobehavioral Reviews* **1998**, 22, (2), 259-73.
39. Fanselow, M. S.; Dong, H. W. *Neuron* **2010**, 65, (1), 7-19.
40. Patel, J.; Fujisawa, S.; Berenyi, A.; Royer, S.; Buzsaki, G. *Neuron* **2012**, 75, (3), 410-7.
41. Hernandez, V. H., Vazquez-Juarez, E., Marquez, M. M., Jauregui Huerta, F., Barrio, R. A. and Zhang, L. *Frontier in Neuroanatomy* **2015**, 9:130.
42. Chen, C.; Diaz Brinton, R. D.; Shors, T. J.; Thompson, R. F. *Hippocampus* **1993**, 3, (2), 193-203.
43. Zhang, L.; Hernandez, V. S. *Neuroscience* **2013**, 228, 139-62.
44. Li, B.; Piriz, J.; Mirrione, M.; Chung, C.; Proulx, C. D.; Schulz, D.; Henn, F.; Malinow, R. *Nature* **2011**, 470, (7335), 535-9.
45. Lecca, S.; Meye, F. J.; Mameli, M. *The European Journal of Neuroscience* **2014**,39, (7), 1170-8.
46. Matsumoto, M.; Hikosaka, O. *Nature* **2007**, 447, (7148), 1111-5.
47. Hikosaka, O. *Nature Reviews. Neuroscience* **2010**, 11, (7), 503-13.
48. Zhang, L.; Hernandez, V. S.; Vazquez-Juarez, E.; Chay, F. K.; Barrio, R. A. *Front Neural Circuits* **2016**, 10, 13.

PROTEIN STRUCTURE ESTIMATION FROM INCOMPLETE NMR DATA*

Z. LI, S. LIN, Y. LI, Q. LEI AND Q. ZHAO

Center for quantum technology research
School of physics
Beijing institute of technology
Beijing, China
E-mail: qzhaoyuping@bit.edu.cn

The knowledge of protein structures is very important to understand their physical and chemical properties. Nuclear Magnetic Resonance (NMR) spectroscopy is one of the main methods to observe the protein structure. In this paper, we propose a two-stage approach to calculate the structure of a protein from a highly incomplete distance matrix, where most of the data is obtained from NMR. We first randomly guess a small part of unobservable distance by utilizing the triangle inequality which is crucial for the second stage. Then we use the Matrix Completion (MC) to calculate the protein structure from the obtained incomplete distance matrix. We also apply the accelerated proximal gradient algorithm (APG) to solve the corresponding optimization problem. Besides, we also analyze the recovery error. We demonstrate the efficiency of our method by several practical examples.

1. Introduction

Knowing the structure of proteins is critical for understanding their physical and chemical properties, which is also useful in drug design. There are two major methods about protein structure determination: X-ray crystallography and protein nuclear magnetic resonance (NMR) [1]. However, the protein NMR method is fundamentally different from the X-ray method: it includes isotope labeling, data collection and processing, peak picking, chemical shift assignment, the Nuclear Overhauser Effect or Enhancement (NOE) peak assignment, and finally, structure calculation. The introduction of protein NMR is a breakthrough, enabling the identification of protein

*Work partially supported by the ministry of science and technology of china (2013yq030595-3, and 2013aa122901), and by nsf of china with the grant (no. 11275024, and 11675014).

structure in the aqueous solutions, which is closer to the native states of proteins. However, the inevitable error of experimental NMR spectra and the intrinsic ambiguity of peak assignments that results from the limited accuracy of frequency measurements turns the tractable problem of finding the chemical shift assignments from ideal spectra into a formidably difficult one under realistic conditions. To elaborate, taking NOE experiments for example, the measurements of long distances (larger than 6Å) are normally unreliable and thus a network of short distances is available for reconstructing the structure of the protein. This process produces an incomplete and noisy Euclidian distance matrix (EDM), from which the short distances of the hydrogen atoms are estimated.

Since the NMR method only provides implicit and indirect information about the protein structure, it relies heavily on complex computational algorithms and methods. Many approaches have been proposed to solve this problem. The existing methods can be categorized into following four techniques:

(1) Euclidean Distance Matrix Completion (EDMC).
(2) Local/global optimization. for authors.
(3) Molecular Dynamics and Simulated Annealing (MD and SA).
(4) Sequence-based Protein Structure Prediction algorithms.

In the early years of protein NMR, many EDMC methods directly worked on distance matrix. The pioneer work of protein NMR by EDMC was realized by Braun *et al.* in 1981[2]. Other efficient ways, such as EMBED [3] and Distance Geometry Algorithms (DISGEO) [4], were developed by Havels group in 1983 and 1984, respectively. Several years later, many researchers thought it was more straight forward to use Gram matrix instead of EDM for solving EDMC problems. Therefore, semidefinite programming (SDP) was chosen to formulate the EDMC problem using Gram matrix. In 2008, Biswas *et al.* proposed Distributed Anchor-Free Graph Localization (DAFGAL) [5], which is based on the idea of divide-and-stitch, to solve EDMC problem. And then, Leung and Toh proposed DISCO method [6], a direct descendant of DAFGAL, in 2009. More recently, Li's group proposed SPROS in 2013[7], which is also a method based on the semi-definite programming-based (SDP) for computing Gram matrix.

In protein NMR methods, the main assumption is that the global minimum of the optimization problem is close to the native structure of the target protein. A effective method, called DGSOL, was presented by Moré and Wu[8,9], in which they solved the problem by a global continuation algorithm. Braun and Go [10] proposed a widely-adapted method called Variable Target Function, which consisted of a group of minimizations. Later on, combining the variable target method with the fast gradient computation technique, Braun and Go proposed DISMAN [10], which was proved to be effective for α-helical proteins due to their mostly short-range restraints.

Running simulated annealing by molecular dynamics (MD) simulation was a breakthrough in protein NMR determination. These methods were able to search the massive conformation space without being trapped in one of the numerous local minima. The first successful and widely-used method was X-plor, which was proposed by Brünger in 1993[11] and consummated by Schwieters *et al.* in 2006[12]. At the same time, torsion angle space was introduced to MD by Jain, introducing the method called torsion angle dynamics (TAD) [13]. The advantage of this approach was that the number of degrees of freedom in the torsion angle place can be reduced nearly 10 times than in Cartesian coordinate space. In 2004, the package CYANA was built by Güntert [14], which was one of the fastest and most widely-used methods.

Another approach for NMR protein determination is Sequence-based Protein Structure Prediction. It has long been known that the 3D structure of a protein is directly related to amino acid sequence [15]. Therefore, there are many other pathways for generating protein structure models, that is structure predictions from the sequence. Among those, ROSETTA is one of the most successful programs for obtaining atomic level 3D structures of small proteins [16]. There are two steps in ROSETTA process: (i) a low-resolution exploration phase using Monte Carlo fragment assembly and a coarse-grained energy function, and (ii) a computationally expensive refinement phase that cycles between combinatorial side-chain optimization and gradient-based minimization of all torsional degrees of freedom in a physically realistic all-atom force field [16]. The structural accuracy of selected fragments can be improved by adding the structural information contained in experimentally determining NMR chemical shifts, and thereby to improve ROSETTA performance without any significant change in the basic structure or functioning of this well established program [17].

2. Two-Stage Method

2.1. *Stage 1: Randomly "guess" a small part of unobservable distances*

The theory of matrix completion (MC) [19-21] requires the sampling pattern of a data matrix should be as random as possible. However, this is not the case in our situation since short distance data may not randomly distribute in the Euclidean distance matrix. In order to utilize the MC theory, we first randomly "guess" a small part of unobservable distances by applying the triangle inequality i.e., the sum of any two sides of a triangle is greater than the third side. The guessed distances are not very accurate in general, but they can be used to recover the original structure matrix by the robust property of matrix completion.

2.2. *Stage 2: Matrix completion*

Matrix completion (MC) based on convex optimization algorithm of compressed sensing [18-19] is a relative frontier area. It mainly studies on the recovery of unknown elements of matrix by inputting a few elements for a low rank matrix. The structure of a protein can be described by a Euclidean distance matrix whose rank is 5 [20]. This makes it possible to apply the MC algorithm to recover the incomplete structure.

Protein distance matrix completion problem can be expressed as follows:

$$\min_{x} \; rank(X) \quad s.t. \; X_{i,j} = M_{i,j}, \quad (i,j) \in \Omega \tag{1}$$

Nevertheless, due to the non-convexity and discontinuity of the objective function, the above problem generally is NP-hard. Candès and his group proposed a widely used method , that is using nuclear norm instead of the rank constraint [21]. Then the rank minimization program can be approximated by seeking nuclear norm minimization, as a convex relaxation of the original problem, as following:

$$\min_{x} \|X\|_* \quad s.t. \; X_{i,j} = M_{i,j}, \quad (i,j) \in \Omega \tag{2}$$

3. Accelerated Proximal Gradient Algorithm

Accelerated Proximal Gradient Algorithm (APG) is a first order algorithm based on Nesterov' technique, whose rate of convergence is very competitive. Toh and Yun [22] converted matrix completion problem into an unconstrained optimization problem given by

$$\min f(X) = \frac{1}{2}\|P_\Omega(X - M)\|_F^2 + \mu\|X\|_*. \tag{3}$$

In each iteration , APG minimizes the second order approximation of Y

$$Q(X,Y) = \frac{1}{2}\|P_\Omega(X - M)\|_F^2 + \langle P_\Omega, (X - Y)\rangle + \frac{L_f}{2}\|X - Y\|_F^2 + \mu\|X\|_*. \tag{4}$$

where L_f is arbitrary constant subjected to

$$\|\nabla f(X_1 - \nabla f(X_2)\| \leq L_f\|X_1 - X_2\|, \tag{5}$$

that is, Lipchitz constant of ∇f

It is easy to see:

$$\underset{X}{\operatorname{argmin}} Q(X,Y) = \underset{X}{\operatorname{argmin}} \frac{L_f}{2}\left\| X - Y + \frac{1}{L_f}P_\Omega(Y)\right\|_F^2 + \mu\|X\|_* \tag{6}$$

For each iteration

$$X_{k+1} = \underset{X}{\operatorname{argmin}} Q(X, Y_k) = \mathcal{D}_{\frac{\mu}{L_f}}\left(Y_k - \frac{1}{L_f}P_\Omega(Y_k)\right). \tag{7}$$

Usually, the rate of convergence can reach $\mathcal{O}(k^{-2})$ when Y_k is the sequence as follows:

$$Y_k = X_k + \frac{t_{k-1} - 1}{t_k}(X_k - X_{k-1}),\ \ t_{k+1}^2 - t_{k+1} \leq t_k^2. \tag{8}$$

The experiments of Toh and Yun[22] show that after APG using the continuation technique and line search technique, the convergence rate of APG is much faster than some other algorithms of matrix completion, such as CVX[23] and SVT[24].

Algorithm 1: APG

1, initializing the X_0, X_{-1},
2, while not converged do,
3, $Y_k = X_k + \frac{t_{k-1}-1}{t_k}(X_k - X_{k-1})$,
4, $X_{k+1} = \mathcal{D}_{\frac{\mu}{L_f}}\left(Y_k - \frac{1}{L_f}P_\Omega(Y_k)\right)$,
5, $t_{k+1} = \frac{1+\sqrt{t_k^2+1}}{2}$, $k = k + 1$
6, end while

---

4. Experimental Results

In our experiments, all of the protein data are obtained from protein data bank (PDB). We establish the initial distance matrix from protein structure model which satisfies the features of NMR data. We set a variety of constraints including chemical bond, rigid plane, and triangle inequality. APG program and RMSD is used to compute the structure and the errors between protein structures, respectively. Visualized results are shown in figures at the end of this section.

We have chosen 4 protein structures from PDB as follows:

ID	Description	Atoms
1G6J	Ubiquitin	1228
1B4R	PKD domain 1 from human polycystein-1	1114
2KT6	Outer membrane usher protein papC	1283
2M5Z	Antimicrobial protein	598

5. Establish the Initial Distance Matrix According to the Features of Protein Structure and NMR

As we all know that proteins are linear hetero polymers, composed of a set of twenty amino acids. By structure, all of the twenty amino acids share a common backbone, Amino acids differ by the side chains attaching to Ca and thus have different chemical properties. Both in backbone and side chain there are partial rigid structures such as peptide plane and stereoscopic structure. Therefore, we can obtain these distances directly from those rigid structures. In addition, generally, chemical bond length is relatively invariant in compound and the distance of atoms through covalent bond can be given directly.

In the backbone of peptides, there are 4 dihedral angles defined by rigid structure, i.e. ϕ, φ, ω and χ. ϕ is the angle of right-hand rotation around $N - C_\alpha$ bond; φ is the angle of right-hand rotation around $C_\alpha - C$ bond; ω is the angle of right-hand rotation around $C - N$ bond. These make 6 neighboring atoms constitute a peptide plane in backbone. In the side chain of an amino acid, taking methionine (MET) as an example, χ is defined by a covalent bond between two heavy atoms (Fig. 1). These rigid structures provide the distance matrix with more distance information. Now considering NMR experiment, according to NOE and isotope labeling, we can get distance less than $6\AA$.

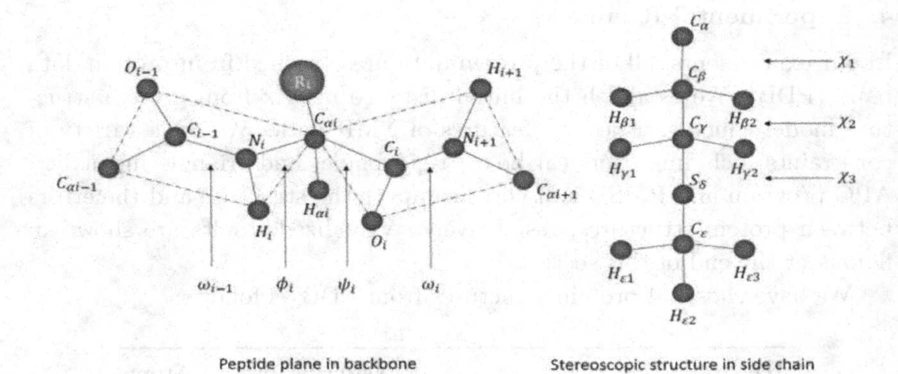

Figure 1. Schematic diagram of the polypeptide backbone and the rigid structure of the branch.

For a stereo chemical structure in space, each inter-atomic distance must satisfy the triangle inequality constraint. We choose the upper bound of the triangle inequality as the estimated value of distance. Due to the inaccuracy of the triangle inequality, only less than 10% values are extracted randomly.

Based on the above, we can establish the initial distance matrix according to the features of protein structure and NMR.

6. Result of APG Algorithm in Calculating Protein NMR Structure

The estimated value may carry some errors. In order to verify the feasibility of our algorithm, we firstly consider an ideal situation that our first estimated values are accurate. The advantage of this hypothesis is that the recovery of protein structure is ideal if accurate data can be obtained, then only improving accuracy of estimated values in the first step can be used to obtain the expected structure. Therefore, firstly, we consider the estimated values computed by the triangle inequality are accurate. The sample rate of these values satisfies $m \geq nr\log_{10}n$, which is the lower bound in matrix completion theory proposed by Candes [18], where n is the dimension of matrix.

In this case, for a specific protein, we need to know how much data weshould obtain at least to recover the protein structure by information retrieval methods. Each protein sample rate can be calculated by the formula $m \geq nr\log_{10}n$ The results are shown in Table 1. Where in Table 1, protein ID: the corresponding code of a protein in the protein database (PDB);

Algorithm 2: Establish the initial distance matrix

```
input: amino acid sequence of a protein, Seq
output: initial distance matrix, D
    1, compute the dimension of a protein, N
    2, D = zeros(N)
      for 1→ Seq
        classify atoms: in backbone and in side chain
        for 1N
          if Seq(i)='certainaminoacid'
            D(j, k) ←distance between atoms in peptide plane and rigid
                      structure
            D(j, k) ←distance between atoms linked by covalent bond
            D(j, k) ←distance obtained by NMR experiment
          end
        end
      end
    3, for 1N
          D(j, k) ← D(j, i) + D(i, k) //upper bound of triangle
  inequality
       end
```

Atoms: the number of atoms in the entire protein molecule, including all the atoms both in the backbone and inside chain; Sampling rate: the ratio of the number of known distance information to the square of matrix dimension; RMSD: root mean square deviation; Time: the total time including the operation of the Fellow Program in Management (FPM) program, the Non-Negative Least-Squares Algorithm (NNLS) program runs as well as the calculation of the RMSD.

In the entire experiment, the realization of protein NMR structure calculation by the two-stage method was tested through 2013b MATLAB software. As the last part of the algorithm, the FPM program package completes the establishment of protein NMR model. For the recovery problem of distance information, we use the FPM algorithm in the APG program package to calculate. This package is designed to solve the least squares problem in the minimization problem of nuclear norm. All tests are completed on the PC Linux computer with 2.8GHz processor and 8GB memory.

Figure 2 shows the recovery structures of the four proteins with all atomic positions calculated by the two-step method. The data of the struc-

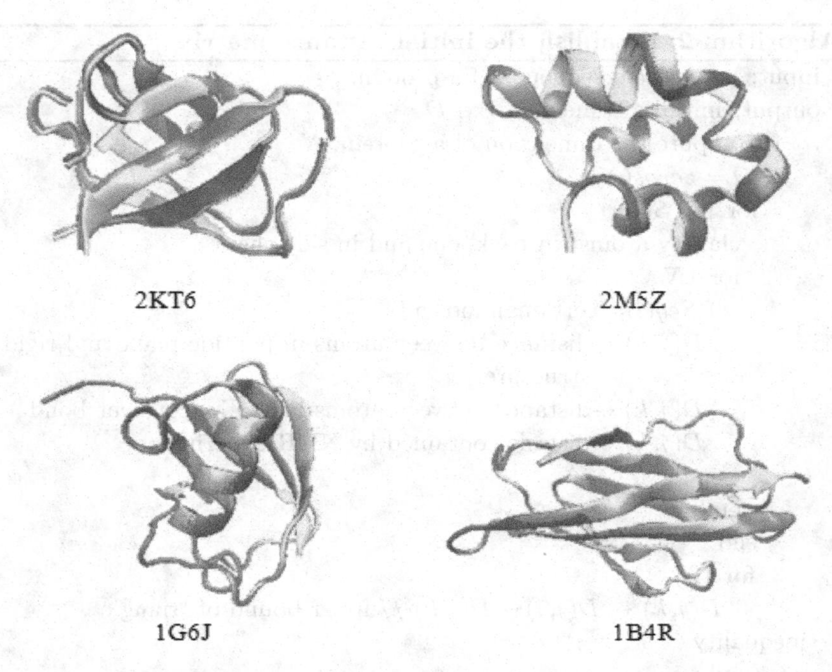

Figure 2. Superimposition of original PDB structures (gray) and structures predicted from ideal distance matrix data (light).

tures for the four proteins all come from the PDB. The RMSD value of this calculation in Figure 2 is so small that we cannot distinguish their structures with the naked eyes. Through these results, we can see that the more the number of atoms in protein molecules and the greater the dimension of the distance matrix, the lower the sampling rate required for restoring the structure. Moreover, the time to run the program will increase if the

Table 1. Calculation results of the NMR structure of accurate estimation.

ID	Description	Atoms	Sampling rate	RMSD	Time
2KT6	Outer membrane usher protein papC	1283	0.0390	2.1E-3	22.23s
2M5Z	Enterocin 7A	743	0.0673	3.4E-4	15.45s
1B4R	PKD domain 1 from human polycystein-1	1114	0.0449	9.6E-4	20.22s
1G6J	Ubiquitin	1228	0.0407	1.7E-3	21.55s

matrix dimension increases.

Previously, we have already concluded that the recovery result is very accurate if the estimate is accurate. However, in general, the estimation results are not very accurate. Now considering the actual situation, we choose triangular constraints as the first step in the two-step estimation method. In the calculation method of triangle constraint, we refer to the calculation process of FW algorithm [25].

In the actual situation, the triangle inequality incorporates errors into the initial distance matrix. We limit the error within 10% of the distance. The calculation results of this situation are listed in Table 2, and their corresponding structures are shown in Figure 3. Since we choose the triangle constraint method as the estimation method, it will bring some errors. The RMSD value of the recovery is larger than that of an accurate estimate of the protein, as shown in Table 2. Although the calculation results are not as accurate as the one obtained by the method with an accurate estimation, this way is closer to the actual situation. If such calculation results are in accord with our expectations, we can conclude that this method is feasible and effective.

Table 2. Calculation results of NMR structure with triangle constraint estimation.

ID	Description	Atoms	Sampling rate	RMSD	Time
2LX6	Lasso Peptide Caulosegnin I	267	0.1873	1.533	60.12s
2M5Z	Enterocin 7A	743	0.0673	1.847	100.76s
1B4R	PKD domain 1 from human polycystein-1	1114	0.0449	1.499	162.53s
1G6J	Ubiquitin	1228	0.0407	1.469	175.89s

As mentioned earlier, some structures whose RMSD values are close to $6\AA$, can only get a little protein structure information. If the RMSD value of protein structure is close to $4\AA$, the position of its amino acids is more accurate, and has more meaning. If the RMSD value of a protein structure is less than $1.5 - 2.0\AA$, its structure is in high resolution, and the most atomic positions of the calculation results are accurate.

Our calculation results of two-step method are listed in Table 2. Their RMSD values are all less than $2\AA$, which are consistent with the results of high resolution standards. In addition, we can clearly see from Figures 2 and 3, the vast majority of the atomic position of the protein are in good

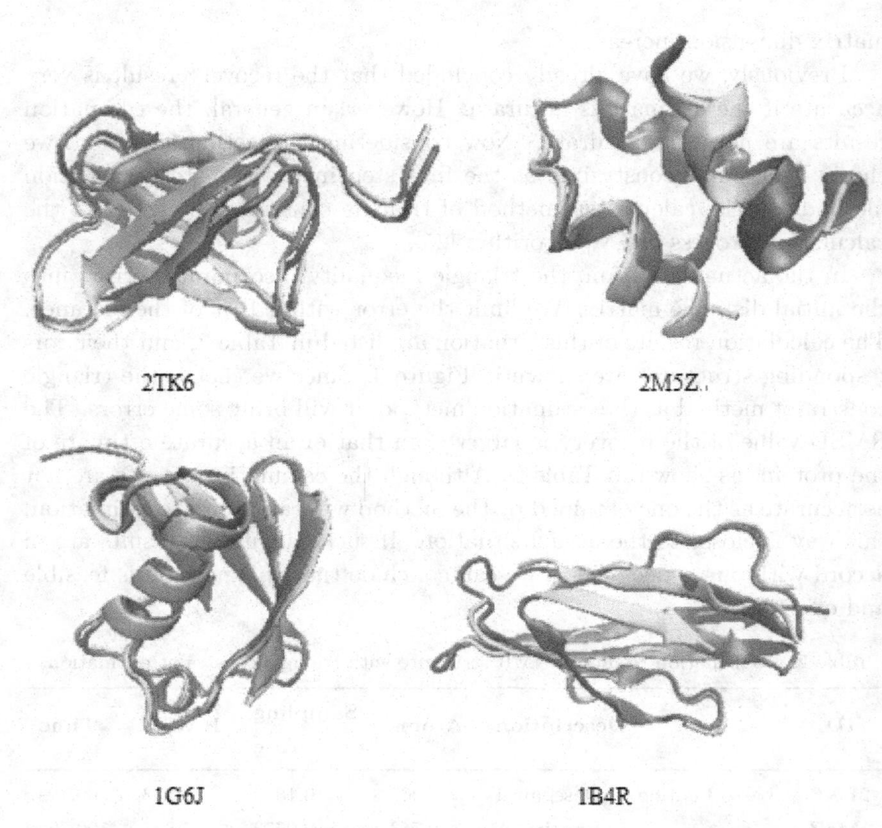

Figure 3. Superimposition of original PDB structures (gray) and structures predicted with the triangle constraint (light).

agreement with accurate location. Therefore, these results have given us a lot of confidence, and we confirm that the two-step method is a very effective way.

7. Error Analysis

Although the recovered structures are acceptable and RMSD values are all less than 2, these results are quite different from the structures recovered by accurate distance data. As we know not only the triangle inequality but also NMR measurement of protein structure, the distances of hydrogen atoms obtained by NOE and angles obtained by J-coupling are inaccurate since they contain noise.

Generally, signal-noise ratio (SNR) is used to evaluate the noise effect on the signal. In order to analyze the relationship between SNR and RMSD, we choose Gaussian white noise as an example. The first step is to establish a noise model, and we assume all the distances obtained through Covalent bonds and rigid structures are set to the exact values. While the distances obtained through NMR and the triangle inequality are set as:

$$\tilde{D}_{ij} = (1 \pm \sqrt{\sigma} R_{ij}) D_{ij} \tag{9}$$

where R_{ij} is a random matrix with rank 1, σ is its squared deviation. SNR is defined by:

$$SNR = 10 \log_{10} \frac{P_s}{P_N} \tag{10}$$

where P_s is the intensity of signal and P_N is the intensity of noise. Generally, we let $P_s = 1$ and $P_N = \frac{1}{\sigma}$. Then the noise model corresponding to this algorithm is established. Firstly we need to analyze the relationship between SNR and RMSD. We choose 1G6J and 2M5Z as examples. SNRs of different magnitudes in the range of (-20, 40) are studied systematically, and each SNR is calculated by APG for 500 times. The results are depictedin the Figure 4.

We can see in both protein 1G6J and 2M5Z, when SNR is less than 10dB, the result is unacceptable. It is easy to conclude that the SNR of the triangle inequality is 14. So APG is a feasible approach for NMR protein calculation.

8. Conclusion

This paper has systematically studied the application of two-step method based on APG algorithm on the protein structure calculation. We mainly took the NMR model of protein structure as a simulation object, and fully considered the various characteristics of the data, such as NOE, small molecule chemistry and disturbance effect. Based on APG, we tested two-step method on 4 proteins with different sizes and topologies, and it determined structures of the test proteins in a short time. The tested proteins generated accurate results. We also studied the noise effect on the recovery result and have given the relationship between SNR and RMSD. Finally, experimental results are promising. Our results demonstrate that we have successfully developed APG in practical application of protein structure, and proved its feasibility.

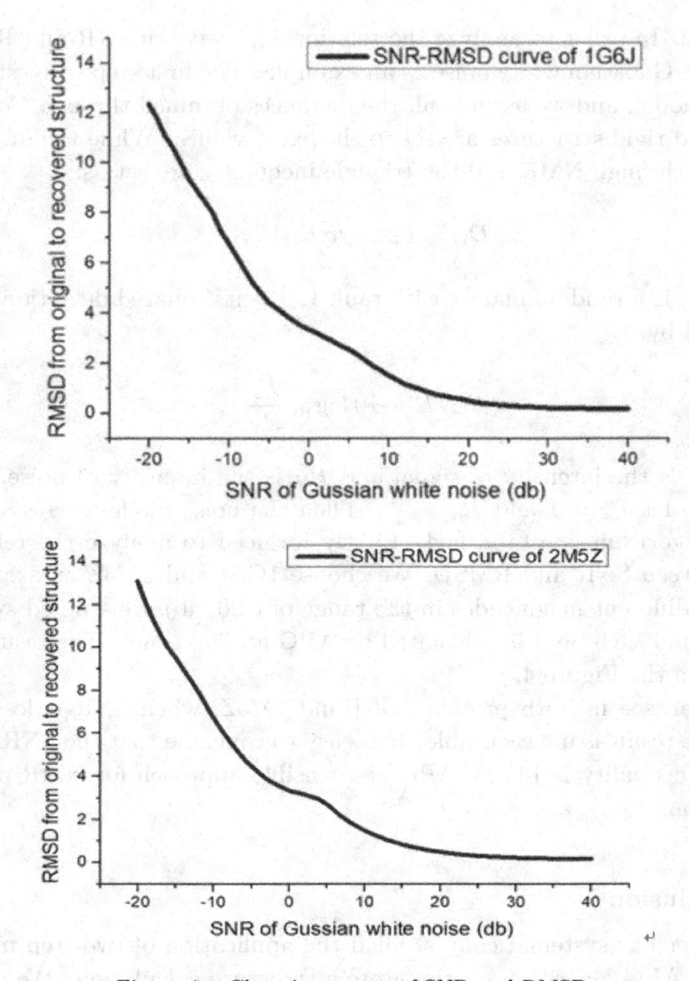

Figure 4. Changing curves of SNR and RMSD.

9. Future Work

Our future work will be to calculate protein structures using Matix Completion. We are also interested in semidefinite programming (SDP). In our numerical tests it appears that standard software packages such as SDPT3 and SeDuMi do not always tackle semidifinite programs, which are not strictly feasible (or nearly so) very well. When the Slaters condition is not satisfied, facial reduction can reduce the problem size and, more importantly, make it strictly feasible.

We want to use partial facial reduction theory [26] to solve this problem, following is the basic flow:

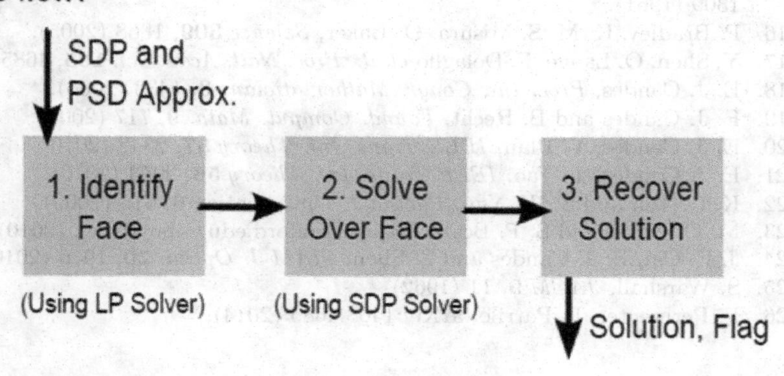

Figure 5. The basic flow of partial facial reduction theory.

Acknowledgments

This work is partly supported by the Ministry of Science and Technology of China (2013YQ030595-3, and 2013AA122901), and by NSF of China with the Grant (No. 11275024, and 11675014).

References

1. M. P. Williamson, T. F. Havel and K. Wüthrich, *J. Mol. Biol.* **182**, 295 (1985).
2. W. Braun, C. Bösch, L. R. Brown, N. Go and K. Wüthrich, *Biochim. Biophys. Acta.* **667**, 377 (1981).
3. T. F. Havel, I. D. Kuntz and G. M. Crippen, *Bull. Math. Biol.* **45**, 665 (1983).
4. T. F. Havel and K. Wüthrich, *Bull. Math. Biol.* **46**, 673 (1984).
5. P. Biswas, K. Toh and Y. Ye, *SIAM J. Sci. Comput.* **30**, 1251 (2008).
6. N. Z. Leung and K. Toh, *SIAM J. Sci. Comput.* **31**, 4351 (1983).
7. B. Alipanahi, N. Krislock, A. Ghodsi, H. Wolkowicz, L. Donaldson and M. Li, *J. Comput. Biol.* **20**, 296 (2013).
8. J. J. Moré and Z. Wu, *SIAM J. Optim.* **7**, 814 (1997).
9. J. J. Moré and Z. Wu, *J. Global Optim.* **15**, 219 (1999).
10. W. Braun and N. Go, *J. Mol. Biol.* **186**, 611 (1985).
11. A. T. Brünger, *Yale University Press* (1993).

12. C. D. Schwieters, J. J. Kuszewski and G. M. Clore, *Prog. Nucl. Magn. Reson. Spectrosc.* **48**, 47 (2006).
13. A. Jain, N. Vaidehi and G. Rodriguez, *J. Comput. Phys.* **106**, 258 (1993).
14. P. Güntert, *Methods Mol. Biol.* **278**, 353 (2004).
15. C. B. Anfinsen, E. Haber, M. Sela, *Proc. Natl. Acad. Scad. Sci. USA* **47**, 1309 (1961).
16. P. Bradley, K. M. S. Misura, D. Baker, *Science* **309**, 1868 (2005).
17. Y. Shen, O. Lange, F. Delaglio *et al.*, *Proc. Natl. Acad. Sci.* **105**, 4685 (2008).
18. E. J. Candès, *Proc. Int. Congr. Mathematicians* **3**, 1433 (2006).
19. E. J. Candès and B. Recht, *Found. Comput. Math.* **9**, 717 (2009).
20. E. J. Candès, Y. Plan, *IEEE Trans. Inf. Theory* **57**, 2342 (2010).
21. E. J. Candès, T. Tao, *IEEE Trans. Inf. Theory* **56**, 2053 (2010).
22. K. C. Toh and S. W. Yun, *Pacific J. Optimization* **6**, 615 (2009).
23. M. C. Grant and S. P. Boyd, http://stanford.edu/~boyd/cvx (2010).
24. J. F. Cai, E. J. Candès and Z. Shen, *SIAM J. Optim.* **20**, 1956 (2010).
25. S. Warshall, *JACM* **9**, 11 (1962).
26. F. Permenter, P. Parrilo, arXiv:1408.4685 (2014).

DECRYPTING HOW PROTEINS MOVE AND CHANGE THEIR SHAPE

ELODIE LAINE

Sorbonne Universités, UPMC-Univ P6, CNRS, IBPS, UMR 7238
Laboratoire de Biologie Computationnelle et Quantitative
4 Place Jussieu, 75005 Paris, France
e-mail: elodie.laine@upmc.fr

Protein structural dynamics is increasingly recognized as important for protein biological functions. The shape and motions of a protein are determined by the physical interactions between the residues of the protein and by the effect of the solvent. Local perturbations can propagate at long distances across a protein structure, resulting in allosteric coupling between distant sites. In recent years, a number of methods have been developed toward describing protein dynamical architectures and identifying communication routes via which signals may be transmitted within proteins. These methods usually represent protein structures as graphs where the nodes are the residues (or the atoms) and the edges encode information about the dynamical behaviour of the protein (persistence or strength of non-covalent interactions and residue cross-correlations). Here, I review some of those methods and the concepts they manipulate, and I present some global properties of proteins that tend to emerge from these analyses.

1. Introduction

A protein can be viewed as a 1-dimensional text (the amino acid sequence of the polypeptide chain), as a 2D arrangement of secondary structures (α-helices, β-strands, loops...), or as an ensemble of points in 3D space whose positions define a tertiary structure. The 3D structure of the protein, encoded by its amino-acid sequence, determines its biological function(s). But proteins are not static, and their dynamical behaviour in solution is increasingly recognized as important for their functions[1,2]. Proteins adapt their shape and motions to environmental conditions. Perturbations such as point mutations, ligand binding or post-translational biochemical modifications[3,4,5] can propagate throughout the protein tertiary structure and impact sites distant from the one where the signal originated. This phenomenon is known as allosteric coupling[6].

Experimental studies have suggested different structural mechanisms underlying allosteric coupling. Hemoglobin gives a typical example where a perturbation signal is propagated via stable non-covalent interactions across the protein structure[7]. Specifically, the binding of oxygen to one hemoglobin subunit induces conformational changes that are relayed to the other subunits, increasing their binding affinity for oxygen. Alternatively, the signal can be transmitted without requiring physical interactions as a support, but simply via local changes in atomic fluctuations[8,9]. These evidence suggest that the nature of allosteric coupling is dual and that the residues comprising a protein structure can "communicate" in different ways.

Over the last decade, computational methods have been developed toward the identification of communication routes across protein structures and the description of protein dynamical architectures. In the following, I review some of those methods and the most important concepts associated to this type of analysis. Most of the methods use graph representations of protein structures, where the nodes of the graph are the residues (or the atoms) and the edges represent contacts and/or dynamical correlations. They can be applied either to a single protein 3D structure or to an ensemble of conformations, determined experimentally or generated by molecular dynamics (MD) simulations.

2. Protein contact network

Adjacent amino acid residues in a polypeptide chain are linked by strong covalent peptidic bonds. The folding of the chain in solution is driven by the hydrophobic effect, that results in the hydrophobic core of the protein being buried from the solvent[10]. The protein residues form secondary structures, such as α-helices and β-sheets, that are stabilized by non-covalent interactions between the atoms of the protein backbone. The 3D arrangement of secondary structure elements is further stabilized by backbone-sidechain and sidechain-sidechain interactions. Hence, the physical contacts between residues are crucial in determining a protein's shape and motions. These contacts can be detected by simply setting a distance cutoff: any two residues closer than that cutoff are considered to be in contact. The distance considered can be the minimum or average interatomic distance, or the Cα-Cα distance, and the cutoff value is typically comprised between 3.5 and 10 Å[11]. More sophisticated geometric criteria can be defined to specifically detect different types of interactions such as hydrogen-bonds, hydrophobic contacts and salt bridges. A protein structure can then be represented by a *protein contact network*, where each node corresponds either to an atom, or a pseudo-atom representing the geometric center of several atoms, or a residue identified with its Cα atom, and the edges represent the interatomic non-covalent interactions[11,12,13,14,15,16]. The edges may be weighted with energy values derived from statistical or physical potentials.

Several online and standalone tools have been designed to construct, visualize and analyse protein contact maps and networks[17,18,19,20,21,22,23,24,25,26]. They allow to analyse one PDB structure at a time and sometimes provide information about the topology of the constructed network (clustering coefficient, average degree, cliques, communities, hubs, shortest paths...). Different types of networks, constructed by using different criteria to define the nodes and the edges, have been designed to perform different types of analyses, such as protein folding analysis, identification of key functional residues, of binding cavities, of domains and structural motifs, and allosteric communication analysis[11]. Some global properties of protein structures have emerged from the characterisation of these network representations. Protein contact networks (PCN) display small-world behaviour[27,28], but this property is conditioned by the protein backbone connectivity[29] and cannot be exploited for 'protein fingerprinting'. The degree of the nodes follows a Poisson distribution[27]. Each amino acid has a characteristic average degree, owing to the limited amino acid sidechain interacting capacity[11]. At most, a residue will physically interact with about 10-15 other residues. Residues with the highest degrees in the network are considered as "hub" nodes. PCNs are assortative (hubs tend to be linked to each other) and hierarchical (there are central and peripheral hubs)[30]. They are formed by several highly connected clusters separated by topological cavities, which seem to correspond to functional binding sites[31]. Key residues governing the folding process were shown to be central residues of the network representing the transition state (which may be very different from the native state)[32].

In a network representing a static structure, each edge encodes the existence, and optionally the strength (energy value), of an interaction. A convenient way to account for the conformational diversity of the protein consists in weighting each edge with the persistence for the pairwise interaction, calculated as the fraction of the number of conformations from an ensemble in which the interaction is present. For example, the PyInteraph tool[33], which is interfaced with PyMOL[34], provides such functionality and can aid the analysis of MD trajectories. In a recent application of PyInteraph to the MZF1 SCAN domain, the authors characterised the properties of cancer-related residues[35]. They showed that these residues tend to display high degree in the network (between 3 and 5 neighbours) and are involved in network paths (chains of residues linked by stable non-covalent interactions) that span long distances across the protein structure.

3. Communication pathways

In addition to pairwise interaction persistence, atomic fluctuations and correlated motions can be computed from protein conformational ensembles. Chennubhotla and Bahar were the first to formally establish a link between signal propagation and fluctuation dynamics in proteins[36]. They showed that the variance of the distance between two residues, which is a measure of their dynamical correlation, is related to the time required for a signal to be transmitted from one to the other residue. Hence, dynamical correlations can be considered as communication propensities between residues. Residues that communicate fast will have a small inter-residue distance variance, *i.e.* they will move together. Under the assumption that non-covalent interactions provide a support for the propagation of the signal, chains of residues linked by stable non-covalent interactions and displaying high dynamical correlations represent good candidates for transmitting information between distant sites of a protein structure. The identification of such chains, called *communication pathways*, has provided mechanistic insights into the allosteric regulation of several proteins and the effects of disease-associated mutations[37,38,39,40,41,42,43,44,45,46,47,48]. The analysis of an oncogenic mutant of KIT receptor tyrosine kinase revealed a mutation-induced disruption of the communication between two distant regions of the protein, and guided the design of a counter-balancing mutation that restored communication[41].

Several methods have been developed to perform such analysis in an automated way[49,50,51,52,53]. They differ mainly by the criteria used to construct the protein structure network and to define communication pathways. In PSN-Ensemble[50], the edges of the network represent non-covalent interactions and are attributed weights that reflect either interaction stabilities (fraction of conformations where the interaction is present), interaction energies (derived from a molecular mechanics potential), or dynamical cross-correlations. Pathways are extracted from the network and a cost is assigned to each pathway, that is equal to the sum of the weights of the edges involved in the pathway. This enables to compare pathways and to identify shortest paths (*i.e.* with the smallest costs) between residues pairs.

Mariani and co-authors[42] extract shortest paths between pairs of residues based on the stability of non-covalent interactions, and then filter them to retain only those that cross at least one residue dynamically correlated with either one of the two extremities of the path. COMMA[53] and MONETA[51] apply much more restrictive criteria: a pathway is defined as a chain of residues linked by stable (frequently observed in a conformational ensemble) interactions and within which each pair of residues communicate fast (small inter-residue variance). The intuition behind is that information is transmitted in the same time between any pair of residues within the pathway. By contrast to other methods, COMMA constructs the protein structure network from the communication pathways, instead of extracting them from the network[53] (Fig. 1). Of note, an original approach was proposed that relies on a simplified representation of the protein using a structural alphabet[54,55]. The protein is decomposed into fragments and each fragment is encoded by a letter that corresponds to a fragment canonical state. Communication pathways are extracted from a network where the nodes represent the protein fragments.

All the published methods and studies use several parameters, *e.g.* cut-off values for distances, interaction frequencies and dynamical correlations. Most often, the parameters are set empirically based on expert knowledge of the system or careful manual analysis of the results. The robustness of the results relative to these parameters is seldom addressed. In COMMA[53], we implemented an automated algorithm to rationally set the thresholds values depending on the system studied and we defined a confidence value for each residue, computed by varying the thresholds. Owing to the noisy and incomplete nature of MD simulation data, this issue becomes particularly relevant.

4. Dynamic domains

The notion of protein domain is one of the most widely used in Biology. Protein domains can viewed as autonomous parts of proteins, that can evolve, fold and function independently from the rest of the protein. Protein domains are the basic units considered in structural, evolutionary and functional classifications of proteins. However, there is no clear mapping between these three levels. Sequence-based domain annotation methods usually restrict themselves to continuous domains, but many structural domains are formed by several segments located far away in the protein sequence. Moreover, domain annotation does not guarantee functional an-

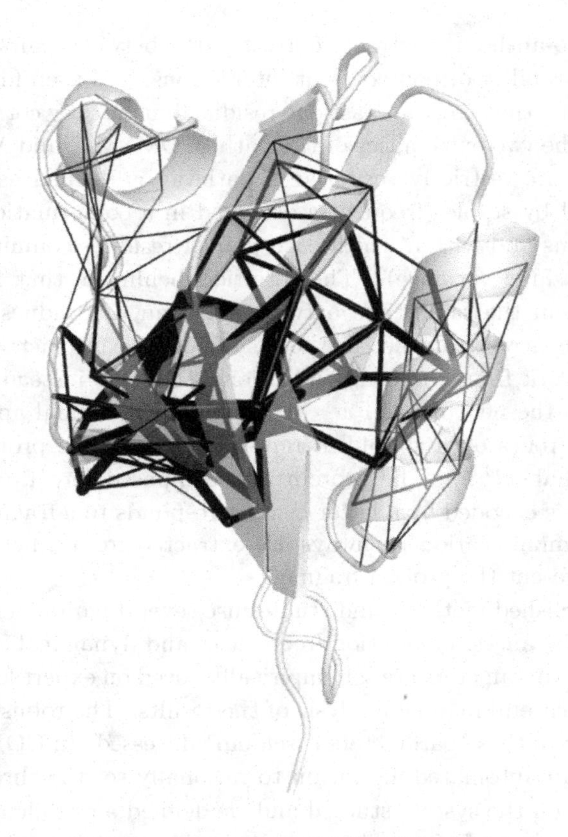

Figure 1. Communication pathways extracted by COMMA for the complex between the third PDZ domain of PSD95 and its cognate ligand[56].

notation. Some domains may exert different functions depending on their structural context, and some functional (*e.g.* catalytic) activities may be ensured by sites lying at the interface between domains.

Structural domains are traditionally defined based on the expert knowledge of structural biologists. Their identification might be eased by the availability of several experimental structures displaying large conformational changes and their functional importance might be supported by biochemical experiments. Protein structure networks provide a convenient mean to automatically partition proteins into connected regions, by extracting either connected components[53], communities[57,58,59] or cliques[45]. If the network is constructed by using information about the dynamical behaviour of the protein, these regions may be referred to as *dynamic domains*. Some studies analysed the functional role of these dynamic domains[58] and suggested that the residues linking them are important for the protein allosteric communication[57].

5. Dynamical correlations

The metric(s) used to quantify correlated motions from protein conformational ensembles directly impact the identification of communication pathways and dynamic domains. The established and commonly used measure is, in analogy to the Pearson correlation coefficient, the normalized covariance of atomic fluctuations, $C_{ij} = \langle \boldsymbol{x}_i \cdot \boldsymbol{x}_j \rangle / (\langle \boldsymbol{x}_i \rangle^2 \langle \boldsymbol{x}_j \rangle^2)^{1/2}$, where $\boldsymbol{x}_i$ and $\boldsymbol{x}_j$ are the positional fluctuation vectors of atoms i and j, respectively, in the molecular fixed frame[60,61]. However, this measure is restricted to linear correlations and therefore cannot capture non-linear correlations, *e.g.* between two atoms oscillating in parallel directions, but with a 90° phase shift. An alternative measure that overcomes this limitation is the generalized correlation coefficient, based on mutual information[62]. Another measure, that is the variance of the inter-residue distance $CP_{ij} = \langle \Delta \boldsymbol{r}_{ij}^T \Delta \boldsymbol{r}_{ij} \rangle$, where $\Delta \boldsymbol{r}_{ij}$ is the normalized vector of distances r_{ij} between atoms i and j, has the advantage of being directly interpretable in terms of communication efficiency[36]. Following another direction, Zhang and Wriggers proposed to apply local feature analysis (LFA) to the covariance matrix[63]. This framework allows to compute residual dynamical correlations that are highly localized (sparse correlation matrix) and that can be used to identify clusters of residues displaying high concerted atomic fluctuations[51,53]. A new method was recently developed, called Principal Feature Decomposition, that also focuses of the protein local dynamics[46]. In GSA Tools, which represent protein structures as strings where each letter stands for a fragment (see above), the dynamical correlation between fragments i and j is determined by calculating the normalized mutual information between the columns i and j of the structural alignment[54,55]. Finally, the NbIt method introduced n-body correlated motions, measured by mutual information terms between three or more residues[43]. This allows to distinguish the residues that stabilize functional sites from those that contribute to allosteric couplings between sites.

Accounting for different measures of dynamical correlations provides a way to identify regions of the protein that behave differently. Among the existing tools for characterising protein dynamical architectures and allosteric communication, COMMA[53] and MONETA[51] are the only ones that integrate two different measures of dynamical correlations within the same framework. On the one hand, communication propensities, computed as the fluctuations of inter-residue distances, are combined with non-covalent interactions to identify communication pathways and "rigid" dynamic domains. On the other hand, residual correlations computed by LFA are combined with inter-residue distances to identify "flexible" dynamic domains, within which residues fluctuate in a concerted way. The idea behind is to represent the dual nature of allosteric communication. We found that the two types of dynamic domains minimally overlap and often fully partition the protein structure[53].

6. Protein kinases

Protein kinases are probably the proteins whose dynamical architectures have been the most extensively characterised (see[40,41,44,45,47,64] for a non-exhaustive list of publications). This owes to several reasons: *(i)* they are ubiquitous, representing about 2% of the proteome, *(ii)* they are involved in signaling pathways crucial for cell survival, growth, differentiation and proliferation, and their deregulation is associated to many cancers, *(iii)* their sequences and their structures are highly conserved through evolution, *(iv)* they display a very high degree of structural plasticity, adopting two clearly distinct conformational states, one inactive and one active, and they are known to be allosterically regulated. Due to their importance for human medicine, protein kinases also present the advantage that there is a very large body of structural data available on them. They have been crystallized in various conditions with different inhibitors. In addition, the kinase domain is relatively small, making it amenable to MD simulations.

Graph representation of protein kinase structures enabled to identify two clusters of hydrophobic residues, namely the regulatory- and catalytic-spines, highly conserved among the protein family[65]. This high conservation suggested an essential role of these clusters for the stability of the kinase active conformation. Recent MD-based studies confirmed their importance in the communication of the protein[64,59]. Residues bridging the two spines were shown to display high node-betweenness centrality (they are crossed by many shortest paths) in a protein structure network constructed based on dynamical correlations[59]. Residues that are indispensable for kinase regulation and catalysis often correspond to the high centrality nodes within the protein structure network[45]. It was also proposed that activating mutations may occur at positions with an average level of network centrality and located at the intersection of high and low stability regions, while inactivating mutations often target catalytically important residues, displaying high centrality in the network[45].

7. Linking structural dynamics with sequence evolution

Residues conserved through evolution are reputed to play important roles for protein structural stability/plasticity, interactions and functions. Evolutionary conservation and coevolution signals have been successfully used to identify protein functional binding sites[66,67,68,69,70] and to determine protein tertiary structures[71,72,73]. A recent study highlighted a set of coevolved residues mediating allosteric communication in the dopamine D2 receptor[74]. Coupling structural dynamics and evolutionary analysis of the rhodopsin-like GPCR family highlighted specific pathways of communication associated to ligand binding and likely generic in the whole protein family[48].

Nevertheless, the evolutionary aspects of protein structural dynamics are not systematically accounted for. This owes in part to the fact that the researchers interested in protein evolution and those studying protein structure and dynamics usually come from different communities and have very different backgrounds. Another reason is that evolutionary signals are sometimes difficult to interpret at the level of the structure. Methods dedicated to protein sequence analysis look at an ensemble of homologous sequences that may cover very long evolutionary times. By contrast, molecular modelling techniques such as MD simulations are limited by the amount of computing time they require. Protein tertiary structures are high-dimensional biological objects, comprising thousands of atoms, and the complete characterisation of the conformational landscape of only one member of a protein family in one organism is still far beyond reach. Until now, no systematic assessment of the impact of sequence variations on protein structural dynamics has been conducted. Tools integrating protein structural dynamics and evolutionary analysis, such as the Bio3D package[52], are needed to perform such analysis.

8. Conclusion

The representation of a protein structure as a graph provides a convenient mean to analyse its global properties and identify residues playing a particular role in the protein structural stability and/or allosteric communication. The growing body of molecular dynamics simulation data has motivated several research groups to develop tools that integrate information recorded in the simulations (*e.g.* interaction persistence/strength and dynamical correlations between residues) in the construction of the protein structure network. These tools proved useful to identify key residues and provide mechanistic insights on the effects of mutations. Although application studies have mainly focused on the detailed description of a particular system, and the tools have not been applied so far at large scale, some global properties of proteins are emerging. This holds promises for the future and we can hope for new exciting developments in the next years toward protein design, functional annotation and mutational effects prediction.

References

1. K. Henzler-Wildman and D. Kern. Dynamic personalities of proteins. *Nature*, 450(7172):964–972, 2007.
2. H. Frauenfelder, S. G. Sligar, and P. G. Wolynes. The energy landscapes and motions of proteins. *Science*, 254(5038):1598–1603, 1991.
3. C.-J. Tsai, A. del Sol, and R. Nussinov. Allostery: Absence of a change in shape does not imply that allostery is not at play. *Journal of Molecular Biology*, 378(1):1 – 11, 2008.
4. D. Kern and E. RP Zuiderweg. The role of dynamics in allosteric regulation. *Current Opinion in Structural Biology*, 13(6):748 – 757, 2003.
5. G. Weber. Ligand binding and internal equilibiums in proteins. *Biochemistry*, 11(5):864–878, 1972.
6. J. Liu and R. Nussinov. Allostery: An Overview of Its History, Concepts, Methods, and Applications. *PLoS Comput. Biol.*, 12(6):e1004966, 2016.
7. J. Monod, J. Wyman, and J. P. Changeux. On the Nature of Allosteric Transitions: A Plausible Model. *J Mol Biol*, 12:88–118, 1965.
8. A. C. Ferreon, J. C. Ferreon, P. E. Wright, and A. A. Deniz. Modulation of allostery by protein intrinsic disorder. *Nature*, 498(7454):390–394, 2013.
9. J. H. Choi, A. H. Laurent, V. J. Hilser, and M. Ostermeier. Design of protein switches based on an ensemble model of allostery. *Nat Commun*, 6:6968, 2015.
10. D. Chandler. Interfaces and the driving force of hydrophobic assembly. *Nature*, 437(7059):640–647, 2005.
11. C. Bode, I. A. Kovacs, M. S. Szalay, R. Palotai, T. Korcsmaros, and P. Csermely. Network analysis of protein dynamics. *FEBS Lett.*, 581(15):2776–2782, 2007.
12. S. Vishveshwara, A. Ghosh, and P. Hansia. Intra and inter-molecular communications through protein structure network. *Curr. Protein Pept. Sci.*, 10(2):146–160, 2009.
13. L. H. Greene. Protein structure networks. *Brief Funct Genomics*, 11(6):469–478, 2012.

14. L. Di Paola, M. De Ruvo, P. Paci, D. Santoni, and A. Giuliani. Protein contact networks: an emerging paradigm in chemistry. *Chem. Rev.*, 113(3):1598–1613, 2013.
15. W. Yan, J. Zhou, M. Sun, J. Chen, G. Hu, and B. Shen. The construction of an amino acid network for understanding protein structure and function. *Amino Acids*, 46(6):1419–1439, 2014.
16. R. K. Grewal and S. Roy. Modeling proteins as residue interaction networks. *Protein Pept. Lett.*, 22(10):923–933, 2015.
17. J. C. Biro and G. Fordos. SeqX: a tool to detect, analyze and visualize residue co-locations in protein and nucleic acid structures. *BMC Bioinformatics*, 6:170, 2005.
18. M. J. Pietal, I. Tuszynska, and J. M. Bujnicki. PROTMAP2D: visualization, comparison and analysis of 2D maps of protein structure. *Bioinformatics*, 23(11):1429–1430, 2007.
19. J. L. Chung, J. E. Beaver, E. D. Scheeff, and P. E. Bourne. Con-Struct Map: a comparative contact map analysis tool. *Bioinformatics*, 23(18):2491–2492, 2007.
20. M. Aftabuddin and S. Kundu. AMINONET – a tool to construct and visualize amino acid networks, and to calculate topological parameters. J proteins as residue interaction networks. *J. Appl. Cryst.*, 43:367–369, 2010.
21. C. Vehlow, H. Stehr, M. Winkelmann, J. M. Duarte, L. Petzold, J. Dinse, and M. Lappe. CMView: interactive contact map visualization and analysis. *Bioinformatics*, 27(11):1573–1574, 2011.
22. M. S. Vijayabaskar, V. Niranjan, and S. Vishveshwara. GraProStr – graphs of protein structures: a tool for constructing the graphs and generating graph parameters for protein structures. *Open Bioinform. J*, 5:53–58, 2011.
23. N. T. Doncheva, K. Klein, F. S. Domingues, and M. Albrecht. Analyzing and visualizing residue networks of protein structures. *Trends Biochem. Sci.*, 36(4):179–182, 2011.
24. D. Kozma, I. Simon, and G. E. Tusnady. CMWeb: an interactive on-line tool for analysing residue-residue contacts and contact prediction methods. *Nucleic Acids Res.*, 40(Web Server issue):W329–333, 2012.
25. D. Piovesan, G. Minervini, and S. C. Tosatto. The RING 2.0 web server for high quality residue interaction networks. *Nucleic Acids Res.*, 44(W1):W367–374, 2016.
26. B. Chakrabarty and N. Parekh. NAPS: Network Analysis of Protein Structures. *Nucleic Acids Res.*, 44(W1):W375–382, 2016.
27. A. R. Atilgan, P. Akan, and C. Baysal. Small-world communication of residues and significance for protein dynamics. *Biophys. J.*, 86(1 Pt 1):85–91, 2004.
28. G. Bagler and S. Sinha. Network properties of protein structures . *Physica A*, 346:27–33, 2005.
29. L. Bartoli, P. Fariselli, and R. Casadio. The effect of backbone on the small-world properties of protein contact maps. *Phys Biol*, 4(4):1–5, 2008.
30. M. Aftabuddin and S. Kundu. Hydrophobic, hydrophilic, and charged amino acid networks within protein. *Biophys. J.*, 93(1):225–231, 2007.

31. E. Estrada. Universality in protein residue networks. *Biophys. J.*, 98(5):890–900, 2010.
32. M. Vendruscolo, N. V. Dokholyan, E. Paci, and M. Karplus. Small-world view of the amino acids that play a key role in protein folding. *Phys Rev E Stat Nonlin Soft Matter Phys*, 65(6 Pt 1):061910, 2002.
33. M. Tiberti, G. Invernizzi, M. Lambrughi, Y. Inbar, G. Schreiber, and E. Papaleo. PyInteraph: a framework for the analysis of interaction networks in structural ensembles of proteins. *J Chem Inf Model*, 54(5):1537–1551, 2014.
34. Warren L DeLano. The pymol molecular graphics system. 2002.
35. M. Nygaard, T. Terkelsen, A. Vidas Olsen, V. Sora, J. Salamanca Viloria, F. Rizza, S. Bergstrand-Poulsen, M. Di Marco, M. Vistesen, M. Tiberti, M. Lambrughi, M. Jäättelä, T. Kallunki, and E. Papaleo. The mutational landscape of the oncogenic mzf1 scan domain in cancer. *Frontiers in Molecular Biosciences*, 3:78, 2016.
36. C. Chennubhotla and I. Bahar. Signal propagation in proteins and relation to equilibrium fluctuations. *PLoS Comput. Biol.*, 3(9):1716–1726, 2007.
37. A. del Sol, H. Fujihashi, D. Amoros, and R. Nussinov. Residues crucial for maintaining short paths in network communication mediate signaling in proteins. *Mol. Syst. Biol.*, 2:2006.0019, 2006.
38. A. Ghosh and S. Vishveshwara. A study of communication pathways in methionyl- tRNA synthetase by molecular dynamics simulations and structure network analysis. *Proc. Natl. Acad. Sci. USA*, 104(40):15711–15716, 2007.
39. G. Morra, G. Verkhivker, and G. Colombo. Modeling signal propagation mechanisms and ligand-based conformational dynamics of the Hsp90 molecular chaperone full-length dimer. *PLoS Comput. Biol.*, 5(3):e1000323, 2009.
40. A. Dixit and G. M. Verkhivker. Computational modeling of allosteric communication reveals organizing principles of mutation-induced signaling in ABL and EGFR kinases. *PLoS Comput. Biol.*, 7(10):e1002179, 2011.
41. E. Laine, C. Auclair, and L. Tchertanov. Allosteric communication across the native and mutated KIT receptor tyrosine kinase. *PLoS Comput. Biol.*, 8(8):e1002661, 2012.
42. S. Mariani, D. Dell'Orco, A. Felline, F. Raimondi, and F. Fanelli. Network and atomistic simulations unveil the structural determinants of mutations linked to retinal diseases. *PLoS Comput. Biol.*, 9(8):e1003207, 2013.
43. M. V. LeVine and H. Weinstein. NbIT–a new information theory-based analysis of allosteric mechanisms reveals residues that underlie function in the leucine transporter LeuT. *PLoS Comput. Biol.*, 10(5):e1003603, 2014.
44. I. Chauvot de Beauchene, A. Allain, N. Panel, E. Laine, A. Trouve, P. Dubreuil, and L. Tchertanov. Hotspot mutations in KIT receptor differentially modulate its allosterically coupled conformational dynamics: impact on activation and drug sensitivity. *PLoS Comput. Biol.*, 10(7):e1003749, 2014.
45. K. A. James and G. M. Verkhivker. Structure-based network analysis of activation mechanisms in the ErbB family of receptor tyrosine kinases: the regulatory spine residues are global mediators of structural stability and allosteric interactions. *PLoS ONE*, 9(11):e113488, 2014.

46. F. Langenfeld, Y. Guarracino, M. Arock, A. Trouve, and L. Tchertanov. How Intrinsic Molecular Dynamics Control Intramolecular Communication in Signal Transducers and Activators of Transcription Factor STAT5. *PLoS ONE*, 10(12):e0145142, 2015.
47. A. Tse and G. M. Verkhivker. Molecular Dynamics Simulations and Structural Network Analysis of c-Abl and c-Src Kinase Core Proteins: Capturing Allosteric Mechanisms and Communication Pathways from Residue Centrality. *J Chem Inf Model*, 55(8):1645–1662, 2015.
48. K. N. Woods, J. Pfeffer, A. Dutta, and J. Klein-Seetharaman. Vibrational resonance, allostery, and activation in rhodopsin-like G protein-coupled receptors. *Sci Rep*, 6:37290, 2016.
49. M. Seeber, A. Felline, F. Raimondi, S. Muff, R. Friedman, F. Rao, A. Caflisch, and F. Fanelli. Wordom: a user-friendly program for the analysis of molecular structures, trajectories, and free energy surfaces. *J Comput Chem*, 32(6):1183–1194, 2011.
50. M. Bhattacharyya, C. R. Bhat, and S. Vishveshwara. An automated approach to network features of protein structure ensembles. *Protein Sci.*, 22(10):1399–1416, 2013.
51. A. Allain, I. Chauvot de Beauchene, F. Langenfeld, Y. Guarracino, E. Laine, and L. Tchertanov. Allosteric pathway identification through network analysis: from molecular dynamics simulations to interactive 2D and 3D graphs. *Faraday Discuss.*, 169:303–321, 2014.
52. L. Skjaerven, X. Q. Yao, G. Scarabelli, and B. J. Grant. Integrating protein structural dynamics and evolutionary analysis with Bio3D. *BMC Bioinformatics*, 15:399, 2014.
53. Y. Karami, E. Laine, and A. Carbone. Dissecting protein architecture with communication blocks and communicating segment pairs. *BMC Bioinformatics*, 17 Suppl 2:13, 2016.
54. A. Pandini, A. Fornili, F. Fraternali, and J. Kleinjung. Detection of allosteric signal transmission by information-theoretic analysis of protein dynamics. *FASEB J.*, 26(2):868–881, 2012.
55. A. Pandini, A. Fornili, F. Fraternali, and J. Kleinjung. GSATools: analysis of allosteric communication and functional local motions using a structural alphabet. *Bioinformatics*, 29(16):2053–2055, 2013.
56. Y. Karami, E. Laine, and A. Carbone. "Infostery" as a new way to look into protein motions and to predict deleterious mutations. *to be submitted.*
57. A. Sethi, J. Eargle, A. A. Black, and Z. Luthey-Schulten. Dynamical networks in tRNA:protein complexes. *Proc. Natl. Acad. Sci. USA*, 106(16):6620–6625, 2009.
58. C. L. McClendon, A. P. Kornev, M. K. Gilson, and S. S. Taylor. Dynamic architecture of a protein kinase. *Proc. Natl. Acad. Sci. USA*, 111(43):E4623–4631, 2014.
59. N. Chopra, T. E. Wales, R. E. Joseph, S. E. Boyken, J. R. Engen, R. L. Jernigan, and A. H. Andreotti. Dynamic Allostery Mediated by a Conserved Tryptophan in the Tec Family Kinases. *PLoS Comput. Biol.*, 12(3):e1004826, 2016.

60. T. Ichiye and M. Karplus. Collective motions in proteins: a covariance analysis of atomic fluctuations in molecular dynamics and normal mode simulations. *Proteins*, 11(3):205–217, 1991.
61. P. H. Hunenberger, A. E. Mark, and W. F. van Gunsteren. Fluctuation and cross-correlation analysis of protein motions observed in nanosecond molecular dynamics simulations. *J. Mol. Biol.*, 252(4):492–503, 1995.
62. O.F. Lange and H. Grubmller. Generalized correlation for biomolecular dynamics. *Proteins*, 62(4):1053–61, 2006.
63. Z. Zhang and W. Wriggers. Local feature analysis: a statistical theory for reproducible essential dynamics of large macromolecules. *Proteins*, 64(2):391–403, 2006.
64. J. R. Ingram, K. E. Knockenhauer, B. M. Markus, J. Mandelbaum, A. Ramek, Y. Shan, D. E. Shaw, T. U. Schwartz, H. L. Ploegh, and S. Lourido. Allosteric activation of apicomplexan calcium-dependent protein kinases. *Proc. Natl. Acad. Sci. USA*, 112(36):E4975–4984, 2015.
65. A. P. Kornev, N. M. Haste, S. S. Taylor, and L. F. Eyck. Surface comparison of active and inactive protein kinases identifies a conserved activation mechanism. *Proc. Natl. Acad. Sci. USA*, 103(47):17783–17788, 2006.
66. O. Lichtarge, H.R. Bourne, and F.E. Cohen. An evolutionary trace method defines binding surfaces common to protein families. *J Mol Biol*, 257:342–358, 1996.
67. O. Lichtarge and M.E. Sowa. Evolutionary predictions of binding surfaces and interactions. *Curr Opin Struct Biol*, 12:21–27, 2002.
68. S. Engelen, L. A. Trojan, S. Sacquin-Mora, R. Lavery, and A. Carbone. Joint evolutionary trees: a large-scale method to predict protein interfaces based on sequence sampling. *PLoS Comput. Biol.*, 5(1):e1000267, 2009.
69. S. Ovchinnikov, H. Kamisetty, and D. Baker. Robust and accurate prediction of residue-residue interactions across protein interfaces using evolutionary information. *Elife*, 3:e02030, 2014.
70. E. Laine and A. Carbone. Local Geometry and Evolutionary Conservation of Protein Surfaces Reveal the Multiple Recognition Patches in Protein-Protein Interactions. *PLoS Comput. Biol.*, 11(12):e1004580, 2015.
71. F. Morcos, A. Pagnani, B. Lunt, A. Bertolino, D. S. Marks, C. Sander, R. Zecchina, J. N. Onuchic, T. Hwa, and M. Weigt. Direct-coupling analysis of residue coevolution captures native contacts across many protein families. *Proc. Natl. Acad. Sci. USA*, 108(49):E1293–1301, 2011.
72. T. A. Hopf, L. J. Colwell, R. Sheridan, B. Rost, C. Sander, and D. S. Marks. Three-dimensional structures of membrane proteins from genomic sequencing. *Cell*, 149(7):1607–1621, 2012.
73. D. S. Marks, T. A. Hopf, and C. Sander. Protein structure prediction from sequence variation. *Nat. Biotechnol.*, 30(11):1072–1080, 2012.
74. Y. M. Sung, A. D. Wilkins, G. J. Rodriguez, T. G. Wensel, and O. Lichtarge. Intramolecular allosteric communication in dopamine D2 receptor revealed by evolutionary amino acid covariation. *Proc. Natl. Acad. Sci. USA*, 113(13):3539–3544, 2016.

DETERMINISTIC AND STOCHASTIC MODEL FOR HTLV-I INFECTION OF CD4$^+$ T CELLS

P. K. SRIVASTAVA

Department of Mathematics
School of Basic Sciences
Indian Institute of Technology Patna
Patna-800013, India
E-mail: pksri@iitp.ac.in

An ODE model of HTLV-I infection with CD4$^+$ T cells is considered and analyzed. The model system has three equilibria: infection free equilibrium, uninfected equilibrium with leukemia cell only and infected equilibrium. It is observed that the infection persists (infected equilibrium is g.a.s.) if basic reproduction number is bigger than one. This model is further modified by introducing discrete time delay to account for the latent period of infection and it is shown that this delay has no destabilizing effect on the local stability of infected equilibrium. Again, this delay model system is extended to corresponding stochastic model considering the randomness in proliferation of leukemia cells. It is done by introducing white noise term in proliferation rate of Leukemia cells and corresponding stochastic delay differential equation model is analyzed using Fourier Transform technique. We observe fluctuations only in Leukemia cells populations and corresponding variations are obtained analytically. A numerical experimentation is performed to analyse and explain the outcomes. We observe that the fluctuations in Leukemia cells' population are contributed by white noise perturbation of proliferation rate and this has no impact of both uninfected and infected CD4$^+$ T cells.

Keywords: HTLV-1 infection; delay; stability; variance; SDDE.

1. Introduction

Human T-cell Lymphotropic Virus-I (HTLV-I) is found to be responsible for the fatal disease Adult T-cell Leukemia/Lymphoma (ATL) and is also related to HTLV-I-Associated Myelopathy/Tropical Spastic Paraparesis (HAM/TSP). As is the case with HIV, its major target is CD4$^+$ T cells. HTLV-I is a retrovirus which contains its genetic information in the form

*Corresponding author.

of RNA. After the infection HTLV-I RNA is reverse transcribed into DNA, which is then integrated to the DNA of host T cell genome. At this stage virus latently persists in the cell for several years without producing virus. Stimulation of these cells activates them to become actively infected. Now these cells can produce virus and they can infect other susceptible cells as well. In very few cases, a fraction of these actively infected T cells is then converted into ATL cells through some mechanism, which is still not understood.

Very few mathematical models have been studied for the case of HTLV-I infection in $CD4^+$ T cells. In 1999 Stilianakis and Seydel proposed a very first model for HTLV-I infection to $CD4^+$ T cells and studied its local dynamics.[1] They considered the interaction amongst four different cell populations: uninfected, latently infected and actively infected $CD4^+$ T cells along with Leukemia cells. The infection in $CD4^+$ T cells spread through cell-to-cell contact of actively infected and uninfected cells. This interaction is taken to be mass action type. Their model describes basic characteristics of HTLV infection, and possible progression to ATL. In 2002 Wang *et al.* studied the complete global dynamics of this model.[2] In 2005 Song and Li considered the Holling type-II interaction for infection and investigated the global stability of the model.[5] In 2011 Cai *et al.* further extended their model considering a more general function response for the interaction and studied global properties of the model.[6] Katri and Ruan modified the model of Stilianakis and Seydel[1] by incorporating discrete time delay to account for the contagion eclipse phase and found that under certain condition the system is stable for all delays.[3] Hence delay in infection does not pose any effect on stability. Gomez-Acevedo and Li in 2005, incorporated both horizontal transmission through cell-to-cell contact as well as vertical transmission through division of infected T cells in the model dynamics of uninfected and infected $CD4^+$ T cells and observed that system undergoes backward bifurcation.[4] Recently some more models considering the dynamics of HTLV, T Cells and CTLs have also been studied.[7,8,10,9]

In this paper, we first modify the model by Stilianakis and Seydel[1] and its modified version by Katri and Ruan.[3] We take only two classes of $CD4^+$ T cells: uninfected and actively infected. It is assumed that the infection develops as mass action type interaction between infected and uninfected $CD4^+$ T cells and then the infected cells immediately become actively infected. The stability of equilibria of this model is investigated. However, it is not a realistic assumption that infected cells become actively infected immediately after infection and there is latency involved which has been accounted in Refs. 1, 3, by taking class of latently infected T cells in the model. Hence to take into account the time lag between the infection of CD4$^+$ T cells and then it becoming actively infected, we introduce a discrete time delay into the ordinary differential equation (ODE) model system, thus obtaining the delay differential equation (DDE) model. This model is then analysed for the stability.

As we know, there is always a randomness involved in all natural systems and this need to be addressed for better understanding of the system as well as for getting an alternate view point of the system. To the best of our knowledge, no model has been studied incorporating randomness in the HTLV-I and CD4$^+$ T cell infection dynamics models. In this paper we attempt to address the inherent randomness of HTLV-I virus by introducing stochastic white noise perturbation through the proliferation rate parameter as the mechanism of development of Leukemia cells in infected patients is still not very clear. We study the effect of stochastic perturbation of proliferation rate by white noise on the dynamics of HTLV-I. The variance of all the populations are found using Fourier Transform technique. We further explored the models numerically to analyse the outcomes.

2. Deterministic Model

In this section we will consider the dynamic behavior of ODE and DDE model for uninfected CD4$^+$ T cells, infected CD4$^+$ T cells and leukemia cells.

2.1. *The basic ODE model*

Here, we take two classes of CD4$^+$ T cells (uninfected (T) and infected (T^*)) along with Leukemia cells (T_M). It is assumed that the infection develops by the interaction between infected and uninfected CD4$^+$ T cells, which is mass action type and the infected cells immediately become actively infected after infection. It is also assumed that Leukemia cells are produced from infected CD4$^+$ T cells and proliferate logistically themselves. Hence we have the following modified form of model[1,3]:

$$\frac{dT}{dt} = \Lambda - \mu T - kTT^*, \tag{1}$$

$$\frac{dT^*}{dt} = k'TT^* - (\rho + \mu_1)T^*, \tag{2}$$

$$\frac{dT_M}{dt} = \rho T^* + \beta T_M \left(1 - \frac{T_M}{T_{M_{max}}}\right) - \mu_M T_M, \tag{3}$$

with $T(0) = T_0 \geq 0$, $T^*(0) = T_0^* \geq 0$, and $T_M(0) = T_{M_0} \geq 0$.

Here Λ is taken to be natural inflow rate of uninfected T cells and μ their natural death rate. k is the rate at which uninfected cells are contacted by infected cells and k' is the rate of infection of T cells with virus from actively infected cells.[3] μ_1 is death rate of infected T cells and ρ is the rate at which Leukemia cells are produced from infected CD4$^+$ T cells. β is the proliferation rate and $T_{M_{max}}$ is carrying capacity of Leukemia cells. μ_M is natural death rate of leukemia cells. T_0, T_0^*, and T_{M_0} are the initial amount of uninfected, infected and CD4$^+$ T cells and Leukemia cells, respectively.

2.1.1. *Positivity and boundedness*

As all variables represent populations it is important to establish their positivity and boundedness at all time for the model system (1)-(3). First we establish positivity. We shall show that if we start with positive initial conditions, the populations will remain positive for all future time. Note that the rate equations give,

$$\left.\frac{dT}{dt}\right|_{T=0,T^*>0,T_M>0} = \Lambda > 0, \quad \left.\frac{dT^*}{dt}\right|_{T^*=0,T>0,T_M>0} = 0,$$

$$\left.\frac{dT_M}{dt}\right|_{T_M=0,,T>0T^*>0} = \rho T^* \geq 0.$$

Thus on each bounding plane of positive octant $\mathbb{R}_+^3$, all the directions are inward. Hence any solution of the model system (1)-(3) initiating in positive octant $\mathbb{R}_+^3$ will always remain there for all future time, ensuring the positivity of populations.

Further, from equations (1) and (2), we have $\frac{dT}{dt} + \frac{k}{k'}\frac{dT^*}{dt} = \Lambda - \mu T - (\rho + \mu_1)\frac{k}{k'}T^* \leq \Lambda - \gamma(T + \frac{k}{k'}T^*)$, where $\gamma = \min\{\mu, \rho + \mu_1\}$. Hence $\limsup_{t\to\infty}(T + \frac{k}{k'}T^*) \leq \frac{\Lambda}{\gamma}$, which along with equation (3) gives $\frac{dT_M}{dt} \leq \frac{\rho\Lambda}{\gamma} + \beta T_M\left(1 - \frac{T_M}{T_{M_{max}}}\right) - \mu_M T_M$. Thus, $\limsup_{t\to\infty} T_M \leq \tilde{T}_M$, where $\tilde{T}_M$ is the positive root of the quadratic equation $\dfrac{\rho\Lambda}{\gamma} + \beta T_M\left(1 - \dfrac{T_M}{T_{M_{max}}}\right) - \mu_M T_M = 0$. Hence, the biological feasible region corresponding to model system is given as

$$\Gamma = \{(T, T^*, T_M) \in R^3 : T, T^*, T_M \geq 0, T + \frac{k}{k'}T^* \leq \frac{\Lambda}{\gamma}, T_M \leq \tilde{T}_M\}$$

which is a positively invariant set.

2.1.2. *Equilibrium points*

The model system (1)-(3) has following three equilibrium points:

(i) the infection free equilibrium point $E_0 = \left(\frac{\Lambda}{\mu}, 0, 0\right)$,

(ii) uninfected equilibrium point with leukemia cells $E_1 = \left(\frac{\Lambda}{\mu}, 0, \hat{T}_M\right)$, where $\hat{T}_M = \frac{\beta - \mu_M}{\beta} T_{M_{max}}$, and

(iii) infected equilibrium point $E_2 = \left(\overline{T}, \overline{T}^*, \overline{T}_M\right)$,

where $\hat{T}_M = \frac{\beta-\mu_M}{\beta} T_{M_{max}}$, $\overline{T} = \frac{\mu_1+\rho}{k'}$, $\overline{T}^* = \frac{k'\Lambda - \mu(\mu_1+\rho)}{k(\mu_1+\rho)}$, and $\overline{T}_M = \frac{T_{M_{max}}}{2\beta}\left[(\beta-\mu_M) + \sqrt{(\beta-\mu_M)^2 + \frac{4\beta\rho\overline{T^*}}{T_{M_{max}}}}\right]$.

Note that the equilibrium point E_0 always exists and E_1 exists if $\beta > \mu_M$. Further E_2 also exists if $\mathcal{R}_0 = \frac{k'\Lambda}{\mu(\mu_1+\rho)} > 1$. Here, $\mathcal{R}_0$ is the basic reproduction number, which is defined as the number of secondary infections from a typical infected individual in virgin population.

2.1.3. *Stability of equilibrium points*

The Jacobian of the system (1)-(3) around an equilibrium point is given by

$$J = \begin{bmatrix} -\mu - k\overline{T}^*_{ep} & -k\overline{T}_{ep} & 0 \\ k'\overline{T}^*_{ep} & k'\overline{T}_{ep} - \mu_1 - \rho & 0 \\ 0 & \rho & \beta - \mu_M - \frac{2\beta\overline{T}_{Mep}}{T_{M_{max}}} \end{bmatrix}$$

where subscript *ep* stands for equilibrium point. For the stability of a system around an equilibrium point, all roots of the characteristic equation of J evaluated at that equilibrium point should have negative real part. In view of this we have following result for local stability:

Theorem 2.1.

(i) E_0 *is locally asymptotically stable (l.a.s.) if* $\mathcal{R}_0 < 1$ *and* $\beta < \mu_M$. *It is unstable otherwise.*

(ii) E_1 *is l.a.s. if* $\mathcal{R}_0 < 1$ *and* $\beta > \mu_M$.

(iii) E_2 *is l.a.s. whenever it exists (i.e.* $\mathcal{R}_0 > 1$ *and* $\beta \neq \mu_M$*).*

To establish the global stability of E_1, we follow the approach of Wang *et al.*[2] Hence we consider the reduced system of equations (1) and (2) in the feasible region $\Omega = \{(T, T^*) \in R^2 : T, T^* \geq 0, T + \frac{k}{k'}T^* \leq \Lambda/\gamma\}$. Using Theorem 3.1 of Ref. 19, it is easy to see that the system (1)-(2) is globally stable around its equilibrium point $(\Lambda/\mu, 0)$. Thus T^* population approaches to zero. Hence the dynamics of leukemia cells satisfy following equation in limiting case:

$$\frac{dT_M}{dt} = \beta T_M \left(1 - \frac{T_M}{T_{M_{max}}}\right) - \mu_M T_M.$$

From above equation it is clear that $T_M \to 0$ as $t \to \infty$ whenever $\beta < \mu_M$ for all non-negative initial conditions. Further, $T_M \to \hat{T}_M$ as $t \to \infty$ for $\beta > \mu_M$. Thus, in Γ, we have the following global stability result:

Theorem 2.2. *For $\mathcal{R}_0 < 1$,*

(i) *E_0 is only equilibrium point and is globally stable in Γ if $\beta < \mu_M$,*
(ii) *E_1 is globally stable in $\Gamma \setminus \{(T, 0, 0) : 0 \leq T \leq \Lambda/\mu\}$ if $\beta > \mu_M$.*

In the following result we establish the global stability of infected equilibrium point E_2.

Theorem 2.3. *The unique non-zero infected equilibrium point E_2 is globally stable in the interior of Γ whenever it exists.*

Proof. Consider the Lyapunov function L_2 for the reduced system of equations (1)-(2):

$$L_2 = \frac{1}{2}(T - \overline{T})^2 + \overline{T}\left(T^* - \overline{T}^* - \overline{T}^* \ln \frac{T^*}{\overline{T}^*}\right).$$

It is easy to see that $\frac{dL_2}{dt} \leq 0$ along the solution trajectories of the reduced system, hence system approaches $(\overline{T}, \overline{T}^*)$ provided $\mathcal{R}_0 > 1$. The equation for T_M reduces to $\frac{dT_M}{dt} \leq \rho \overline{T}^* + \beta T_M \left(1 - \frac{T_M}{T_{M_{max}}}\right) - \mu_M T_M$. Thus we conclude that $T_M \to \overline{T}_M$ as $t \to \infty$ for all non-negative initial conditions. Hence the theorem. □

Remark 2.1. The system (1)-(2) is uniformly persistent[16] in the interior of Ω if and only if $\mathcal{R}_0 > 1$. Using result from Ref. 17 and following proof of Proposition 3.3 of Ref. 18, the instability of $(\Lambda/\mu, 0)$ implies uniform persistence when $\mathcal{R}_0 > 1$.

2.2. *The delay differential equation model*

In model (1)-(3) it was assumed that infected T cells become actively infected immediately after infection. As it has been discussed in introduction, there is a latency involved for infected cells to become actively infected which has also been accounted in Refs. 1, 2, 3, 5 by taking class of latently infected T cells in the model. Here, we shall model this latent period (the time of infection of T cell and it becoming actively infected) by incorporating discrete time lag into infection rate term. Taking time lag τ into consideration the following model is proposed:

$$\frac{dT(t)}{dt} = \Lambda - \mu T(t) - kT(t)T^*(t), \tag{4}$$

$$\frac{dT^*(t)}{dt} = k'T(t-\tau)T^*(t-\tau) - (\rho + \mu_1)T^*(t), \tag{5}$$

$$\frac{dT_M(t)}{dt} = \rho T^*(t) + \beta T_M(t)\left(1 - \frac{T_M(t)}{T_{M_{max}}}\right) - \mu_M T_M(t), \tag{6}$$

with initial conditions $T(\theta) = \varphi_1(\theta) > 0$, $T^*(\theta) = \varphi_2(\theta) > 0$, and $T_M(\theta) = \varphi_3(\theta) > 0$ for $\theta \in [-\tau, 0]$. Here $\varphi(\theta) = (\varphi_1(\theta), \varphi_2(\theta), \varphi_3(\theta)) \in C([-\tau, 0], \mathbb{R}^3_{+0})$, the Banach space of continuous functions mapping the interval $[-\tau, 0]$ into $\mathbb{R}^3_{+0} = \{(T, T^*, V) : T, T^*, V > 0\}$. Here all the parameters have same meaning as in previous section.

2.2.1. *Boundedness*

Assuming T, T^* and T_M are all positive for all t. Now define a function: $\Phi(t) = T(t) + \frac{k}{k'}T^*(t+\tau)$, so that we get

$$\begin{aligned}\frac{d\Phi}{dt} &= \Lambda - \mu T(t) - kT(t)T^*(t) + kT(t)V(t) - \frac{k(\rho + \mu_1)}{k'}T^*(t+\tau)\\ &= \Lambda - \mu T(t) - \frac{k(\rho + \mu_1)}{k'}T^*(t+\tau)\\ &\leq \Lambda - \gamma\Phi\end{aligned}$$

where $\gamma = \min\{\mu, \rho + \mu_1\}$. Hence $\Phi(t) \leq \frac{\Lambda}{\gamma}$ for all large t.

2.2.2. *Stability of equilibria*

The delay model system has same equilibria as were in case of ODE model. In order to perform the local stability analysis of the delay model (4)-(6), we linearize the system.The linearized system is given by

$$Y' = AY(t) + BY(t-\tau) \tag{7}$$

where

$$A = \begin{bmatrix} -\mu - kT^*_{ep} & -kT_{ep} & 0 \\ 0 & -\mu_1 - \rho & 0 \\ 0 & \rho & \beta - \mu_M - \frac{2\beta T_{M\,ep}}{T_{M\,max}} \end{bmatrix} \; \& \; B = \begin{bmatrix} 0 & 0 & 0 \\ k'T^*_{ep} & k'T_{ep} & 0 \\ 0 & 0 & 0 \end{bmatrix}.$$

It can be easily seen by finding characteristic equation of linearized system that equilibrium point E_0, which was stable when $\tau = 0$, remains stable for all $\tau > 0$ provided the conditions $\beta < \mu_M$ and $\mathcal{R}_0 < 1$ are satisfied. Similarly, E_1 also remains stable for all $\tau > 0$ if $\beta > \mu_M$ and $\mathcal{R}_0 < 1$. Further we investigate whether delay has any effect on stability of infected infected equilibrium E_2.

The characteristic equation of the linearized system (7) at E_2 is given by

$$det(A + Be^{-\lambda\tau} - \lambda I) = 0, \tag{8}$$

which gives

$$\left(\beta - \mu_M - \frac{2\beta\overline{T}_M}{T_{M_{max}}} - \lambda\right)[\lambda^2 + p_1\lambda + q_1 - k'\overline{T}(\lambda + \mu)e^{-\lambda\tau}] = 0,$$

where $p_1 = \mu + k\overline{T}^* + \mu_1 + \rho$ and $q_1 = (\mu_1 + \rho)(\mu + k\overline{T}^*)$. Clearly, $\lambda = \beta - \mu_M - \frac{2\beta\overline{T}_M}{T_{M\,max}} < 0$ whenever E_2 exists. Hence the stability of E_2 for delayed model system will depend upon the nature of roots of following transcendental equation,

$$\lambda^2 + p_1\lambda + q_1 - k'\overline{T}(\lambda + \mu)e^{-\lambda\tau} = 0 \tag{9}$$

Since this is transcendental equation, this will have infinitely many roots in $\mathbb{C}$. We are interested to find if there exist any root with positive real part. This is done by investigating the existence of purely imaginary roots, if any. To do so, putting $\lambda = i\omega$ in (9) and separating real and imaginary parts, we have

$$-\omega^2 + q_1 = k'\overline{T}(\mu\cos\omega\tau + \omega\sin\omega\tau) \tag{10}$$

$$p_1\omega = k'\overline{T}(\omega\cos\omega\tau - \mu\sin\omega\tau) \tag{11}$$

squaring and adding equations (10) and (11), we get the following bi-quadratic equation in ω^2:

$$\omega^4 + (\mu + k\overline{T}^*)^2\omega^2 + (\mu_1 + \rho)^2(k^2\overline{T}^{*2} + 2\mu k\overline{T}^*) = 0.$$

The above equation has no real root for ω^2 and hence no real ω satisfy above equation. Therefore, the characteristic equation (8) has roots with negative real part only for all $\tau > 0$. Finally we put the result in following theorem:

Theorem 2.4. *For the delay system (4)-(6), the infected equilibrium E_2, provided it exists, is locally asymptotically stable for all $\tau \geq 0$ i.e. delay has no effect on the local stability of E_2.*

3. Stochastic Delay Differential Equation Model

In this section we shall extend the delay differential equation model of previous section to a stochastic differential equation model. As we know that biological systems are always prone to stochastic perturbations and hence incorporating it in the model will give more realistic picture. There are many ways to consider the stochasticity in the modeling. We shall use the idea of parameter perturbation using white noise introduced in[12] and successfully applied in Refs. 13, 14. As mentioned in introduction, here we consider stochastic perturbation to account for the uncertainty of the development of ATL cells. Thus we introduce noise in the proliferation rate of ATL parameter β by considering $\beta \to \beta + \sigma_1 \xi(t)$ in DDE model system, where $\xi(t)$ is Gaussian white noise with zero mean and unit variance, σ_1 is the intensity of the stochastic perturbation. Hence we get the following system of stochastic delay differential equations:

$$\frac{dT(t)}{dt} = \Lambda - \mu T(t) - kT(t)T^*(t), \tag{12}$$

$$\frac{dT^*(t)}{dt} = k'T(t-\tau)T^*(t-\tau) - (\rho + \mu_1)T^*(t), \tag{13}$$

$$\begin{aligned}\frac{dT_M(t)}{dt} &= \rho T^*(t) + \beta T_M(t)\left(1 - \frac{T_M(t)}{T_{Mmax}}\right) - \mu_M T_M(t)\\ &\quad + \sigma_1 T_M\left(1 - \frac{T_M(t)}{T_{M_{max}}}\right)\xi(t).\end{aligned} \tag{14}$$

To analyse the impact of stochastic perturbation on populations, we shall compute the intensity of fluctuation around the infected equilibrium point E_2 due to the noise using the method introduced by Nisbet & Gurney.[12] To linearize, substitute $T(t) = \overline{T} + x(t)$, $T^*(t) = \overline{T^*} + y(t)$ and $T_M(t) = \overline{T_M} + z(t)$ in system (12)-(14) we get (see details in Refs. 13, 14, 12),

$$\frac{dX(t)}{dt} = AX(t) + BX(t-\tau) + \Xi(t) \tag{15}$$

where $X(\cdot) = (x(\cdot), y(\cdot), z(\cdot))^T$, A and B same as defined in equation (7) and $\Xi(t) = (0, 0, \sigma_1 \alpha \xi(t))^T$, where $\alpha = \overline{T}_M \left(1 - \frac{\overline{T}_M}{T_{M\,max}}\right)$. Now we are in position to compute intensity of fluctuations in different populations around the infected equilibrium E_2 due to noise following Ref. 12.

Assume a continuous function $X(t)$, known within the interval $-\frac{\omega}{2} \leq t \leq \frac{\omega}{2}$ and define another function $\widetilde{X}(s)$ which is related to it by

$$\widetilde{X}(s) = \int_{-\frac{\omega}{2}}^{\frac{\omega}{2}} X(t) e^{-ist} dt. \tag{16}$$

Taking inverse Fourier transform of $\widetilde{X}(s)$ we get

$$X(t) = \frac{1}{2\pi} \int_{-\infty}^{\infty} \widetilde{X}(s) e^{ist} ds. \tag{17}$$

Equation (17) implies that $\frac{1}{2\pi}\widetilde{X}(s)$ is the amplitude density of the components of $X(t)$ in the angular frequency interval s to $s + ds$. Hence $\frac{1}{2\pi}\widetilde{X}(s)$ is a crude estimate of the amplitude of the component of $X(t)$ with angular frequency 's' (see Ref. 12 for details). If ω is sufficiently large to assume $\omega \to +\infty$ and if $X(t) = \frac{d^n}{dt^n} Y(t)$, then

$$\widetilde{X}(s) = (is)^n \widetilde{Y}(s),$$

where $\widetilde{Y}(s)$is the Fourier transform of $Y(t)$. If $\omega \to +\infty$ and if $X(t) = Y(t - \tau)$, then

$$\widetilde{X}(s) = e^{-is\tau} \widetilde{Y}(s).$$

Using results mentioned above and taking Fourier transform of equation (15), we get

$$\widetilde{A}(s) \widetilde{X}(s) = \widetilde{\Xi}(s) \tag{18}$$

where

$$\begin{aligned} \widetilde{A}(s) &\equiv (a_{ij})_{3\times 3} \\ &= \begin{pmatrix} is + \mu + k\overline{T}^* & k\overline{T} & 0 \\ -k'\overline{T}^* e^{-is\tau} & is + \mu_1 + \rho - k'\overline{T} e^{-is\tau} & 0 \\ 0 & -\rho & is - \beta + \mu_M + \frac{2\beta \overline{T}_M}{T_{M_{max}}} \end{pmatrix}, \end{aligned}$$

$\widetilde{X}(s) = (\widetilde{x}(s), \widetilde{y}(s), \widetilde{z}(s))^T$ and $\widetilde{\Xi}(s) = (0, 0, \sigma_1 \alpha \widetilde{\xi}(s))$.

Solving the algebraic system (18), we get

$$\widetilde{X}(s) = \widetilde{A}^{-1}(s) \widetilde{\Xi}(s) \tag{19}$$

where $\widetilde{A}^{-1}(s) = \frac{1}{\Delta}\begin{pmatrix} a_{22}a_{33} & k\overline{T}a_{33} & 0 \\ -a_{21}a_{33} & a_{11}a_{33} & 0 \\ \rho k'\overline{T^*}e^{-is\tau} & 0 & a_{11}a_{22} + kk'\overline{TT^*}e^{-is\tau} \end{pmatrix}$ and Δ is given by: $\Delta = (R_1 + iR_2)(is - \beta + \mu_M + \frac{2\beta\overline{T}_M}{T_{Mmax}})$ where $R_1 = (\mu + k\overline{T}^*)(\mu_1 + \rho) - s^2 - k'\overline{T}(\mu\cos s\tau + s\sin s\tau)$, and $R_2 = s(\mu + k\overline{T}^* + \mu_1 + \rho) + k'\overline{T}(\mu\sin s\tau - s\cos s\tau)$.

If the function $X(t)$ has zero mean value then the fluctuation intensity (variance) of the components within the frequency band s to $s + ds$ is denoted by $S_X(s)ds$ and $S_X(s)$ is defined by (for details see Ref. 12),

$$S_X(s) = \lim_{\omega\to\infty}\frac{|\widetilde{X}(s)|^2}{\omega}$$

Also, if $X(t)$ has zero mean value, the inverse transform of $S_X(s)$ is the 'auto-covariance function'

$$C_X(\tau) = \frac{1}{2\pi}\int_{-\infty}^{\infty} S_X(s)e^{is\tau}ds$$

The corresponding 'variance of fluctuations' in $X(t)$ is therefore given by

$$\sigma_X^2 = C_X(0) = \frac{1}{2\pi}\int_{-\infty}^{\infty} S_X(s)ds$$

For a Gaussian white noise process it is given by

$$\begin{aligned} S_{\xi_i\xi_j}(s) &= \lim_{\omega\to\infty}\frac{E\left[\widetilde{\xi}_i(s)\widetilde{\xi}_j(s)\right]}{\omega} \\ &= \lim_{\omega\to\infty}\frac{1}{\omega}\int_{\frac{-\omega}{2}}^{\frac{\omega}{2}}\int_{\frac{-\omega}{2}}^{\frac{\omega}{2}} E\left[\widetilde{\xi}_i(t)\widetilde{\xi}_j(t')\right]e^{is(t'-t)}dtdt' \\ &= \delta_{ij} \end{aligned}$$

Hence solving system (19) we get,

$$\widetilde{X}(s) = \left(0, 0, \sigma_1\left(\frac{\overline{T}_M\left(1 - \frac{\overline{T}_M}{T_{Mmax}}\right)}{is - \beta + \mu_M + \frac{2\beta\overline{T}_M}{T_{Mmax}}}\right)\widetilde{\xi}(s)\right).$$

Thus $S_{\widetilde{x}} = 0$, $S_{\widetilde{y}} = 0$, and

$$S_{\widetilde{z}} = \left|\sigma_1\frac{\overline{T}_M\left(1 - \frac{\overline{T}_M}{T_{Mmax}}\right)}{is - \beta + \mu_M + \frac{2\beta\overline{T}_M}{T_{Mmax}}}\right|^2 = \frac{\sigma_1^2\overline{T}_M^2\left(1 - \frac{\overline{T}_M}{T_{Mmax}}\right)^2}{s^2 + \left(-\beta + \mu_M + \frac{2\beta\overline{T}_M}{T_{Mmax}}\right)^2}.$$

Hence the variance of all the tree populations are given as,

$$\sigma_x^2 = \sigma_y^2 = 0,$$

$$\sigma_z^2 = \frac{1}{2\pi}\int_{-\infty}^{\infty} |S_{\tilde{z}}(s)| ds = \frac{\sigma_1^2 \overline{T}_M^2 \left(1 - \frac{\overline{T}_M}{T_{Mmax}}\right)^2}{2\left(-\beta + \mu_M + \frac{2\beta \overline{T}_M}{T_{Mmax}}\right)} > 0.$$

Hence we got explicit expressions for the spectral densities of all the three cell populations, two of which are zero. It can be easily seen that the non-zero spectral density for T_M population is delay independent. Hence we conclude this section with the comment that only stochastic fluctuation in parameter shows fluctuations in T_M population and these fluctuations are independent of the magnitude of time delay τ.

4. Numerical Simulation Results

In this section we numerically investigate the models studied and also verify the results obtained. For this we choose following set of the parameters: $\Lambda = 6\ mm^{-3}\ day^{-1}$, $\mu = 0.006\ day^{-1}$, $k = 0.001\ mm^3\ day^{-1}, k' = 0.00007\ mm^3\ day^{-1}, \rho = 0.00004\ day^{-1}$, $\mu_1 = 0.05\ day^{-1}$, $\beta = 0.0003\ day^{-1}$, $T_{Mmax} = 2200$, $\mu_M = 0.0005\ day^{-1}$ with initial condition $T(0) = 1000\ mm^{-3}$, $T^*(0) = 10\ mm^{-3}$, $T_M(0) = 0\ mm^{-3}$. Most of the parameter values are chosen from Ref. 1. For this choice of parameters, $\mathcal{R}_0 = 1.4 > 1$ and $\beta < \mu_M$, hence E_0 and E_2 exist and E_0 will be unstable while E_2 will be stable. The simulation result is plotted in Fig. 1 in solid curve. One can see that both uninfected and infected T cell populations reach to equilibrium value but leukemia cells are increasing for 1000 days. Next we simulate the delay model system (4)-(6) for 1000 days with same values of parameters and same initial point and with $\tau = 10$ days. The corresponding solution trajectories of all the three populations are plotted in Fig. 1 in dashed curve. It is noted that the solution trajectory again approaches towards E_2 and follow similarly as in case of ODE model. The result corresponds to Theorem 2.4, that delay has no impact on the local stability of E_2.

Further, to see the impact of stochasticity, we simulate the SDDE model system (12)-(14) with parameter values same as earlier, $\tau = 10$ along with $\sigma_1 = 0.05$, following the technique given in Ref. 14 and Ref. 15. The initial values are chosen as constant function $T_0 = 1000\ mm^{-3}$, $T_0^* = 10\ mm^{-3}$, $T_{M0} = 5\ mm^{-3}$ for $[0, \tau]$. In Fig. 2 we plotted numerical results of all the populations. We can see that effect of noise is only in leukemia cell population. Thus introduction of noise term into proliferation rate β is solely responsible for the stochastic oscillations of T_M cell populations and does not produce any kind of stochastic fluctuations in other cell populations.

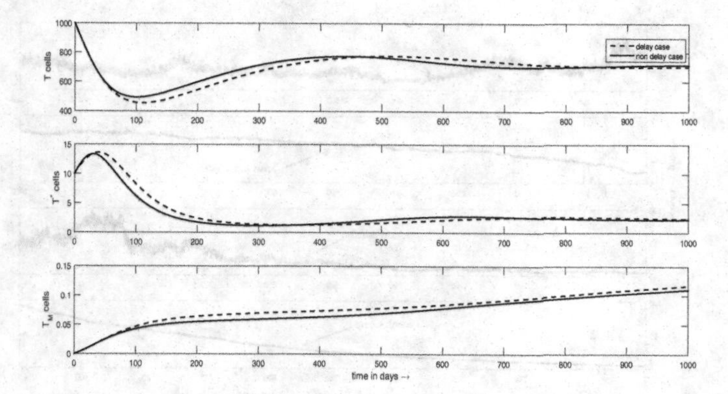

Figure 1. Population profiles for ODE model system (1)-(3) (in solid curve) and DDE model system (4)-(6) (in dashed curve for $\tau = 10$) for $\Lambda = 6, \mu = 0.006, k = 0.001, k' = 0.00007, \rho = 0.00004, \mu_1 = 0.05, \beta = 0.0003, T_{Mmax} = 2200, \mu_M = 0.0005, \mathcal{R}_0 = 1.4$.

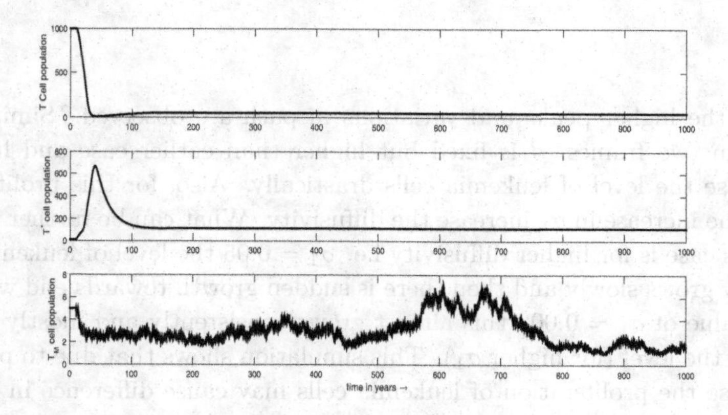

Figure 2. Population profile of SDDE model for parameter values $\Lambda = 6, \mu = 0.006, k = 0.001, k' = 0.00007, \rho = 0.00004, \mu_1 = 0.05, T_{Mmax} = 2200, \mu_M = 0.0005, \beta = 0.0003, \tau = 10, \sigma_1 = 0.05$. showing effect of stochastic perturbation in leukemia cell population only.

Since the effect of noise is only on T_M population, we consider the time evolution of this population for a combination of $\sigma_1 = 0.003, 0.05$ and $\beta = 0.003, 0.0003$. The numerical simulation results are plotted in Fig. 3. The top two frames show that if diffusivity is increased fixing proliferation, then the leukemia cell population follow similar pattern but for higher value

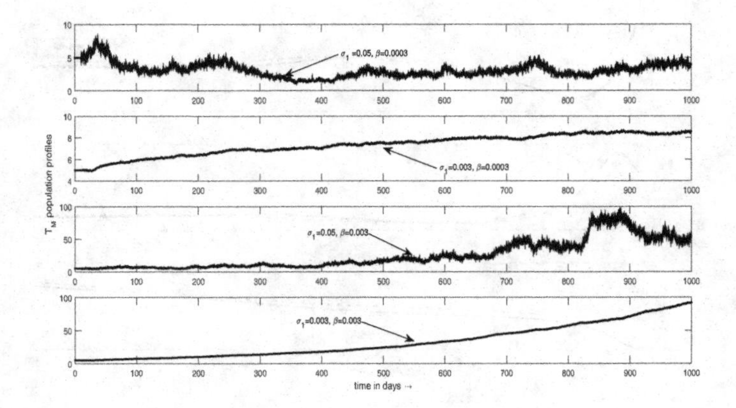

Figure 3. T_M population for various values of β, σ_1 and other parameters as $\Lambda = 6, \mu = 0.006, k = 0.001, k' = 0.00007, \rho = 0.00004, \mu_1 = 0.05, \beta = 0.0003, T_{Mmax} = 2200, \mu_M = 0.0005, \tau = 10$.

of σ_1 the higher peaks and variations in path are observed. Similarly in bottom two frames, β is fixed but higher than earlier case and hence it increase the level of leukemia cells drastically. Also, for this proliferation rate the increase in σ_1 increase the diffusivity. What can be further noticed in this case is for higher diffusivity i.e. $\sigma_1 = 0.05$ the level of leukemia cells initaly grows slowly and then there is sudden growth towards end while for low value of $\sigma_1 = 0.003$ this almost grows consistently and mostly remain above the level (for higher σ_1). This simulation shows that due to presence of noise the proliferation of leukemia cells may cause difference in level of leukemia cell population attained. Hence noise has significant impact on level of leukemia cells for this particular simulation run.

Further in Fig. 4 we plotted the 5 simulations of the SDDE model system for the parameters $\Lambda = 6$, $\mu = 0.006$, $k = 0.001$, $k' = 0.00007$, $\rho = 0.00004$, $\mu_1 = 0.05 T_{Mmax} = 2200$, $\mu_M = 0.0005$, $\tau = 10$, $\sigma_1 = 0.05$ and $\beta = 0.0003$, & 0.003. The effect of increase in β can be seen easily. In case of low proliferation rate the diffusion is more but the population is low so variation is significant, but in case of high proliferation rate the population level is high towards the end of simulation period. Due to high proliferation the population of leukemia cell is also higher in this case.

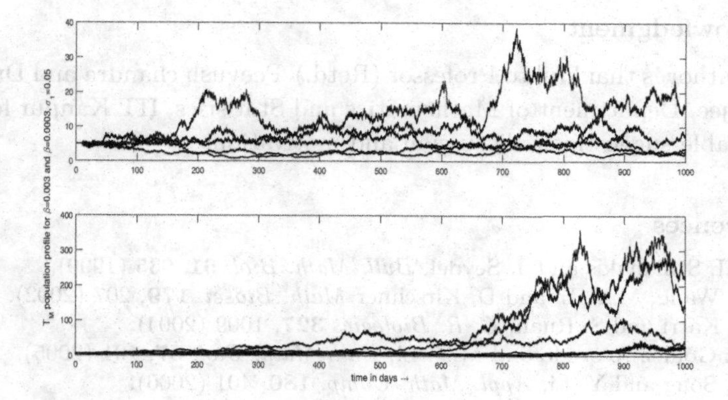

Figure 4. Five numerical paths for T_M population for varying β. $\beta = 0.0003$ (top panel) and $\beta = 0.003$ (bottom panel) and other parameters are $\Lambda = 6$, $\mu = 0.006$, $k = 0.001$, $k' = 0.00007$, $\rho = 0.00004$, $\mu_1 = 0.05$, $T_{Mmax} = 2200$, $\mu_M = 0.0005$, $\tau = 10$, $\sigma_1 = 0.05$.

5. Conclusion

In this paper, we considered a three dimensional model for the infection of CD4$^+$ T cells by HTLV-I virus. First we considered an ODE model for the uninfected CD4$^+$ T cells, infected CD4$^+$ T cells and leukemia cells. We performed the local and global stability analysis of the ODE model to show that infected cells die out if $\mathcal{R}_0 < 1$ and if $\mathcal{R}_0 > 1$ then the primary infection always lead to an infected equilibrium E_2. If $\beta > \mu_M$, then in former case ATLs saturate to a level $\widehat{T}_M$ and no chronic infection is possible while in later case, this leads to chronic infection with ATLs level $\overline{T}_M$ higher than $\widehat{T}_M$. These results are similar to those obtained in Refs. 1, 2. Then we accounted for the latency by introducing discrete time delay in ODE model. For the DDE model we established that the discrete time delay has no effect on the local stability of infected equilibrium point E_2. This is important observation that delay due to latency cause no destabilizing effect and hence it is justified to consider only productively infected CD4$^+$ T cell population. Finally, we studied the effect of stochastic perturbation by introducing white noise in proliferation rate β, which results in the SDDE model corresponding to the DDE model. Analytical as well as numerical simulation revealed that the stochastic oscillation in T_M population is arising due to white noise perturbation in proliferation rate and had no impact on time evolution of uninfected and infected CD4$^+$ T cell population.

Acknowledgment

The author is thankful to Professor (Retd.) Peeyush chandra and Dr Malay Banerjee, Department of Mathematics and Statistics, IIT Kanpur for their invaluable suggestions, discussion and motivation.

References

1. N.I. Stilianakis and J. Seydel, *Bull. Math. Biol.* **61**, 935 (1999).
2. L. Wang, M.Y. Li and D. Kirschner *Math. Biosci.* **179**, 207 (2002).
3. P. Katri and S. Ruan, *C. R. Biologies* **327**, 1009 (2004).
4. H. Gómez-Acevedo and M.Y. Li, *Bull. Math. Biol.* **67**, 101 (2005).
5. X. Song and Y. Li, *Appl. Math. Comp.* **180**, 401 (2006).
6. L. Cai, X. Li and M. Ghosh, *Appl. Math. Modell.* **35**, 3587 (2011).
7. H. Gómez-Acevedo, M.Y. Li and S. Jacobson, *Bull. Math. Biol.* **72**, 681 (2010).
8. M.Y. Li and H. Shu, *Bull. Math. Biol.* **73**, 1774 (2011).
9. M.Y. Li and H. Shu, *Nonlin. Anal.: RWA* **13**, 1080 (2012).
10. J. Lang and M.Y. Li, *J. Math. Biol.* **65**, 181 (2012).
11. J. Hale, *Theory of functional differential equation*, Springer Verlag, New York (1977).
12. R.M. Nisbet and G.C. Gurney, *Modelling Fluctuating Populations*, Wiley Interscience, New York (1982).
13. M. Bandyopadhyay, T. Saha, R. Pal, *Nonlin. Anal. Hyb. Sys.* **2**, 958 (2008).
14. M. Carletti, *Math. Med. Biol.* **23**, 297 (2006).
15. C.T.H. Baker and E. Buckwar, *LMS J. Comput. Math.* **3**, 315 (2000).
16. P. Waltman, *A brief survey of persistence, in: Busenberg, S., Martelli, M. (Eds.), Delay differential equations and dynamical systems* Springer, New York, (1991).
17. H.I. Freedman, M.X. Tang and S.G. Ruan, *J. Dyna. Diff. Equ.* **6**, 583 (1994).
18. M.Y. Li, J.R. Graef, L.C. Wang and J. Karsai, *Math. biosc.* **160**, 191 (1999).
19. S. Iwami, Y. Takeuchi and X. Liu, *Math. Biosci.* **207**, 1 (2007).

STUDY OF GLOBAL STABILITY AND OPTIMAL TREATMENT FOR AN INFECTIOUS DISEASE MODEL

A. KUMAR, P. K. SRIVASTAVA AND A. YADAV

Department of Mathematics
Indian Institute of Technology Patna, Patna-801103, India
E-mail: anujdubey17@gmail.com, pksri@iitp.ac.in, anuyadav@iitp.ac.in

In this paper we considered a nonlinear compartmental $SIRS$ model that accounts for the impact of awareness effect in population. We analyzed and established the global stability of unique infected equilibrium by constructing a suitable Lyapunov function. We found that the basic reproduction number for the model system is independent of awareness parameter and hence eradication of the disease cannot be achieved through awareness only, though it reduces disease prevalence in initial phase. Further, by introducing treatment to infective, we formulate and study corresponding optimal control problem to find optimal treatment policy for providing treatment with minimum cost. With the help of Pontryagin's Maximum Principle the optimal treatment path is obtained analytically. We performed numerical simulations to analyze the results obtained for a representative set of parameters. It is observed that the awareness reduces the infective population during initial phase but is not effective if treatment is not available. While optimal treatment in aware population is very effective in keeping disease load significantly low throughout the period and reducing economical cost. Thus the combined impact of awareness and optimal treatment can bring down disease prevalence in population and also minimize cost incurred.

Keywords: Global Stability; Lyapunov Function; Optimal Treatment; Awareness.

1. Introduction

Infectious diseases are one of the major hurdle in social and economical development of many countries. It is a challenging task for the policy makers worldwide to plan for controlling the spread of diseases. Mathematical models have been found useful and hence are used to predict disease prevalence, to decide control measure and also to formulate policies so that either disease is eradicated or at least it can be contained to a significantly low

*Corresponding author.

level.[4,11,16,24,29,30] Using mathematical models, in general, we seek the pattern of the spread of a disease so that a control measure can be devised and analysed. For example, control aspects such as isolation, vaccination and treatment etc. have been studied in.[11,18,36] Behavioural changes of individuals have been observed and studied in presence of information of the disease in population (see Refs. 1, 5, 6, 7, 26, 28, 31, 32). Though, this particular aspect has not been explored much in epidemiological models in past.

Recently there is growing interest among researchers to study the impact of various behaviour influencing factors such as educational campaigns, awareness programs and other social programs on the spread of infectious diseases.[2,7,26,27,28,35] It is observed that the impact of media coverage plays an important role in reducing the disease transmission in population.[26,27,32]

In epidemiological modeling, vaccination, treatment and isolation have been used as mitigation strategies[2,11,36] but there is always a trade off to apply a suitable strategy. It is also challenging to find out the optimum response which minimizes the growth of disease and also reduces the cost burden incurred in any applied mitigation strategy. Cost optimizing for any implemented intervention in modeling process is always a challenging problem. For last two decades, many researchers focused on these types of mitigation strategies to reduce cost incurred in such policies. A substantial work has been done using optimal control in epidemiology.[3,11,13,17,21,22,32,34,37]

Horst Behncke proposed general epidemiological model with different control regimes such as vaccination, quarantine and health promotion campaigns etc. and explored optimum control strategies theoretically.[2] In Ref. 13, authors proposed infectious disease model with treatment and discussed a cost optimizing treatment policy for the model. A useful theoretical background with numerical simulation of optimal control theory in various epidemiological models with multiple controls strategies found in Refs. 18, 24. In 2008, Zaman *et al.* proposed an SIR model with optimal vaccination policy and explained the impact of vaccination in controlling the spread of disease with variable population size.[36] Goff *et al.* used both treatment and vaccination as control measure for various epidemiological models and explained their importance in the duration of epidemic and obtained various patterns of these policies.[11] Recently social and educational programs such as awareness programs and health promoting campaigns etc. have also been used as control in models to reduce the incidence of epidemic.[2,7,19]

In 2008, Liu *et al.*[26] proposed following nonlinear SIRS model where the incidence rate was corrected to include the effect of information through media due to prevalence of the disease:

$$\begin{aligned}\frac{dS}{dt} &= r - dS - \left(\beta_1 - \beta_2 \frac{I}{m+I}\right) SI + \delta R,\\ \frac{dI}{dt} &= \left(\beta_1 - \beta_2 \frac{I}{m+I}\right) SI - (d+\gamma+\alpha)I, \qquad (1)\\ \frac{dR}{dt} &= \gamma I - (d+\delta)R,\end{aligned}$$

with initial conditions $S(0) \geq 0$, $I(0) \geq 0$ and $R(0) \geq 0$. Here they also assumed that individuals in recovered class revert back to susceptible class with loss of immunity.

Here $S(t)$, $I(t)$, and $R(t)$ are respective individuals, at any given time t, in susceptible, infective, and recovered compartments of the total population $N(t)$. The parameter r represents growth rate of susceptible population, γ is a constant recovery rate of infected population, d is natural mortality rate and α is disease related death rate. β_1 is transmission rate of the disease under consideration in absence of awareness and $\left(\beta_1 - \beta_2 \frac{I}{m+I}\right)$ is effective contact rate after getting influenced by information through media or other social program.[26] The details on choosing the function $\left(\beta_1 - \beta_2 \frac{I}{m+I}\right)$ can be found in Ref. 26. Also note that $\beta_1 \geq \beta_2$ as media or other social program can not prevent the spreading of disease completely. m represents the number of infective with respect to media program or other social program to the disease. Parameter δ is the rate constant of immunity loss.

Liu *et al.*[26] observed that model system has a disease free equilibrium, which always exists, and another unique endemic equilibrium which exists whenever a threshold condition on parameters is satisfied. They discussed stability of disease free equilibrium and also established the global stability of the endemic equilibrium for the special case when $\delta = 0$ by showing non-existence of periodic orbits. Note that in this case original model system (1) reduces to an SIR model.

In this paper, we first establish the global stability of endemic equilibrium of model system (1) for the general case $\delta \neq 0$. This is done by constructing a Lyapunov function and using LaSalle's Theorem.[23] Further, it is considered that treatment is available and is provided to the infective. To understand and obtain optimal treatment, the corresponding optimal control problem is formulated with aim to minimize the cost of treatment as well as the cost due to disease burden. We study the corresponding optimal control problem using Pontryagin's Maximum Principle and optimal treatment profile is obtained. We analysed our results numerically for a certain set of parameters with emphasis on comprehensive effect of information of disease and optimal treatment, on the dynamics of disease as well as on disease prevalence.

2. Stability Analysis of Model (1)

In this section we first reproduce stability results obtained in Ref. 26. The model system (1) has following two equilibria:

(i) the disease free equilibrium $E_1 = (\frac{r}{d}, 0, 0)$, and
(ii) the infected equilibrium $E_2 = (S_*, I_*, R_*)$, which exists whenever $R_0 > 1$, where R_0 is basic reproduction number given as $R_0 = \frac{r\beta_1}{d(d+\alpha+\gamma)}$. Components $S_* = \frac{(d+\alpha+\gamma)}{(\beta_1 - \beta_2 \frac{I_*}{m+I_*})}$, $R_* = \frac{\gamma I_*}{d+\delta}$ and I_* is the positive root of following equation:

$$f(I) = AI^2 + BI + C = 0,$$

where $A = -\frac{(\beta_1-\beta_2)}{(d+\delta)}(\delta(d+\alpha)+d(d+\alpha+\gamma))$, $B = -\frac{d\beta_1 m\gamma}{(d+\delta)} - \beta_1 m(d+\delta) + r\beta_1(1-\frac{1}{R_0}) - r\beta_2$ and $C = dm(d+\alpha+\gamma)(R_0 - 1)$.

In Ref. 26, authors showed that if $R_0 > 1$, $C > 0$ and $A < 0$, then $f(I) = 0$ has a unique positive root and system (1) has unique infected equilibrium point E_2. Note that $A < 0$ only when $\beta_1 > \beta_2$ and in that case $f(I) = 0$ has a unique positive root $I_* = \frac{-B-\sqrt{B^2-4AC}}{2A}$. Further, we note that if $\beta_1 = \beta_2$ and $R_0 > 1$ then $A = 0$, $C > 0$ and $B < 0$. In this case also $f(I) = 0$ has a unique positive root $I_* = -\frac{C}{B}$ and system (1) has unique infected equilibrium point.

The biologically feasible region of the model system (1) is the following positive invariant set[26]:

$$\Gamma = \{(S, I, R) \in \mathbb{R}^3_+ \mid 0 < S + I + R \leq \frac{r}{d}, S \geq 0, I \geq 0, R \geq 0\}.$$

Theorem 2.1.[26]

(1) The disease free equilibrium E_1 of system (1) is globally asymptotically stable if $R_0 < 1$ and is unstable if $R_0 > 1$.
(2) If $R_0 > 1$ then system (1) has unique infected equilibrium point E_2 and in this case E_2 is locally asymptotically stable.
(3) For $\delta = 0$, the unique infected equilibrium point E_2 is globally asymptotically stable when $R_0 > 1$.

2.1. *Global Stability of Infected Equilibrium E_2*

In this section we prove the global stability of unique infected equilibrium point E_2 for $\delta \neq 0$ by constructing a suitable Lyapunov function[9] for model system (1) and using Lyapunov LaSalle's theorem.[23]

Consider the following Lyapunov function in Γ,

$$V(S,I,R) = \frac{1}{2}[(S-S_*)+(I-I_*)+(R-R_*)]^2 + m_1\left(I - I_* - I_* \ln\frac{I}{I_*}\right) + \frac{m_2}{2}(R-R_*)^2.$$

Here m_1 and m_2 are positive constants and will be chosen suitably later. Clearly V is a positive definite function. Differentiating V with respect to t along the solution trajectories of system (1) and using following relations:

$$d(S_* + I_* + R_*) + \alpha I_* = r,$$
$$\left(\beta_1 - \beta_2\frac{I_*}{m+I_*}\right)S_* = d + \alpha + \gamma,$$
$$(d+\delta)R_* = \gamma I_*,$$

we get,

$$\begin{aligned}
\dot{V} &= [(S-S_*)+(I-I_*)+(R-R_*)]\frac{d(S+I+R)}{dt} + m_1\frac{(I-I_*)}{I}\frac{dI}{dt} \\
&\quad + m_2(R-R_*)\frac{dR}{dt} \\
&= [(S-S_*)+(I-I_*)+(R-R_*)][r - d(S+I+R) - \alpha I] + m_1\frac{(I-I_*)}{I} \\
&\quad \left(\left(\beta_1 - \beta_2\frac{I}{m+I}\right)SI - (d+\alpha+\gamma)I\right) + m_2(R-R_*)[\gamma I - (d+\delta)R] \\
&= -d[(S-S_*)+(R-R_*)]^2 - \left((d+\alpha) + \frac{m_1 m\beta_2 S}{(m+I)(m+I_*)}\right)(I-I_*)^2 \\
&\quad - m_2(d+\delta)(R-R_*)^2 + \left(m_1\left(\beta_1 - \frac{\beta_2 I_*}{(m+I_*)}\right) - (2d+\alpha)\right) \\
&\quad (S-S_*)(I-I_*) + (m_2\gamma - (2d+\alpha))(R-R_*)(I-I_*).
\end{aligned}$$

Now choose $m_1 = \frac{(2d+\alpha)S_*}{(d+\alpha+\gamma)}$ and $m_2 = \frac{2d+\alpha}{\gamma}$, so that $\dot{V} \leq 0$. Hence $\dot{V}$ is negative and $\dot{V} = 0$ if and only if $S = S_*, I = I_*$ and $R = R_*$ in Γ. Hence the singleton set $\{E_2\}$ is the largest positively invariant set contained in $\{(S,I,R) \in \Gamma : \dot{V} = 0\}$. Then by Lyapunov LaSalle's theorem[23] E_2 is globally asymptotically stable in the interior of Γ. Hence we have following result:

Theorem 2.2. *If $R_0 > 1$ then the unique infected equilibrium E_2 is globally asymptotically stable.*

Remark 2.1. The same Lyapunov function $V(S, I, R)$ will work for the case $\delta = 0$ as well and hence the results obtained in Ref. 26 follows from here immediately.

Here, we establish the global stability result numerically, for this we use a hypothetical set of parameter values as: $r = 5, d = 0.02, \gamma = 0.025, \delta = 0.01, \alpha = 0.1, \beta_1 = 0.002, \beta_2 = 0.0016$ and $m = 30$. In this case, model system (1) has two equilibria (*i*) disease free equilibrium $E_1 = (250, 0, 0)$ and (*ii*) unique infected equilibrium $E_2 = (107.825, 20.806, 17.338)$ and basic reproduction number $R_0 = 3.448$. As in this case $R_0 > 1$, from Theorem 2.2 E_2 is globally asymptotically stable. To show this we solve system (1) using four different initial conditions: $I_1 = (500, 20, 3), I_2 = (250, 9, 10), I_3 = (150, 30, 60), I_4 = (50, 60, 90)$ and note that all solution trajectories approach to infected equilibrium E_2 (Fig. 1). This mimics the results obtained in Theorem 2.2.

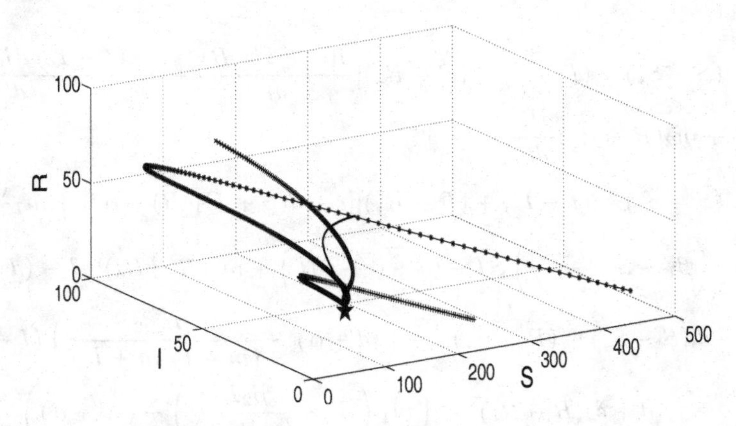

Figure 1. Solutions of model system (1) starting from different initial conditions: $I_1 = (500, 20, 3), I_2 = (250, 9, 10), I_3 = (150, 30, 60), I_4 = (50, 60, 90)$, approach the infected equilibrium $E_2 = (107.825, 20.806, 17.338)$.

3. Optimal Control Problem

In this section we discuss optimal control strategy for controlling disease prevalence in model system (1) and formulate the corresponding optimal control problem. In model (1) infected individuals are recovering with a constant recovery rate γ through immunity. In order to control the spread of disease we need to provide additional treatment to the infected individuals as recovery through immunity may not be sufficient. Hence we consider the per capita treatment rate as u in model system (1). As there is always cost involved in providing treatment and other related efforts, we consider $u(t)$ as control variable and study the corresponding optimal control problem. Our aim is to find an optimal treatment function $u^*(t)$ in presence of information about the disease that minimizes infective as well as the cost incurred in treatment and related efforts.

As we can not provide unlimited treatment due to restriction on available resources, in practice, the admissible set for the treatment function $u(t)$ is

$$U = \{u(t) : u(t) \in [0, u_{max}], t \in [0, T]\},$$

where $u(t)$ is measurable and bounded and T is final time for applied treatment policy.

In view of above let us consider following control problem which minimizes the cost functional

$$J[u(t)] = \int_0^T [AI(t) + Bu^2(t)]dt, \tag{2}$$

subject to

$$\begin{aligned} \frac{dS}{dt} &= r - dS - \left(\beta_1 - \beta_2 \frac{I}{m+I}\right) SI + \delta R, \\ \frac{dI}{dt} &= \left(\beta_1 - \beta_2 \frac{I}{m+I}\right) SI - (d + \alpha + \gamma)I - u(t)I, \\ \frac{dR}{dt} &= \gamma I + u(t)I - (d + \delta)R, \end{aligned} \tag{3}$$

with initial conditions $S(0) \geq 0$, $I(0) \geq 0$ and $R(0) \geq 0$.

The cost functional J chosen above represents the total cost incurred in time $[0, T]$ corresponding to model system (3). The integrand $L(S, I, R, u) = AI(t) + Bu^2(t)$ represents cost at any time t. The first term $AI(t)$ in $L(S, I, R, u)$ represents cost associated with infected population and which usually occur because of potential man hour loss due to illness etc. and the second term $Bu^2(t)$ accounts for the cost incurred in treatment of infected individuals, which also includes the costs associated with treatment, diagnosis and hospitalization etc.[11,22] Cost corresponding to treatment is taken as nonlinear because such costs are not linear due to the requirement of increasing cost for reaching higher fraction of the population. Quadratic nonlinearity is chosen here because of simplicity and has been used in literature.[11,22] Constants A and B are positive weights that are used to balance units of measurement as well as to emphasize relative trade off amongst the arguments of cost functional. When there are more than one control functions in a model, such as vaccination, treatment and awareness etc., these constants represent relative balancing factor amongst them.[11,18]

Now our problem is to find control function $u(t)$ in U that minimizes the cost functional J, *i.e.* to find $u^*(t) \in U$ such that $J[u^*(t)] = min[J[u(t)] : u(t) \in U]$. The existence of such $u^*(t)$ is guaranteed in the following.

For the existence of optimal control $u^*(t)$ corresponding to optimal control problem (2)-(3) the following conditions must satisfied:[10,11,12,18]

(*a*) The set of solutions to system (3) with control variable in U is non empty.

(*b*) U is closed and convex and state system can be written as linear function of control variable with coefficients depending on time and state variables.

(*c*) Integrand L of Eq. (2) is convex on U and $L(S, I, R, u) \geq c_1 \mid u \mid^\rho - c_2$ where $c_1 > 0$ and $\rho > 1$.

Note that total population $N = S + I + R$ is given by following differential equation:

$$\frac{dN}{dt} = r - dN - \alpha I.$$

Thus $\frac{dN}{dt} \leq r - dN$ which implies $N \leq \frac{r}{d} \Rightarrow S + I + R \leq \frac{r}{d}$. Hence solutions of system (3) are bounded for each bounded control variable in U. Clearly the right hand side functions in system (3) satisfy Lipschitz condition with respect to state variable. Hence using Picard-Lindelöf Theorem[8] condition (*a*) satisfied. Also U is closed and convex by definition and system (3) is linear in control $u(t)$ and hence condition (*b*) is satisfied too. Moreover L being quadratic in control variable $u(t)$, it is convex. Further, $L(S, I, R, u) = AI(t) + Bu^2(t) \geq B \mid u \mid^2$. Now taking $B = c_1 > 0$ and $\rho = 2 > 1$, with these choices condition (*c*) is also satisfied. Now, using results from Refs. 10, 11, 18, we conclude that there is a control $u^*(t)$ such that we get $J[u^*(t)] = min[J[u(t)]]$. We summarize above in the following theorem:

Theorem 3.1. *There exists an optimal control $u^*(t)$ in U such that*

$J[u^*(t)] = min[J[u(t)]]$ corresponding to the control system (2)-(3).

3.1. *Characterization of Optimal Control*

In this section we characterize the optimal control function $u^*(t)$ for system (2)-(3) using the Pontryagin's Maximum Principle[10,14] and also derive necessary conditions for the optimal control function through Hamiltonian. Pontryagin's Maximum Principle characterizes the optimal control function using adjoint variables which connect cost functional and state system.

Theorem 3.2. *Let $u^*(t)$ (hereafter referred as u^*) be optimal control variable and S^*, I^* and R^* be corresponding state variables of the optimal control problem (2)-(3). Then there exist adjoint variables λ_1, λ_2 and λ_3 which satisfy*

$$\frac{d\lambda_1}{dt} = \left(d + \left(\beta_1 - \beta_2\frac{I}{m+I}\right)I\right)\lambda_1 - \left(\beta_1 - \beta_2\frac{I}{m+I}\right)I\lambda_2,$$

$$\frac{d\lambda_2}{dt} = -A + \left(\beta_1 - \beta_2\frac{I(2m+I)}{(m+I)^2}\right)S\lambda_1$$
$$-\left\{\left(\beta_1 - \beta_2\frac{I(2m+I)}{(m+I)^2}\right)S - (d+\alpha+\gamma) - u\right\}\lambda_2 - (\gamma+u)\lambda_3,$$

$$\frac{d\lambda_3}{dt} = -\delta\lambda_1 + (d+\delta)\lambda_3,$$

with transversality conditions $\lambda_1(T) = 0, \lambda_2(T) = 0$ and $\lambda_3(T) = 0$. The optimal control u^* is characterized as

$$u^* = min\left\{max\left\{0, \frac{I^*}{2B}(\lambda_2 - \lambda_3)\right\}, u_{max}\right\}. \tag{4}$$

Proof. Consider u^* is given optimal control and S^*, I^* and R^* are corresponding optimal state variables of system (3) which minimize the cost functional (2). Now we define the corresponding Hamiltonian H as,

$$H(S,I,R,u,\lambda_1,\lambda_2,\lambda_3) = L(S,I,R,u) + \lambda_1\dot{S} + \lambda_2\dot{I} + \lambda_3\dot{R}$$
$$= AI + Bu^2 + \lambda_1\left(r - dS - \left(\beta_1 - \beta_2\frac{I}{m+I}\right)SI + \delta R\right)$$
$$+ \lambda_2\left(\left(\beta_1 - \beta_2\frac{I}{m+I}\right)SI - (d+\alpha+\gamma)I - uI\right)$$
$$+ \lambda_3(\gamma I + uI - (d+\delta)R). \tag{5}$$

Then from Pontryagin's Maximum Principle these adjoint variables λ_1, λ_2 and λ_3 satisfy following canonical equations

$$\frac{d\lambda_1}{dt} = -\frac{\partial H}{\partial S}, \quad \frac{d\lambda_2}{dt} = -\frac{\partial H}{\partial I} \text{ and } \frac{d\lambda_3}{dt} = -\frac{\partial H}{\partial R}, \tag{6}$$

with transversality conditions $\lambda_1(T) = 0, \lambda_2(T) = 0$ and $\lambda_3(T) = 0$.

From Eq. (6), we get

$$\begin{aligned}
\frac{d\lambda_1}{dt} &= \left(d + \left(\beta_1 - \beta_2 \frac{I}{m+I}\right) I\right) \lambda_1 - \left(\beta_1 - \beta_2 \frac{I}{m+I}\right) I \lambda_2, \\
\frac{d\lambda_2}{dt} &= -A + \left(\beta_1 - \beta_2 \frac{I(2m+I)}{(m+I)^2}\right) S \lambda_1 \\
&\quad - \left\{\left(\beta_1 - \beta_2 \frac{I(2m+I)}{(m+I)^2}\right) S - (d + \alpha + \gamma) - u\right\} \lambda_2 - (\gamma + u)\lambda_3, \\
\frac{d\lambda_3}{dt} &= -\delta\lambda_1 + (d + \delta)\lambda_3,
\end{aligned}$$

with transversality conditions $\lambda_1(T) = 0, \lambda_2(T) = 0$ and $\lambda_3(T) = 0$.

Further, from the optimality condition, we have

$$\frac{\partial H}{\partial u} = 0, \text{ at } u = u^*$$

which gives $2Bu^* - \lambda_2 I^* + \lambda_3 I^* = 0$, *i.e.* $u^* = \frac{I^*}{2B}(\lambda_2 - \lambda_3)$.

We note from property of control space U and above discussion that the optimal control u^* has following form:

$$u^* = \begin{cases} 0 & \text{if } \frac{I^*}{2B}(\lambda_2 - \lambda_3) < 0, \\ \frac{I^*}{2B}(\lambda_2 - \lambda_3) & \text{if } 0 \le \frac{I^*}{2B}(\lambda_2 - \lambda_3) \le u_{max}, \\ u_{max} & \text{if } \frac{I^*}{2B}(\lambda_2 - \lambda_3) > u_{max}. \end{cases}$$

This can be equivalently written as

$$u^* = min\left\{max\left\{0, \frac{I^*}{2B}(\lambda_2 - \lambda_3)\right\}, u_{max}\right\}.$$

Hence by Pontryagin's Maximum Principle, we characterize $u^* \in U$ such that

$$J[u^*] = min[J[u]].$$

□

3.2. *Optimality System*

Here, we use the optimal control characterized in previous section to get the following optimality system:

$$
\begin{aligned}
\frac{dS^*}{dt} &= r - dS^* - \left(\beta_1 - \beta_2 \frac{I^*}{m+I^*}\right) S^* I^* + \delta R^*,\\
\frac{dI^*}{dt} &= \left(\beta_1 - \beta_2 \frac{I^*}{m+I^*}\right) S^* I^* - (d+\alpha+\gamma) I^* - u^* I^*,\\
\frac{dR^*}{dt} &= \gamma I^* + u^* I^* - (d+\delta) R^*,
\end{aligned}
\tag{7}
$$

with initial conditions $S^*(0) \geq 0, I^*(0) \geq 0$ and $R^*(0) \geq 0$. The corresponding adjoint system with Hamiltonian H^* at $(S^*, I^*, R^*, u^*, \lambda_1, \lambda_2, \lambda_3)$ is given as,

$$
\begin{aligned}
\frac{d\lambda_1}{dt} &= \left(d + \left(\beta_1 - \beta_2 \frac{I^*}{m+I^*}\right) I^*\right) \lambda_1 - \left(\beta_1 - \beta_2 \frac{I^*}{m+I^*}\right) I^* \lambda_2,\\
\frac{d\lambda_2}{dt} &= -A + \left(\beta_1 - \beta_2 \frac{I^*(2m+I^*)}{(m+I^*)^2}\right) S^* \lambda_1\\
&\quad - \left\{\left(\beta_1 - \beta_2 \frac{I^*(2m+I^*)}{(m+I^*)^2}\right) S^* - (d+\alpha+\gamma) - u^*\right\} \lambda_2 - (\gamma + u^*) \lambda_3,\\
\frac{d\lambda_3}{dt} &= -\delta \lambda_1 + (d+\delta) \lambda_3,
\end{aligned}
\tag{8}
$$

with transversality conditions $\lambda_1(T) = 0, \lambda_2(T) = 0$ and $\lambda_3(T) = 0$ and u^* is as in Eq. (4).

4. Numerical Experimentations and Discussion

In this section we shall analyze and discuss the results, obtained in previous section, numerically using MATLAB®. We consider the set of parameters as given in Table 1 with initial population size $S(0) = 990, I(0) = 9$ and $R(0) = 1$. Numerically, we intend to see the impact of optimal treatment on the disease dynamics in absence as well as presence of effect of awareness driven by media. The effect of awareness on population is also investigated when no treatment is available.

We solve model system (3) with no treatment ($u = 0$), to understand the effect of awareness on disease progression. First we choose $\beta_2 = 0$ *i.e.* awareness has no impact on disease incidence and plot the corresponding solution trajectories in Fig. 2 in continuous curves. We observe from Fig. 2 that infective first increase sharply and then reach to its maximum within

Table 1. Description of parametric values used in model system.

Parameter	Description	Value	Source
r	Growth rate	5 person day^{-1}	Ref. 26
d	Natural mortality rate	0.00004 day^{-1}	Refs. 11, 25
γ	Recovery rate	0.05 day^{-1}	Ref. 26
δ	Loss of immunity rate	0.01 day^{-1}	Ref. 26
α	Disease related death rate	0.008 day^{-1}	Refs. 11, 15
β_1	Transmission rate before awareness	0.002 person^{-1} day^{-1}	Ref. 26
β_2	Reduction in contact rate due to awareness	0.0018 person^{-1} day^{-1}	Ref. 26
m	Constant which reflects the impact of awareness	30	Ref. 26

7-8 days and then decrease gradually before settling to its equilibrium level. We also note that there is sharp decay in susceptible population before settling to equilibrium level and corresponding lower level of recovered population. As plotted in Fig. 2 high disease prevalence is observed in absence of awareness effect for around 40 days whereas after that less prevalence is observed. This high prevalence not only drastically affect the social well being but also contribute to higher productivity loss due to disease burden during the disease outbreaks. Further, to see the effect of awareness, we choose $\beta_2 = 0.0018$, and solve model system (3). The corresponding solution trajectories are plotted in Fig. 2 in doted curves. In this case, due to awareness there is a slow growth in infective in comparison to previous case as seen in Fig. 2. Also the peak of infective during first 15-18 days is far less smaller than in case of $\beta_2 = 0$. Thus awareness influences a significant delay in infective's peak. We also note in this case that though there is decay in susceptible population, they settle to higher level than the earlier one as in Fig. 2. Hence we observe, from Fig. 2, that the effect of awareness reduces disease prevalence within the population under constant recovery. But $R_0 = \frac{r\beta_1}{d(d+\alpha+\gamma)}$ is independent of β_2, so eradication of the disease can not be achieved through awareness only. Thus impact of awareness plays an important role in reducing the disease prevalence but not effective in eradication of disease.

We now solve the optimal problem (2)-(3) numerically to obtain optimal treatment corresponding to minimum cost functional J. For this we solve the optimality system (7)-(8) using numerical iterative scheme. We use the forward-backward sweep method starting with an initial guess for optimal control and solve the optimal state system forward in time and after that solve the adjoint state system backward in time using ode45 in

MATLAB. Then the optimal control u_* is updated using the Hamiltonian of the optimal system and continue the same process till a pre-defined convergence criteria is met (for details see Refs. 20, 24, 33). For this purpose, we have used the steepest descent method of optimization that is discussed in Refs. 20, 33.

We choose weight constants $A = 1.64$ and $B = 200$. As cost involved for treatment may be far more higher than losses due to disease burden, this motivated us to choose A and B as above.[11,19,36] As has been mentioned in Sec. 3, the weight constants A and B are associated with infected population and treatment respectively which balance disease burden and cost incurred in treatment. Since $u \in [0, u_{max}]$, where u_{max} represents the 100% treatment for treatment policy at time t and therefore in our problem we consider $u_{max} = 1$.[24,36] The corresponding optimal treatment path so obtained is plotted in Fig. 2 (bottom figure). The optimal states corresponding to this optimal treatment are given in Fig. 2 in dashed curves. On comparing from Fig. 2, we observe that peak of disease prevalence due to combined effect of awareness and optimal treatment is very low than in case of only awareness effect. Also infective decreases very fast before settling at lower level in this case.

From the above numerical experimentations, we observe that effect of awareness (β_2), in absence of treatment, only reduces the epidemic peak. But the combined effect of both awareness and optimal treatment not only reduces the epidemic peak but also reduces the disease prevalence. This is happening because awareness reduces the contacts and thus decreases the new infections whereas treatment reduces the infectious period of infective. Thus the optimal treatment and awareness together help to reduce the disease prevalence in very short time.

To vet the criticality, suitability and cost-effectiveness for applied treatment policy, costs design analysis has been performed and made comparative study of costs numerically for various cases. Optimal cost profiles for different scenario along with profiles of infective population in respective cases are given in Fig. 3. Note that in the absence of any control measure the cost is via opportunity loss (productivity loss) which is contributed by infective population. This is significantly high and is so due to the reason that in absence of any control disease will spread and infective will be higher and hence opportunity loss will increase and add to the economic burden. It is important to note here the optimal cost, when both effect of awareness and optimal treatment are used, is remarkably low in comparison with others. The count of infective in this case significantly reduces

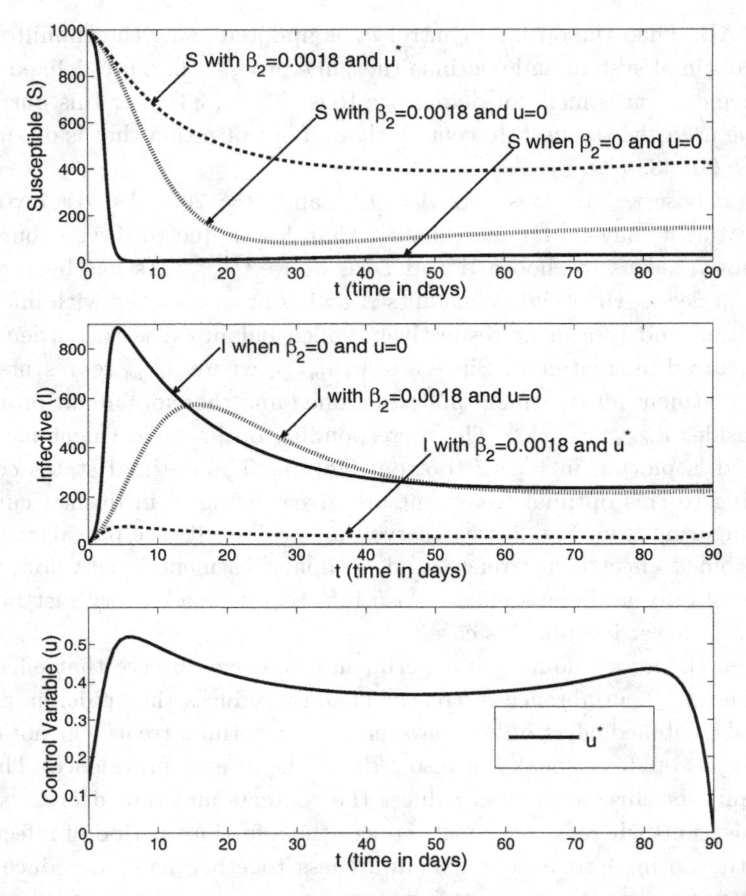

Figure 2. (Upper) Profiles of susceptible populations under various control scenario. (Middle) Profiles of infective populations under control scenario. (Bottom) Optimal path of treatment u^*. Parameters are as in Table 1.

the total cost via opportunity loss. Whereas effect of awareness alone also reduces economic burden significantly but not up to with combination of treatment.

In the following, we see the impact of various level of awareness effect on optimal treatment and infective population. As has been noted in Ref. 26, $\beta(I) = \beta_1 - \beta_2 \frac{I}{m+I}$ and $\frac{\partial \beta(I)}{\partial m} > 0$, thus disease transmission becomes lower by reducing m. We varied the saturation constant m from 0 to 10 and obtained its impact on populations and optimal treatment profile and cost functional. The corresponding population profiles are plotted in

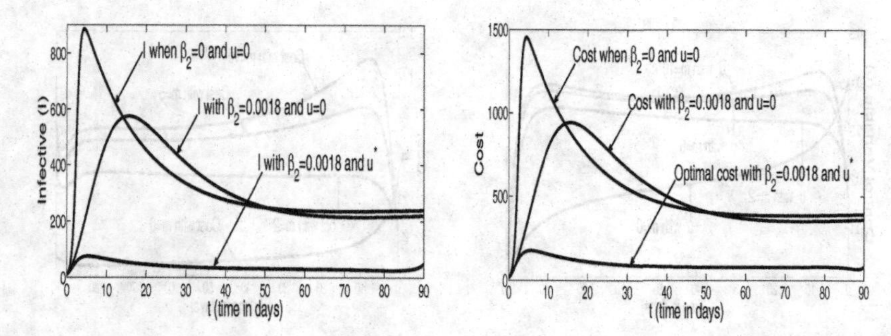

Figure 3. (a) Various profiles of infective. (b) Various profiles of cost. Other parameters are as in Table 1.

Fig. 4 and optimal treatment and cost profiles are plotted Fig. 5. No saturation in awareness effect *i.e.* $m = 0$ lowers the disease transmission and in this case the count of infective is found lower Fig. 4(b). We also observe that a small potential on optimal treatment is required for disease control along with minimal cost Fig. 5. Whereas for larger m, one has to employ higher efforts on treatment to lower the disease transmission and prevalence causing significantly higher cost.

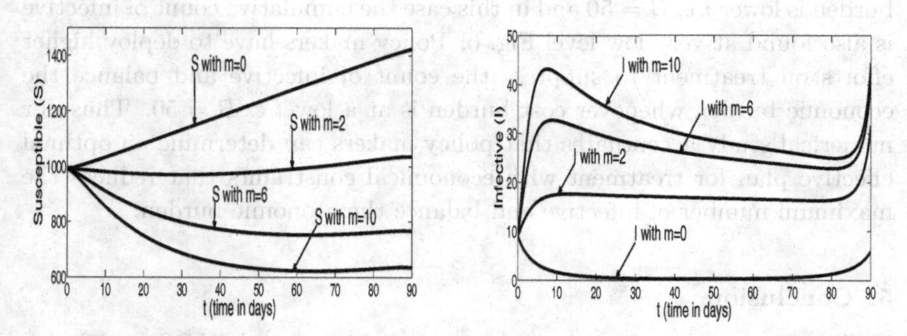

Figure 4. (a) Profiles of susceptible for various values of m under optimal treatment u^*. (b) Profiles of infective for various values of m under optimal treatment u^*. Other parameters are as in Table 1.

4.1. *Effect of Weight Constant on Optimal Control Problem*

Actual estimation of total cost associated due to disease burden and applied control interventions is one the critical issue for policy makers and health

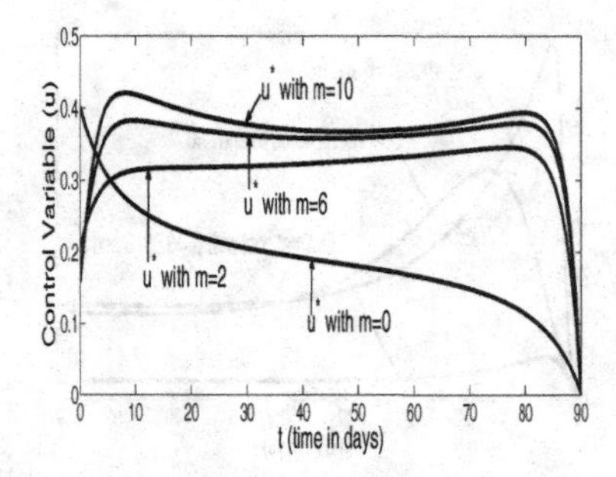

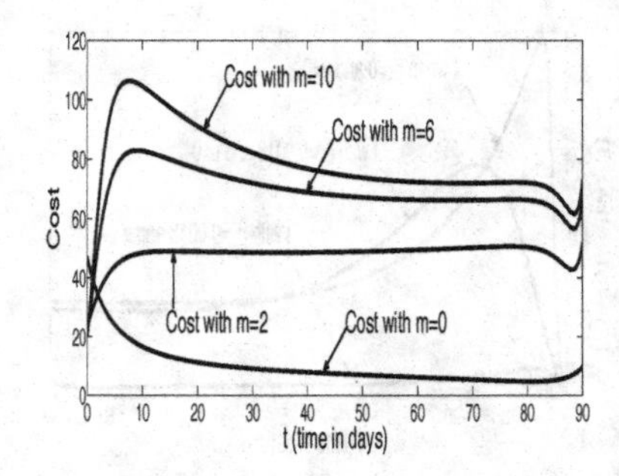

Figure 5. (a) Profiles of optimal treatment u^* for various values of m. (b) Profiles of cost for various values of m under optimal treatment u^*. Other parameters are as in Table 1.

agencies. To see the effect of variation of costs associated in treatment process, we vary the corresponding weight constant B from 50 to 1000 and observe the pattern of various treatment plans along with the corresponding prevalence of disease. The corresponding numerical results are given in Fig. 6 and Fig. 7. We note that as B increases the corresponding cost involved also increases and the count of infective also increases.

Count of susceptible population is found at a higher level when cost burden is lower *i.e.* $B = 50$ and in this case the cumulative count of infective is also found at very low level Fig. 6. Policy makers have to deploy higher efforts on treatment to suppress the count of infective and balance the economic burden whenever cost burden is at a low *i.e.* $B = 50$. Thus our numerical study accentuates that policy makers can determine an optimal effective plan for treatment with economical constraints that reduces the maximum number of infective and balance the economic burden.

5. Conclusion

In this paper we considered a nonlinear compartmental $SIRS$ model proposed by Liu *et al.*[26] which account for the impact of awareness effect in population. We analyzed and established the global stability of unique infected equilibrium by constructing a suitable Lyapunov function. As the basic reproduction number for the model system, $R_0 = \frac{r\beta_1}{d(d+\alpha+\gamma)}$ is independent of awareness parameter β_2, hence eradication of the disease can not be achieved through awareness only. Though the impact of awareness plays an important role in reducing the disease prevalence.

Further, by introducing treatment to infective, we formulate and study

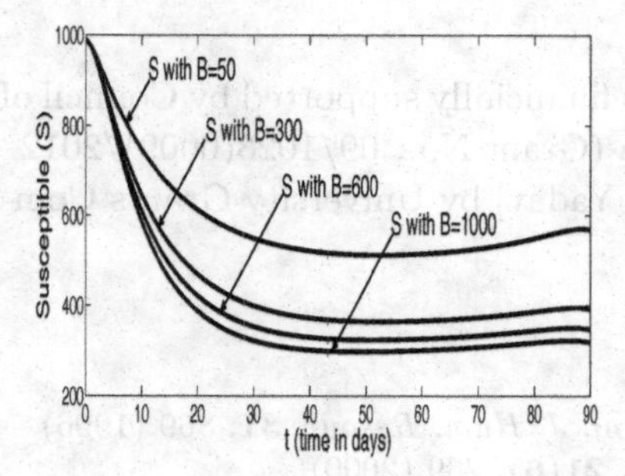

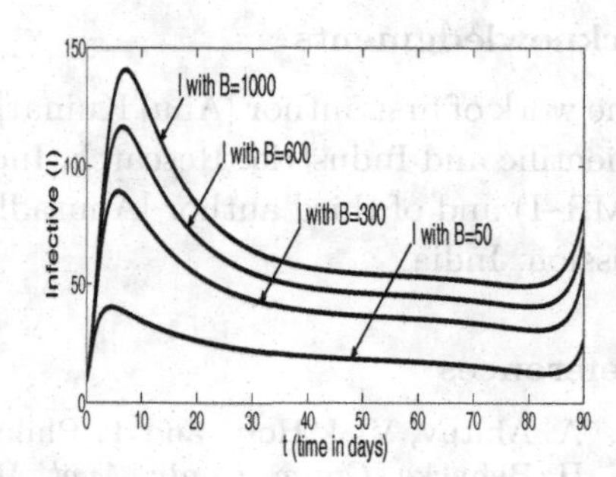

Figure 6. (a) Profiles of susceptible for various values of B under optimal treatment u^*. (b) Profiles of infective for various values of B under optimal treatment u^*. Other parameters are as in Table 1.

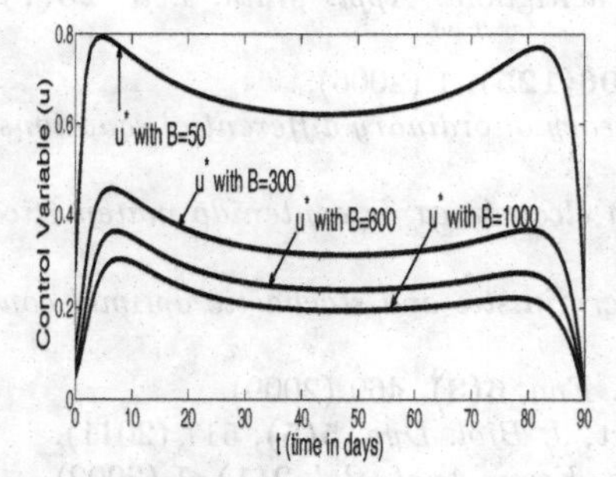

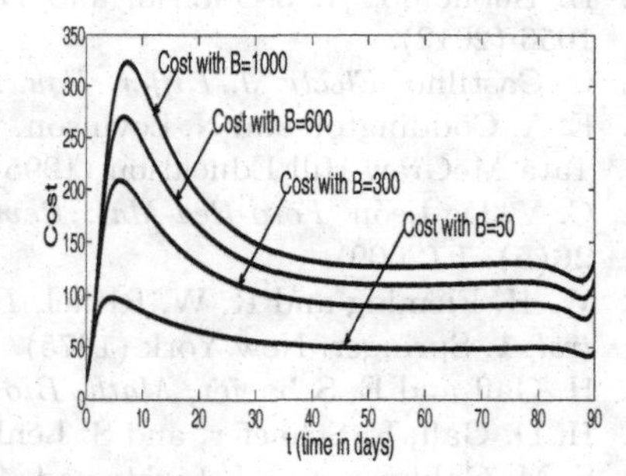

Figure 7. (a) Profiles of optimal treatment u^* for various values of B. (b) Profiles of cost for various values of B under optimal treatment u^*. Other parameters are as in Table 1.

corresponding optimal control problem to find optimal treatment policy for providing treatment with minimum cost. With the help of Pontryagin's Maximum Principle we obtained the optimal treatment path analytically for the finite time period. We performed numerical simulations to analyse the results obtained. It is observed that the awareness reduces the infective population during initial phase but is not effective if treatment is not available. While optimal treatment in aware population is very effective in keeping disease load significantly low throughout the period and reducing economical cost. Thus the combined impact of awareness and optimal treatment can bring down disease prevalence in population and also minimize cost incurred.

Acknowledgments

The work of first author [Anuj Kumar] is financially supported by Council of Scientific and Industrial Research, India (Grant No.: 09/1023(0009)/2012–EMR–I) and of third author [Anuradha Yadav] by University Grants Commission, India.

References

1. A. Ahituv, V. J. Hotz, and T. Philipson, *J. Hum. Resour.* **31**, 869 (1996).
2. H. Behncke, *Optim. Contr. Appl. Met.* **21(6)**, 269 (2000).
3. L. E. Bobisud, *Math. Biosci.* **35(1)**, 165 (1977).
4. F. Brauer and C. Castillo-Châavez, *Mathematical models in population biology and epidemiology*, **2**, Springer (2012).
5. B. Buonomo, A. d'Onofrio, and D. Lacitignola, *Math. Biosci.* **216(1)**, 9 (2008).
6. B. Buonomo, A. d'Onofrio, and D. Lacitignola, *Appl. Math. Lett.* **25(7)**, 1056 (2012).
7. C. Castilho, *Electr. J. Differ. Equ.* **2006(125)**, 1 (2006).
8. E. A. Coddington and N. Levinson, *Theory of ordinary differential equations*, Tata McGraw-Hill Education (1995).
9. C. V. De León, *Foro-Red-Mat: Revista electrónica de contenido matemático* **26(5)**, 1 (2009).
10. W. H. Fleming and R. W. Rishel, *Deterministic and stochastic optimal control*, **1**, Springer, New York (1975).
11. H. Gaff and E. Schaefer, *Math. Biosci. Eng.* **6(3)**, 469 (2009).
12. H. D. Gaff, E. Schaefer, and S. Lenhart, *J. Biol. Dyn.* **5(5)**, 517 (2011).
13. S. M. Goldman and J. Lightwood, *Top. Econ. Anal. Pol.* **2(1)**, 1 (2002).
14. D. Grass, *Optimal control of nonlinear processes: With applications in drugs, corruption and terror*, Springer (2008).
15. A. B. Gumel, S. Ruan *et al.*, *Proc. R. Soc. Lond., B, Biol. Sci.* **271(1554)**, 2223 (20004).
16. H. W. Hethcote, *SIAM Rev.* **42(4)**, 599 (2000).
17. H. R. Joshi, *Optim. Contr. Appl. Met.* **23(4)**, 199 (2002).
18. H. R. Joshi, S. Lenhart, M. Y. Li, and L. Wang, *Contemp. Math.* **410**, 187 (2006).
19. S. M. Kassa and A. Ouhinou, *J. Math. Biol.* **70(1-2)**, 213 (2015).
20. D. E. Kirk, *Optimal control theory: an introduction*, Dover Publications (2012).
21. D. Kirschner, S. Lenhart, and S. Serbin, *J. Math. Biol.* **35(7)**, 775 (1997).
22. Q. Kong, Z. Qiu, Z. Sang, and Y. Zou, *Math. Control Relat. Fields* **1(4)**, 493 (2011).
23. J. P. LaSalle, *The stability of dynamical systems, regional conference series in applied mathematics*, SIAM Philadelphia (1976).
24. S. M. Lenhart and J. T. Workman, *Optimal control applied to biological models*, **15**, CRC Press (2007).

25. X. Liu, Y. Takeuchi, and S. Iwami, *J. Theor. Biol.* **253(1)**, 1 (2008).
26. Y. Liu and J. Cui, *Int. J. Biomath.* **1(1)**, 65 (2008).
27. A. K. Misra, A. Sharma, and J. Li, *Discrete Continuous Dyn. Syst. Ser. B* **18(7)**, 1909 (2013).
28. A. K. Misra, A. Sharma, and J. B. Shukla, *Math. Comput. Model.* **53(3)**, 1221 (2011).
29. D. Moulay, M. A. Aziz-Alaoui, and H. Kwon, *Math. Biosci. Eng.* **9(2)**, 369 (2012).
30. A. S. Perelson and P. W. Nelson, *SIAM Rev.* **41(1)**, 3 (1999).
31. T. Philipson, *J. Hum. Resour.* **31**, 611 (1996).
32. J. M. Tchuenche, N. Dube, C. P. Bhunu, C. T. Bauch *et al.*, *BMC Public Health* **11(Suppl 1)**, S5 (2011).
33. X. Wang, *Solving optimal control problems with MATLAB: Indirect methods*, Technical report, ISE Dept., NCSU (2009).
34. K. H. Wickwire, *Math. Biosci.* **30(1)**, 129 (1976).
35. S. Xia and J. Liu, *J. R. Soc. Interface* **11(94)**, 20140013 (2014).
36. G. Zaman, Y. H. Kang, and I. H. Jung, *Biosystems* **93(3)**, 240 (2008).
37. I. Zeiler, J. P. Caulkins, D. Grass, and G. Tragler, *Siam. J. Control. Optim.* **48(6)**, 3698 (2010).

GLOBAL STABILITY OF A MODIFIED HIV INFECTION MODEL WITH SATURATION INCIDENCE*

JUNXIAN YANG
School of Science
Anhui Agricultural University
Hefei 230036, PR China
E-mail: yangjunxian1976@126.com

QINGGUO ZHANG
School of Science
Anhui Agricultural University
Hefei 230036, PR China
E-mail: qgzhang@ahau.edu.cn

LEIHONG WANG
School of Forestry and Landscape Architecture
Anhui Agricultural University
Hefei 230036, PR China
E-mail: 75455127@qq.com

In this paper, a modified HIV infection model with saturation incidence is investigated. By analyzing the corresponding characteristic equations, the local stability of an infection-free equilibrium and an endemic infection equilibrium are discussed. By using suitable Lyapunov functions and the LaSalle invariant principle, it is proved that if the basic reproductive number is less than 1, the infection-free equilibrium is globally asymptotically stable. If the basic reproductive number is larger than 1, by means of the second additive compound matrix, the globally asymptotical stability of the endemic equilibrium is obtained. Numerical simulations are carried out to illustrate the main theoretical results.

*This work was supported by the National Natural Science Foundation of China (No. 11201002)

1. Introduction

HIV is the human immunodeficiency virus that targets the CD4+ T lymphocytes, which are the most abundant white blood cells of the immune system. Chronic HIV infection wreaks the serious damage on the CD4+ T cells by causing their decline and destruction, and leads to Acquired Immunodeficiency Syndrome (AIDS) [1].

Mathematical modeling, especial with saturated infection rate, has become an important tool to analyze the spread and control of infection diseases. Since the early 1980s there has been tremendous effort made in the mathematical modelling of HIV/AIDS [2-4]. In recent years, the study of HIV/AIDS has made greater progress [5-9]. The development of such models is aimed at both better understanding of the observed epidemiological patterns and providing short and long term prediction of HIV/AIDS incidence. The basic mathematical model widely used for studying the dynamics of HIV infection has the following form [10, 11]:

$$\begin{cases} \dot{T}(t) = A - d_1T(t) - \beta T(t)V(t), \\ \dot{I}(t) = \beta T(t)V(t) - d_2I(t), \\ \dot{V}(t) = kI(t) - d_3V(t), \end{cases} \tag{1}$$

where $T(t), I(t), V(t)$ are the number of uninfected cells, infected cells, and free virus at time t, respectively. The constant $A(A > 0)$ is the rate at which new target cells are generated. The constants d_1, d_2, d_3 are the death rate of uninfected cells, infected cells, free virus, respectively. The constant β is the rate describing the infection process, and infected cells are produced at rate $\beta T(t)V(t)$ at time t. Free virus is produced from infected cells at rate $kI(t)$ at time t.

In the 1990s, there has been a discussion about the process of the HIV RNA transcribing into DNA: when an HIV enters a resting CD4+ T cell, the HIV RNA may not be completely reverse transcribed into DNA [12]. A proportion of resting infected cells can revert to the uninfected state before the viral genome is integrated into the genome of the lymphocyte [13]. Recently, some mathematical models of HIV infection have been proposed based on the assumption that a fraction of infected CD4 + T cells return to the uninfected class [14-16]. Srivastava and Chandra [15] have considered a model with three populations:

$$\begin{cases} \dot{T}(t) = A - d_1T(t) - \beta T(t)V(t) + pI(t), \\ \dot{I}(t) = \beta T(t)V(t) - d_2I(t) - pI(t), \\ \dot{V}(t) = kI(t) - d_3V(t), \end{cases} \tag{2}$$

where the meanings of the variables $T(t), I(t), V(t)$ and the parameters $A, d_1, d_2, d_3, \beta, k$ are the same as those given in (1). The term $pI(t)$ is the rate of infected cells in the latent stage reverting to the uninfected class.

In (1) and (2), the rate of HIV infection is assumed to be bilinear by the term $\beta T(t)V(t)$. However, the actual incidence rate is probably not linear over the entire range of virus $V(t)$ and uninfected cells $T(t)$. Qilin Sun and Lequan Min [17] have considered the following model:

$$\begin{cases} \dot{T}(t) = A - d_1 T(t) - \dfrac{\beta T(t)V(t)}{T(t)+V(t)} + pI(t), \\ \dot{I}(t) = \dfrac{\beta T(t)V(t)}{T(t)+V(t)} - d_2 I(t) - pI(t), \\ \dot{V}(t) = kI(t) - d_3 V(t), \end{cases} \tag{3}$$

In this model, they have used a saturated infection rate $\dfrac{\beta T(t)V(t)}{T(t)+V(t)}$ to replace the mass action term $\beta T(t)V(t)$ in (2). At this time, the actual incidence rate is not linear over the entire range of virus $V(t)$ and uninfected cells $T(t)$ any more.

In this paper, we consider a modified HIV-1 infection model with saturated infection rate $\dfrac{\beta T(t)V(t)}{1+\alpha V(t)}$:

$$\begin{cases} \dot{T}(t) = A - d_1 T(t) - \dfrac{\beta T(t)V(t)}{1+\alpha V(t)} + pI(t), \\ \dot{I}(t) = \dfrac{\beta T(t)V(t)}{1+\alpha V(t)} - d_2 I(t) - pI(t), \\ \dot{V}(t) = kI(t) - d_3 V(t), \end{cases} \tag{4}$$

We assume that constant $A, d_1, d_2, d_3, \beta, p, k, \alpha$ are positive and $d_1 \leqslant d_2$.

It is easy to show that the solutions of (4) with initial conditions $T(0) > 0, I(0) > 0$,and $V(0) > 0$ have all positive components for $t > 0$. Hence, we begin the analysis of (4) by observing the nonnegative octant

$$D = \{(T(t), I(t), V(t)) \in R^3_{+0} | t \geqslant 0\},$$

where $R^3_{+0} = \{(x_1, x_2, x_3) | x_i \geqslant 0, i = 1, 2, 3\}$.

According to the first two equations of (4), we can get

$$\dot{T}(t) + \dot{I}(t) = A - d_1 T(t) - d_2 I(t) \leqslant A - d_1(T(t) + I(t)),$$

and then

$$T(t) + I(t) \leqslant \frac{A}{d_1}. \tag{5}$$

So $T(t)$ and $I(t)$ are bounded. From the last equation of (4), it follows that

$$\dot{V}(T) = kI(t) - d_3V(t) \leqslant \frac{kA}{d_1} - d_3V(t),$$

and then

$$V(t) \leqslant \frac{kA}{d_1 d_3}. \tag{6}$$

So $V(t)$ are bounded. Hence there is a bounded set

$$\Omega = \{(T(t), I(t), V(t)) | 0 \leqslant T(t) + I(t) \leqslant \frac{A}{d_1}, 0 \leqslant V(t) \leqslant \frac{kA}{d_1 d_3}\}, \tag{7}$$

such that any solution trajectory $(T(t), I(t), V(t))$ of (4) with initial value $(T(0), I(0), V(0))$ will keep in the set Ω.

In this paper, our primary goal is to carry out a complete mathematical analysis of system (4) and establish its global dynamics. The organization of this paper is as follows. In the section 2, by analyzing the corresponding characteristic equations, the local asymptotic stability of an infection-free equilibrium and an endemic infection equilibrium of model (4) were studied. In the section 3, the global stability of the infection-free equilibrium and the endemic infection equilibrium were discussed by means of suitable Lyapunov functional and LaSalle's invariant principle, respectively. In the section 4, some numerical simulations are performed to illustrate the analytical results. In the last section, a brief discussion and the summary of the main results were provided.

2. Local Asymptotical Stability of Equilibrium Point

In this section, we study the local stability of each equilibrium point of system (4). Clearly, system (4) always has an infection-free equilibrium point $E_0(T_0, 0, 0)$, where $T_0 = \dfrac{A}{d_1}$.

Denote

$$R_0 = \frac{A\beta k}{d_1(d_2 + p)d_3}.$$

Here, R_0 is called the basic reproduction number of system (4). It is easy to show that if $R_0 > 1$, system (4) has a unique endemic infection equilibrium point $E^*(T^*, I^*, V^*)$, where

$$T^* = \frac{(d_2 + p)d_3(1 + \alpha V^*)}{\beta k}, \ \ I^* = \frac{d_3 V^*}{k}, \ \ V^* = \frac{d_1(d_2 + p)}{d_1(d_2 + p)\alpha + \beta d_2}(R_0 - 1).$$

According to $T^* > 0, I^* > 0, V^* > 0$, we can get that the endemic infection equilibrium point $E^*(T^*, I^*, V^*)$ exists in the interior of Ω:

$$\begin{aligned}\Omega^0 = \{(T, I, V) \in \Omega | T > 0, I > 0, V > 0, \\ T + I < \frac{A}{d_1}, V < \frac{kA}{d_1 d_3}\},\end{aligned} \tag{8}$$

Therefore, the stability of $E^*(T^*, I^*, V^*)$ only needs to be discussed in Ω^0.

2.1. *Locally asymptotical stability of the infection-free equilibrium point* E_0

Theorem 2.1 *If* $R_0 < 1$, *then the infection-free equilibrium point* $E_0(T_0, 0, 0)$ *of system* (4) *is locally asymptotically stable; If* $R_0 > 1$, *then* $E_0(T_0, 0, 0)$ *is unstable.*

Proof The Jacobian matrix of system (4) at an arbitrary point is given by

$$J(T, I, V) = \begin{pmatrix} -d_1 - \dfrac{\beta V(t)}{1 + \alpha V(t)} & p & -\dfrac{\beta T(t)}{(1 + \alpha V(t))^2} \\ \dfrac{\beta V(t)}{1 + \alpha V(t)} & -(d_2 + p) & \dfrac{\beta T(t)}{(1 + \alpha V(t))^2} \\ 0 & k & -d_3 \end{pmatrix} \tag{9}$$

Then

$$J(E_0) = \begin{pmatrix} -d_1 & p & -\dfrac{\beta A}{d_1} \\ 0 & -(d_2 + p) & \dfrac{\beta A}{d_1} \\ 0 & k & -d_3 \end{pmatrix} \tag{10}$$

The corresponding characteristic equation of matrix $J(E_0)$ is

$$|\lambda E - J(E_0)| = (\lambda + d_1)\left[(\lambda + d_2 + p)(\lambda + d_3) - \frac{A\beta k}{d_1}\right] = 0. \tag{11}$$

Clearly, Equation (11) always has a negative real root $\lambda_1 = -d_1$. Other roots of (11) are determined by the following equation

$$(\lambda + d_2 + p)(\lambda + d_3) - \frac{A\beta k}{d_1} = 0. \tag{12}$$

Equation (12) can be written as

$$\lambda^2 + (d_2 + p + d_3)\lambda + (d_2 + p)d_3 - \frac{A\beta k}{d_1} = 0. \tag{13}$$

Clearly, $d_2 + p + d_3 > 0$.

If $R_0 < 1, (d_2+p)d_3 - \dfrac{A\beta k}{d_1} > 0$, all roots of equation (13) are negative real roots, then the infection-free equilibrium point E_0 is locally asymptotically stable.

If $R_0 > 1, (d_2+p)d_3 - \dfrac{A\beta k}{d_1} < 0$, equation (13) has one positive root, hence, the infection-free equilibrium point E_0 is unstable. The proof of Theorem 2.1 is completed.

2.2. *Locally asymptotical stability of the endemic infection equilibrium point $E^*(T^*, I^*, V^*)$*

Theorem 2.2 *If $R_0 > 1$, then the endemic infection equilibrium point $E^*(T^*, I^*, V^*)$ is locally asymptotically stable.*

Proof The Jacobian matrix of system (4) at $E^*(T^*, I^*, V^*)$ is given by

$$J(E^*) = \begin{pmatrix} -d_1 - \dfrac{\beta V^*}{1+\alpha V^*} & p & -\dfrac{\beta T^*}{(1+\alpha V^*)^2} \\ \dfrac{\beta V^*}{1+\alpha V^*} & -(d_2+p) & \dfrac{\beta T^*}{(1+\alpha V^*)^2} \\ 0 & k & -d_3 \end{pmatrix} \tag{14}$$

The corresponding characteristic equation of matrix $J(E^*)$ is

$$\begin{aligned} |\lambda E - J(E^*)| = & \left(\lambda + d_1 + \frac{\beta V^*}{1+\alpha V^*}\right)(\lambda + d_2 + p)(\lambda + d_3) \\ & + \frac{k\beta T^*}{(1+\alpha V^*)^2} \cdot \frac{\beta V^*}{1+\alpha V^*} - \frac{p\beta V^*}{1+\alpha V^*}(\lambda + d_3) \\ & - \left(\lambda + d_1 + \frac{\beta V^*}{1+\alpha V^*}\right)\frac{k\beta T^*}{(1+\alpha V^*)^2} = 0. \end{aligned} \tag{15}$$

Reorganizing equation (15), here is

$$\lambda^3 + a_1\lambda^2 + a_2\lambda + a_3 = 0,$$

where

$$a_1 = d_1 + \frac{\beta V^*}{1+\alpha V^*} + d_2 + p + d_3 > 0,$$

$$\begin{aligned} a_2 &= \left(d_1 + \frac{\beta V^*}{1+\alpha V^*}\right)(d_2+p+d_3) + (d_2+p)d_3 \\ &\quad - \frac{p\beta V^*}{1+\alpha V^*} - \frac{k\beta T^*}{(1+\alpha V^*)^2} \\ &= d_1(d_2+p+d_3) + \frac{\beta V^*}{1+\alpha V^*}(d_2+d_3) + (d_2+p)d_3 - \frac{(d_2+p)d_3}{1+\alpha V^*} \\ &= d_1(d_2+p+d_3) + \frac{\beta V^*}{1+\alpha V^*}(d_2+d_3) + \frac{\alpha V^*(d_2+p)d_3}{1+\alpha V^*} > 0, \end{aligned}$$

$$\begin{aligned} a_3 &= \left(d_1 + \frac{\beta V^*}{1+\alpha V^*}\right)(d_2+p)d_3 + \frac{k\beta T^*}{(1+\alpha V^*)^2} \cdot \frac{\beta V^*}{1+\alpha V^*} - \frac{p\beta V^* d_3}{1+\alpha V^*} \\ &\quad - \left(d_1 + \frac{\beta V^*}{1+\alpha V^*}\right)\frac{k\beta T^*}{(1+\alpha V^*)^2} \\ &= \left(d_1 + \frac{\beta V^*}{1+\alpha V^*}\right)(d_2+p)d_3 - \frac{p\beta V^* d_3}{1+\alpha V^*} - \frac{d_1 k\beta T^*}{(1+\alpha V^*)^2} \\ &= d_1(d_2+p)d_3 + \frac{\beta V^* d_2 d_3}{1+\alpha V^*} - \frac{d_1(d_2+p)d_3}{1+\alpha V^*} \\ &= \frac{\alpha V^* d_1(d_2+p)d_3}{1+\alpha V^*} + \frac{\beta V^* d_2 d_3}{1+\alpha V^*} > 0, \end{aligned}$$

$$\begin{aligned} a_1 a_2 - a_3 &= \left(d_1 + \frac{\beta V^*}{1+\alpha V^*} + d_2 + p + d_3\right) \\ &\quad \cdot \left[d_1(d_2+p+d_3) + \frac{\beta V^*}{1+\alpha V^*}(d_2+d_3) + \frac{\alpha V^*(d_2+p)d_3}{1+\alpha V^*}\right] \\ &\quad - \frac{\alpha V^* d_1(d_2+p)d_3}{1+\alpha V^*} - \frac{\beta V^* d_2 d_3}{1+\alpha V^*} > 0. \end{aligned}$$

By the Routh-Hurwitz criterion, the endemic infection equilibrium point $E^*(T^*, I^*, V^*)$ is locally asymptotically stable.

3. Global Asymptotical Stability of Equilibrium Point

3.1. *Globally asymptotical stability of the infection-free equilibrium point E_0*

Theorem 3.1 *If $R_0 < 1$, then the infection-free equilibrium $E_0(T_0, 0, 0)$ of system (4) is globally asymptotically stable in Ω.*

Proof Let $(T(t), I(t), V(t))$ be any positive solution of system (4). Define

$$V_1(t) = I(t) + \frac{(d_2+p)}{k}V(t). \tag{16}$$

Calculating the derivative of $V_1(t)$ along the positive solutions of system (4), it follows that

$$\begin{aligned}
\dot{V}_1(t) &= \dot{I}(t) + \frac{d_2+p}{k}\dot{V}(t) \\
&= \frac{\beta T(t)V(t)}{1+\alpha V(t)} - d_2 I(t) - pI(t) + \frac{d_2+p}{k}(kI(t) - d_3 V(t)) \\
&= \left(\frac{\beta T(t)}{1+\alpha V(t)} - \frac{(d_2+p)d_3}{k}\right)V(t) \\
&\leqslant \left(\frac{A\beta}{d_1} - \frac{(d_2+p)d_3}{k}\right)V(t) \\
&= \frac{(d_2+p)d_3}{k}(R_0-1)V(t).
\end{aligned}$$

If $R_0 < 1$, then $\dot{V}_1(t) \leqslant 0$ holds in Ω. Moreover, $\dot{V}_1(t) = 0$ if and only if $V(t) = 0$. Hence, the largest compact invariant set in Ω is

$$M_1 = \{(T, I, V) \in \Omega | V(t) = 0\}. \tag{17}$$

By the LaSalle's invariance principle, $\lim\limits_{t\to+\infty} V(t) = 0$. Then we can get limit equations:

$$\begin{cases} \dot{T}(t) = A - d_1 T(t) + pI(t), \\ \dot{I}(t) = -(d_2+p)I(t). \end{cases} \tag{18}$$

Define

$$V_2(t) = T(t) - T_0 - T_0 \ln\frac{T(t)}{T_0} + I(t), \tag{19}$$

where $T_0 = \dfrac{A}{d_1}$.

Calculating the derivative of $V_2(t)$ along the positive solutions of system

(18), it follows that

$$
\begin{aligned}
\dot{V}_2(t) &= \left(1 - \frac{T_0}{T(t)}\right)\dot{T}(t) + \dot{I}(t) \\
&= \left(1 - \frac{T_0}{T(t)}\right)(A - d_1T(t) + pI(t)) - (d_2 + p)I(t) \\
&= A - d_1T(t) - \frac{T_0A}{T(t)} + d_1T_0 - \frac{T_0pI(t)}{T(t)} - d_2I(t) \\
&= d_1T_0 - d_1T(t) - \frac{d_1T_0^2}{T(t)} + d_1T_0 - \frac{T_0pI(t)}{T(t)} - d_2I(t) \\
&= d_1T_0\left(2 - \frac{T(t)}{T_0} - \frac{T_0}{T(t)}\right) - \left(\frac{T_0p}{T(t)} + d_2\right)I(t) \\
&\leqslant -\left(\frac{T_0p}{T(t)} + d_2\right)I(t),
\end{aligned}
\tag{20}
$$

here, we use inequality $\frac{T(t)}{T_0} + \frac{T_0}{T(t)} \geqslant 2$, and the equality is established if and only if $T(t) = T_0$. Then it follows from (20) that $\dot{V}_2(t) \leqslant 0$ holds in M_1. By Theorem 5.3.1 in [18], the solutions limit to M_2, the largest invariant subset of $\{\dot{V}_2(t) = 0\}$. Clearly, it follows from (20) that $\{\dot{V}_2(t) = 0\}$ if and only if $T(t) = T_0, I(t) = 0$. Accordingly, the global asymptotic stability of $E_0(T_0, 0, 0)$ follows from LaSalle's invariance principle. This completes the proof.

3.2. *Globally asymptotical stability of the endemic infection equilibrium point E^**

In this subsection, we firstly introduce a lemma outlined by Li and Wang [19]. The lemma is briefly summarized as follows:

Let $x \mapsto f(x) \in R^n$ be a C^1 function for x in an open set $\Gamma \subset R^n$. Consider the differential equation

$$\dot{x} = f(x) \tag{21}$$

Denote by $x(t, x_0)$ the solution to (21) such that $x(0, x_0) = x_0$. Let $\bar{x}$ be an equilibrium point of (21). Li and Wang [19] have made the following two basic assumptions:

(H_1) There exists a compact absorbing set $K \subset \Gamma$;

(H_2) Equation (21) has a unique equilibrium $\bar{x}$ in Γ.

They (Theorem 2.5 in [19]) have given the following result.

Lemma 3.1 (see [19]). *Assume that*

(1) *assumptions* (H_1) *and* (H_2) *hold;*

(2) *equation* (21) *satisfies the Poincaré – Bendixson Property;*

(3) *for each periodic solution* $x = p(t)$ *to* (21) *with* $p(0) \in \Gamma$, *the linear system (the second additive compound system)*

$$\dot{z}(t) = \frac{\partial f^{[2]}}{\partial x}(p(t))z(t), \tag{22}$$

is asymptotically stable, where $\frac{\partial f^{[2]}}{\partial x}$ *is the second additive compound matrix of the Jacobian matrix* $\frac{\partial f}{\partial x}$ *of* f;

(4) $(-1)^n \det\left(\frac{\partial f}{\partial x}(\bar{x})\right) > 0.$

Then the unique equilibrium $\bar{x}$ *is globally asymptotically stable in* Γ.

Theorem 3.2 *If* $R_0 > 1$, *then the endemic infection equilibrium point* $E^*(T^*, I^*, V^*)$ *of* (4) *is globally asymptotically stable in* Ω^0.

Proof Based on Lemma, the proof of Theorem 3.2 has been implemented by the following four steps:

(1) (H_1) is equivalent to the uniform persistence of the system (4)[20]. By (8), Ω^0 is bounded, so the necessary and sufficient condition for the uniform persistence of (4) is equivalent to the equilibrium point E_0 being unstable [21]. Theorem 2.1 has shown that E_0 is unstable if $R_0 > 1$. Therefore, the system (4) is uniformly persistent if $R_0 > 1$ so that (H_1) holds if $R_0 > 1$.

Meanwhile, because $E^*(T^*, I^*, V^*)$ is the unique equilibrium point of the system (4) in Ω^0, so that (H_2) holds.

The results above verify the condition (1) of Lemma 3.1.

(2) By (9), the Jacobian matrix of (4) is

$$J(T,I,V) = \begin{pmatrix} -d_1 - \dfrac{\beta V(t)}{1+\alpha V(t)} & p & -\dfrac{\beta T(t)}{(1+\alpha V(t))^2} \\ \dfrac{\beta V(t)}{1+\alpha V(t)} & -(d_2+p) & \dfrac{\beta T(t)}{(1+\alpha V(t))^2} \\ 0 & k & -d_3 \end{pmatrix}$$

If $H = \mathrm{diag}(1, -1, 1)$, then

$$HJH = \begin{pmatrix} -d_1 - \dfrac{\beta V(t)}{1+\alpha V(t)} & -p & -\dfrac{\beta T(t)}{(1+\alpha V(t))^2} \\ -\dfrac{\beta V(t)}{1+\alpha V(t)} & -(d_2+p) & -\dfrac{\beta T(t)}{(1+\alpha V(t))^2} \\ 0 & -k & -d_3 \end{pmatrix}$$

and we can obtain that HJH has non-positive off-diagonal elements in Ω^0. Therefore (4) satisfies the *Poincaré – Bendixson* property. This verifies condition (2) of Lemma 3.1.

(3) Let $(T(t), I(t), V(t))$ be a periodic solution in Ω^0. The second additive compound matrix of the Jacobian matrix of (4) is given by

$$J^{[2]} = \begin{pmatrix} -(d_1+d_2+p) - \dfrac{\beta V(t)}{1+\alpha V(t)} & \dfrac{\beta T(t)}{(1+\alpha V(t))^2} & \dfrac{\beta T(t)}{(1+\alpha V(t))^2} \\ k & -(d_1+d_3) - \dfrac{\beta V(t)}{1+\alpha V(t)} & p \\ 0 & \dfrac{\beta V(t)}{1+\alpha V(t)} & -(d_2+p+d_3) \end{pmatrix} \tag{23}$$

and the second additive compound system of (4) along the periodic solution $(T(t), I(t), V(t))$ is

$$\begin{aligned} \dot{\omega}_1 &= \left(-d_1 - d_2 - p - \frac{\beta V(t)}{1+\alpha V(t)}\right)\omega_1 + \frac{\beta T(t)}{(1+\alpha V(t))^2}\omega_2 + \frac{\beta T(t)}{(1+\alpha V(t))^2}\omega_3, \\ \dot{\omega}_2 &= k\omega_1 + \left(-d_1 - d_3 - \frac{\beta V(t)}{1+\alpha V(t)}\right)\omega_2 + p\omega_3, \\ \dot{\omega}_3 &= \frac{\beta V(t)}{1+\alpha V(t)}\omega_2 - (d_2 + p + d_3)\omega_3. \end{aligned} \tag{24}$$

Define a global Lyapunov function by

$$V_3(\omega_1, \omega_2, \omega_3, T(t), I(t), V(t)) = \left\| \left(\omega_1, \frac{I(t)}{V(t)}\omega_2, \frac{I(t)}{V(t)}\omega_3\right) \right\|,$$

where $\|\cdot\|$ is the norm in set D defined by

$$\|(\omega_1, \omega_2, \omega_3)\| = \sup\{|\omega_1|, |\omega_2| + |\omega_3|\}.$$

Along a solution $(\omega_1, \omega_2, \omega_3)$ of (24), $V_3(\omega_1, \omega_2, \omega_3, T(t), I(t), V(t))$ becomes

$$V_3(\omega_1, \omega_2, \omega_3, T(t), I(t), V(t)) = \sup\left\{|\omega_1|, \frac{I(t)}{V(t)}(|\omega_2| + |\omega_3|)\right\}.$$

The orbit of the periodic solution $(T(t), I(t), V(t))$ is at a positive distance from the boundary of Ω by the uniform persistence, thus there exists a constant $\mu > 0$ such that $T(t) > \mu, I(t) > \mu, V(t) > \mu$ for large enough t.

Since $V(t) < \dfrac{kA}{d_1 d_3}$ by (8),

$$V_3(\omega_1, \omega_2, \omega_3, T(t), I(t), V(t)) \geqslant \sup\left\{|\omega_1|, \frac{\mu d_1 d_3}{kA}(|\omega_2| + |\omega_3|)\right\}. \tag{25}$$

for all $(\omega_1, \omega_2, \omega_3) \in R^3$.

The right derivative of V_3 along a solution $(\omega_1, \omega_2, \omega_3)$ to (24) and $(T(t), I(t), V(t))$ can be estimated as follows:

$$\begin{aligned} D_+|\omega_1| &\leqslant \left(-d_1 - d_2 - p - \frac{\beta V(t)}{1+\alpha V(t)}\right)|\omega_1| \\ &\quad + \frac{\beta T(t)}{(1+\alpha V(t))^2}(|\omega_2| + |\omega_3|), \\ D_+|\omega_2| &\leqslant k|\omega_1| + \left(-d_1 - d_3 - \frac{\beta V(t)}{1+\alpha V(t)}\right)|\omega_2| + p|\omega_3|, \\ D_+|\omega_3| &\leqslant \frac{\beta V(t)}{1+\alpha V(t)}|\omega_2| - (d_2 + p + d_3)|\omega_3|. \end{aligned} \tag{26}$$

Therefore

$$\begin{aligned} & D_+\frac{I(t)}{V(t)}(|\omega_2| + |\omega_3|) \\ &= \frac{\dot{I}(t)V(t) - I(t)\dot{V}(t)}{V^2(t)}(|\omega_2| + |\omega_3|) + \frac{I(t)}{V(t)}D_+(|\omega_2| + |\omega_3|) \\ &\leqslant \frac{I(t)}{V(t)}\left(\frac{\dot{I}(t)}{I(t)} - \frac{\dot{V}(t)}{V(t)}\right)(|\omega_2| + |\omega_3|) \\ &\quad + \frac{I(t)}{V(t)}[k|\omega_1| - (d_1 + d_3)|\omega_2| - (d_2 + d_3)|\omega_3|] \\ &\leqslant \frac{I(t)}{V(t)}k|\omega_1| + \left(\frac{\dot{I}(t)}{I(t)} - \frac{\dot{V}(t)}{V(t)} - d_1 - d_3\right)\frac{I(t)}{V(t)}(|\omega_2| + |\omega_3|), \end{aligned} \tag{27}$$

here, we use conditions $d_1 \leqslant d_2$.

Relations (26) and (27) lead to

$$D_+V_3(t) \leqslant \sup\{g_1(t), g_2(t)\}V_3(t), \tag{28}$$

where

$$\begin{aligned} g_1(t) &= -d_1 - d_2 - p - \frac{\beta V(t)}{1+\alpha V(t)} + \frac{\beta T(t)V(t)}{I(t)(1+\alpha V(t))^2} \\ &\leqslant -d_1 - d_2 - p + \frac{\beta T(t)V(t)}{I(t)(1+\alpha V(t))}, \\ g_2(t) &= \frac{kI(t)}{V(t)} + \frac{\dot{I}(t)}{I(t)} - \frac{\dot{V}(t)}{V(t)} - d_1 - d_3. \end{aligned} \tag{29}$$

Rewriting (4), we find that

$$\frac{\beta T(t)V(t)}{I(t)(1+\alpha V(t))} = \frac{\dot{I}(t)}{I(t)} + d_2 + p, \quad \frac{kI(t)}{V(t)} = \frac{\dot{V}(t)}{V(t)} + d_3. \tag{30}$$

Combining (29) and (30), it follows that

$$\sup\{g_1(t), g_2(t)\} \leqslant \frac{\dot{I}(t)}{I(t)} - d_1. \tag{31}$$

By (28),(31) and Gronwall's inequality, we obtains

$$V_3(t) \leqslant \frac{V_3(0)}{I(0)} I(t) e^{-d_1 t} \leqslant \frac{V_3(0)}{I(0)} \frac{A}{d_1} e^{-d_1 t}. \tag{32}$$

It shows that $V_3(t) \to 0$ as $t \to +\infty$, and in turn that $(\omega_1, \omega_2, \omega_3) \to 0$ as $t \to +\infty$ by (25). The second additive compound system is asymptotically stable. This verifies the condition (3) of Lemma 3.1.

(4) The Jacobi matrix of system (4) at the endemic infection equilibrium $E^*(T^*, I^*, V^*)$ is

$$J(E^*) = \begin{pmatrix} -d_1 - \dfrac{\beta V^*}{1+\alpha V^*} & p & -\dfrac{\beta T^*}{(1+\alpha V^*)^2} \\ \dfrac{\beta V^*}{1+\alpha V^*} & -(d_2+p) & \dfrac{\beta T^*}{(1+\alpha V^*)^2} \\ 0 & k & -d_3 \end{pmatrix}$$

and then

$$\begin{aligned} \det(J(E^*)) &= \begin{vmatrix} -d_1 - \dfrac{\beta V^*}{1+\alpha V^*} & p & -\dfrac{\beta T^*}{(1+\alpha V^*)^2} \\ \dfrac{\beta V^*}{1+\alpha V^*} & -(d_2+p) & \dfrac{\beta T^*}{(1+\alpha V^*)^2} \\ 0 & k & -d_3 \end{vmatrix} \\ &= -\left(d_1 + \frac{\beta V^*}{1+\alpha V^*}\right)(d_2+p)d_3 - \frac{k\beta^2 T^* V^*}{(1+\alpha V^*)^3} + \frac{d_3 p \beta V^*}{1+\alpha V^*} \\ &\quad + \frac{k\beta T^*}{(1+\alpha V^*)^2}\left(d_1 + \frac{\beta V^*}{1+\alpha V^*}\right) \\ &= -d_1(d_2+p)d_3 - \frac{\beta V^*}{1+\alpha V^*}(d_2+p)d_3 + \frac{d_3 p \beta V^*}{1+\alpha V^*} + \frac{d_1 k \beta T^*}{(1+\alpha V^*)^2} \\ &= \frac{d_1 k \beta T^*}{(1+\alpha V^*)^2} - d_1(d_2+p)d_3 - \frac{d_2 d_3 \beta V^*}{1+\alpha V^*} \\ &= \frac{d_1(d_2+p)d_3}{1+\alpha V^*} - d_1(d_2+p)d_3 - \frac{d_2 d_3 \beta V^*}{1+\alpha V^*} \\ &= \left(\frac{1}{1+\alpha V^*} - 1\right) d_1(d_2+p)d_3 - \frac{d_2 d_3 \beta V^*}{1+\alpha V^*} < 0. \end{aligned}$$

Since $J(E^*)$ is a 3×3 matrix, we gets $n = 3$, and then

$$(-1)^3 |J(E^*)| = -|J(E^*)| > 0.$$

This verifies condition (4) of Lemma 3.1.

Hence, if $R_0 > 1$, then the endemic infection equilibrium point $E^*(E^*, T^*, I^*)$ is globally asymptotically stable in Ω^0 by Lemma 3.1.

4. Numerical Simulation

In this section, we perform numerical simulation to support the theoretical analysis. In system (4), let $A = 5, k = 10, \beta = 0.0002, \alpha = 0.5, p = 0.5, d_1 = 0.1, d_2 = 0.5, d_3 = 0.3$ and initial values $T(0) = 20, I(0) = 12, V(0) = 7$. It is easy to show that $R_0 = \frac{1}{3} < 1$. By Theorem 3.1, we see that the disease-free equilibrium $E_0(50, 0, 0)$ of system (4) is globally asymptotically stable. Numerical simulation illustrates this result(see, Figure 1).

In system (4), Let $A = 5, k = 10, \beta = 0.02, \alpha = 0.5, p = 0.5, d_1 = 0.1, d_2 = 0.5, d_3 = 0.3$ and initial values $T(0) = 7, I(0) = 0.2, V(0) = 0.05$. It is easy to show that $R_0 = \frac{100}{3} > 1$, system (4) has a unique endemic equilibrium $E^*(41.92, 1.62, 53.89)$. By Theorem 3.2, the endemic equilibrium E^* of system (4) is globally asymptotically stable. Numerical simulation illustrates this fact (see, Figure 2).

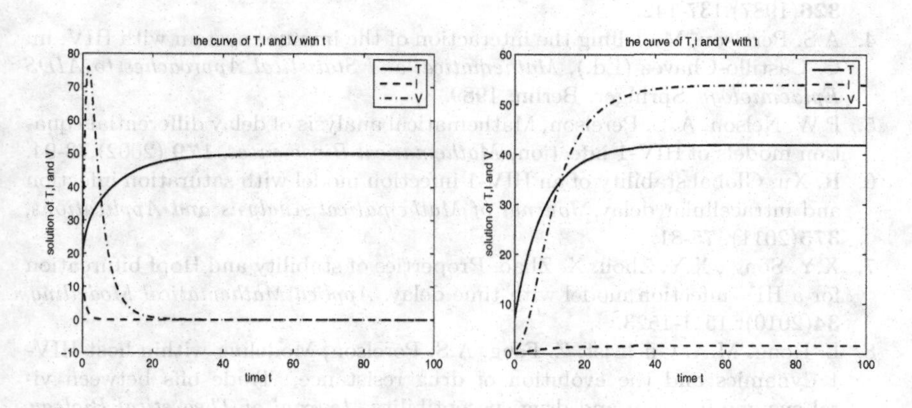

Figure 1. $R_0 < 1$, stability of E_0 Figure 2. $R_0 > 1$, stability of E^*

5. Discussion

In this paper, we formulated a modified HIV infection model with saturation incidence $\frac{\beta T(t)V(t)}{1 + \alpha V(t)}$ and the proportion of infected cells reverting to the uninfected state. The stabilities of the disease-free equilibrium $E_0(T_0, 0, 0)$ and the endemic equilibrium $E^*(T^*, I^*, V^*)$ of system (4) were studied by characteristic equation, Lyapunov function and the second additive compound matrix. It is found that the infection is cleared out when $R_0 < 1$, namely the uninfected steady state is globally stable, while the endemic equilibrium is not feasible. When $R_0 > 1$, the infection persists and the steady state is globally stable. By the definition of R_0, it is clear that R_0 decreases as the reverting rate p of infected cells increases, hence R_0 can be low for a high parametric value of p. In the further work, we will incorporate a delay into the model to describe the time delay between the infection of CD4 + T cell and the emission of virus particles on a cellular level.

References

1. L. Wang and M. Y. Li, Mathematical analysis of the global dynamics of a model for HIV infection of CD4+ T cells, *Mathematical Biosciences*, **200**(2006), 44-57.
2. R. M. Anderson, The role of mathematical models in the study of HIV transmission and the epidemiology of AIDS, *Journal of acquired immune deficiency syndromes*, **1**(1988), 241-256.
3. R. M. May, R. M. Anderson, Transmission dynamics of HIV infection, *Nature*, **326**(1987),137-142.
4. A.S. Perelson, Modelling the interaction of the immune system with HIV, in: C. Castillo-Chavez (Ed.), *Mathematical and Statistical Approaches to AIDS Epidemiology*, Springer, Berlin, 1989.
5. P.W. Nelson, A. S. Perelson, Mathematical analysis of delay differential equation models of HIV-1 infection, *Mathematical Biosciences*, **179** (2002),73-94.
6. R. Xu, Global stability of an HIV-1 infection model with saturation infection and intracellular delay, *Journal of Mathematical Analysis and Applications*, **375**(2011), 75-81.
7. X.Y. Song , X.Y. Zhou, X. Zhao, Properties of stability and Hopf bifurcation for a HIV infection model with time delay, *Applied Mathematical Modelling*, **34**(2010), 1511-1523.
8. L. Rong, M.A. Gilchrist, Z. Feng, A.S. Perelson, Modeling within host HIV-1 dynamics and the evolution of drug resistance: Trade offs between viral enzyme function and drug susceptibility, *Journal of Theoretical Biology*, **247**(2007), 804-818.
9. J. L. Wang, R. Zhang, T. Kuniyab, Global dynamics for a class of age-infection HIV models with nonlinear infection rate, *Journal of Mathematical Analysis and Applications*, **432**(2015), 289-313.
10. M. A. Nowak, C. R. Bangham, Population dynamics of immune responses to persistent viruses, *Science*, **272**(1996), 74-79.
11. P. W. Nelson, A. S. Perelson, Mathematical analysis of delay differential equation models of HIV-1 infection, *Mathematical Biosciences*, **179** (2002), 73-94.

12. J. A. Zack, S. J. Arrigo, S. R. Weitsman *et al.*, HIV-1 entry into quiescent primary lymphocytes: molecular analysis reveals a labile, latent viral structure, *Cell*, **61**(1990), 213-222.
13. P. Essunger, A. S. Perelson, Modeling HIV infection of CD4+ t-cell subpopulations, *Journal of Theoretical Biology*, **170**(1994), 367-391.
14. L. Rong, M. A. Gilchrist, Z. Feng *et al.*, Modeling within-host HIV-1 dynamics and the evolution of drug resistance: trade-offs between viral enzyme function and drug susceptibility, *Journal of Theoretical Biology*, **247**(2007), 804-818.
15. P. K. Srivastava, P. Chandra, Modeling the dynamics of HIV and CD4+ T cells during primary infection, *Nonlinear Analysis: Real World Applications*, **1**(2010), 612-618.
16. B. Buonomo, C. Vargas-De-Leon, Global stability for an HIV-1 infection model including an eclipse stage of infected cells, *Journal of Mathem- atical Analysis and Applications*, **385**(2012), 709-720.
17. Q. L. Sun, L.Q. Min, Dynamics Analysis and Simulation of a Modified HIV Infection Model with a Saturated Infection Rate, *Computational and Mathematical Methods in Medicine*, 2014, Article ID **145162**, 14 pages.
18. J. Hale, Theory of Functional Differential Equations, *Springer-Verlag, Heidelberg*, 1977.
19. M. Y. Li , L. Wang, Global stability in some SEIR epidemic models, *Mathematical Approaches for Emerging and Reemerging Infectious Diseases: Models, Methods, and Theory*, 2002, 295-311.
20. P. Waltman, A brief survey of persistence, *In Proceedings of a Conference in Honor of Kenneth Cooke*, 1991, 31-40.
21. M. Y. Li, J. R. Graef, L. Wang *et al.*, Global dynamics of a SEIR model with varying total population size, *Mathematical Biosciences*, **160**(1999), 191-213.

A SURVEY OF GEOMETRIC TECHNIQUES FOR PATTERN RECOGNITION OF PROBABILITY OF OCCURRENCE OF AMINO ACIDS IN PROTEIN FAMILIES

R. P. MONDAINI

Federal University of Rio de Janeiro
Centre of Technology, COPPE, 21947-972
Rio de Janeiro - RJ, Brazil
E-mail: Rubem.Mondaini@ufrj.br

A Converse of the Steiner's theorem on straigthedge geometry is used to introduce Steiner Loci curves (SLC). A motivation from molecular architecture is used to derive a generic pattern recognition method based on segments of the SLC. An application to the probability of occurrence of amino acids in protein families is then shown in order to testify to the efficiency and robustness of these Steiner Loci techniques.

1. Introduction

A Pattern Recognition process will start from identified functions of the representations associated to the behaviour of the variables of a natural process. Usually these representations are given in the form of curves of the space which has been chosen as the scenario of the natural process. Steiner Loci curves are a restriction on this representation, since they are arc circles associated to sets of curves of the representation. In section 2, we revisit the method of describing the molecular structure in terms of generic Fermat points and we derive the fundamental equation which characterize this description, by emphasising the usefulness of direction cosines. An Ansatz is introduced to solve these problems with an Euclidean distance. In section 3, the same fundamental equations of section 2 are derived from the concept of potential energy minimization. In section 4, by inspiration of the ideas of Schrödinger and Anfinsen[1,2], we introduced an Ansatz for calculating the direction cosines for a class of non-Euclidean distances in terms of Chebyshev polynomials. In section 5, we present the concept of Steiner Loci and its advantage for transforming the results obtained as a dense set of curves into a sparse set. This is then proposed as a new pattern recognition method of their images and we present its usefulness on a problem of level curves associated to the probability occurrence of amino acids in protein families.

2. Generalized Fermat Points in $\mathbb{R}^n$ with Euclidean Distance

Given p points in $\mathbb{R}^n$ space with the adoption of an Euclidean metric to calculate distances, the generalized Fermat problem is to determine the coordinates of a $(p+1)$th point such that the sum of distances to the p given points is a minimum.

This Euclidean distance between the F point and a generic jth point is given by

$$R_{jF}^{(2)} = \left(\sum_{k=1}^{n} (x_{kF} - x_{kj})^2 \right)^{1/2}, \quad k = 1, 2, \ldots, n \qquad (1)$$

$$j = 1, 2, \ldots, p$$

The Fermat problem is then posed as:

$$\sum_{j=1}^{p} R_{jF}^{(2)} = \text{minimum} \qquad (2)$$

The condition (2) will lead to:

$$0 = \frac{\partial}{\partial x_{kF}} \sum_{j=1}^{p} R_{jF}^{(2)} = -\sum_{j=1}^{p} \frac{(x_{kF} - x_{kj})}{R_{jF}^{(2)}} \qquad (3)$$

After multiplying the n equations (3) by $\hat{\imath}_k$, the unit vector of the k-direction, we have:

$$0 = \sum_{k=1}^{n} \hat{\imath}_k \sum_{j=1}^{p} \frac{(x_{kF} - x_{kj})}{R_{jF}^{(2)}} = \sum_{j=1}^{p} \sum_{k=1}^{n} \hat{\imath}_k \frac{(x_{kF} - x_{kj})}{R_{jF}^{(2)}} \qquad (4)$$

or

$$\sum_{j=1}^{p} \hat{r}_{jF} = 0 \tag{5}$$

where $\hat{r}_{jF}$ is the unit vector

$$\hat{r}_{jF} = \frac{\vec{r}_j - \vec{r}_F}{R_{jF}^{(2)}} \tag{6}$$

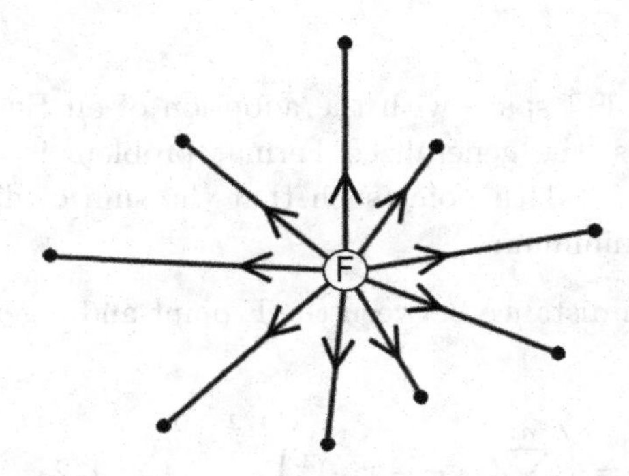

Figure 1. The problem of Eq. (5) — A Fermat problem to be solved in terms of unit vectors.

The last term in Eq. (4) can be also written as

$$\sum_{k=1}^{n} \sum_{j=1}^{p} \cos\alpha_{jk}\, \hat{\imath}_k = 0 \tag{7}$$

where $\cos\alpha_{jk}$ is the direction cosine of the vector $\vec{r}_F - \vec{r}_j$ w.r.t. the k-axis of coordinates.

From linear independence, we can also write from Eq. (7),

$$\sum_{j=1}^{p} \cos\alpha_{jk} = 0\,, \quad k = 1, 2, \ldots, n \tag{8}$$

Eq. (1) can be also written

$$\sum_{k=1}^{n} \cos^2\alpha_{jk} = 1\,, \quad j = 1, 2, \ldots, p \tag{9}$$

Eqs. (8), (9) are $(n + p)$ fundamental constraints on the direction cosines $\cos \alpha_{jk}$.

From these equations we see that there is a maximum number of $n.p - (n + p)$ direction cosines to be chosen as free parameters of the modelling process to be introduced in the following sections.

Let us now write Eq. (5) as:

$$\hat{r}_l + \sum_{\substack{j=1 \\ j \neq l}}^{p} \hat{r}_j = 0 \tag{10}$$

We can then write,

$$1 + \sum_{\substack{j=1 \\ j \neq l}}^{p} \hat{r}_j \cdot \hat{r}_l = 1 + \sum_{\substack{j=1 \\ j \neq l}}^{p} \cos \gamma_{jl} = 0 \tag{11}$$

There are p Eqs. (9) and $\binom{p}{2}$ unknowns, corresponding to $\binom{p}{2}$ planar angles γ_{jl}.

The solution for $p = 3$ is unique. It corresponds to the vectors connecting the centre to the vertices of an equilateral triangle since

$$\cos \gamma_{jk} = -\frac{1}{2}, \quad \forall j \neq k, \quad j, k = 1, 2, 3 \tag{12}$$

A special class of solutions of Eq. (11) can be written[5]

$$\cos \gamma_{jk} = \frac{p\,\delta_{jk} - 1}{p - 1} \tag{13}$$

where δ_{jk} is the Kronecker symbol.

For γ_{jk}, $j \neq k$, they will include solution (12). An example is given by the solution for $p = 4$ which corresponds to vectors connecting the centre of a regular tectrahedron to its vertices since

$$\cos \gamma_{jk} = -\frac{1}{3}, \quad \forall j \neq k, \quad j, k = 1, 2, 3, 4$$

The idealistic representations of the solutions for $p = 3, 4$ presented above do correspond to the angles of 120° among the "edges" of C–N, C–H, C–O ($p = 3$) as well as the angles of 109.47° among C_α–N, C_α–H, C_α–R

($p = 4$) where C, N, H, C_α, R stand for carbon, nitrogen, hydrogen, oxygen, α-carbon and lateral chain of a specific amino acid on an ideal protein structure, respectively.

After putting emphasis on the information to be obtained from the angles γ_{jk} above, we go back to the characterization of special solutions through the direction cosines $\cos\alpha_{jk}$, which satisfy Eqs. (8), (9).

We start from the identity

$$\sum_{j=1}^{p} e^{i\frac{2\pi j}{p}} \equiv 0 \tag{14}$$

and we choose the parameterization of the α_{jk} angle as

$$\alpha_{jk} = \theta_k + \frac{2\pi j}{p}, \quad j = 1, \ldots, p \quad k = 1, \ldots, n \tag{15}$$

From Eq. (14), we have:

$$\sum_{j=1}^{p} \cos\left(\theta_k + \frac{2\pi j}{p}\right) \equiv 0, \quad k = 1, \ldots, n \tag{16}$$

$$\sum_{j=1}^{p} \sin\left(\theta_k + \frac{2\pi j}{p}\right) \equiv 0, \quad k = 1, \ldots, n \tag{17}$$

Eqs. (15)–(17) will lead to introduce the following $p\,\mathrm{x}\,n$ matrix as a class of solutions with $(n-1)$ free parameters

$$\cos\alpha = \begin{pmatrix} \cos\left(\theta_1 + \frac{2\pi}{p}\right) & \frac{\cos\theta_2}{\sin\theta_1}\sin\left(\theta_1 + \frac{2\pi}{p}\right) & \cdots & \frac{\cos\theta_n}{\sin\theta_1}\sin\left(\theta_1 + \frac{2\pi}{p}\right) \\ \cos\left(\theta_1 + \frac{4\pi}{p}\right) & \frac{\cos\theta_2}{\sin\theta_1}\sin\left(\theta_1 + \frac{4\pi}{p}\right) & \cdots & \frac{\cos\theta_n}{\sin\theta_1}\sin\left(\theta_1 + \frac{4\pi}{p}\right) \\ \vdots & \vdots & \ddots & \vdots \\ \cos\left(\theta_1 + \frac{2\pi(p-1)}{p}\right) & \frac{\cos\theta_2}{\sin\theta_1}\sin\left(\theta_1 + \frac{2\pi(p-1)}{p}\right) & \cdots & \frac{\cos\theta_n}{\sin\theta_1}\sin\left(\theta_1 + \frac{2\pi(p-1)}{p}\right) \\ \cos\theta_1 & \cos\theta_2 & \cdots & \cos\theta_n \end{pmatrix} \tag{18}$$

since $np - (n+p)$ is the maximum number of free parameters. This class of solutions is characterized by

$$(n-1) \leq np - (n+p)$$

or,

$$p \geq \frac{2n-1}{n-1} \Rightarrow p \geq 3 \tag{19}$$

Eq. (18) stands for generalized Ansatz for the solution of Euclidean conditions on the direction cosines, (8), (9). In section 3 we will present the treatment of the most elementary non-trivial Euclidean solution, with $p = 3$, $n = 2$ which corresponds to the Euclidean Steiner problem in $\mathbb{R}^2$.

Before leaving this section let us emphasize the possibility of constructing diagrams corresponding to an structure to be obtained by grouping those which satisfy Eq. (5). We have: Each group has p_s unit vectors and

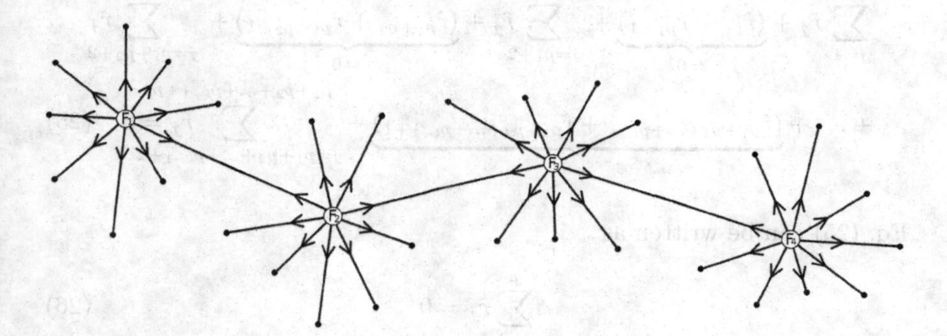

Figure 2. The junction of sets of unit vectors.

there are l sets of unit vectors, $s = 1, 2, \ldots, l$. The Fermat problem can be extended to the entire structure above if a partial Fermat problem is solved on each set, or:

$$\sum_{j=1}^{p_1} \hat{r}_j = 0 \tag{20}$$

$$\sum_{j=p_1+1}^{p_1+p_2} \hat{r}_j = 0 \tag{21}$$

$$\vdots$$

$$\sum_{j=p_1+p_2+\ldots+p_{l-1}+1}^{p_1+p_2+\ldots+p_{l-1}+p_l} \hat{r}_j = 0 \tag{22}$$

From Eqs. (20)–(22), we can write,

$$\sum_{j=1}^{p_1+p_2+\ldots+p_{l-1}+p_l} \hat{r}_j = 0 \tag{23}$$

The entire structure has p external unit vectors and

$$p = p_1 + p_2 + \ldots + p_l - 2(l-1) \tag{24}$$

where $(l-1)$ is the number of pairs of opposite internal unit vectors, since we can write from Eq. (23),

$$\sum_{j=1}^{p_1-1} \hat{r}_j + \underbrace{(\hat{r}_{p_1} + \hat{r}_{p_1+1})}_{=0} + \sum_{j=p_1+2}^{p_1+p_2-1} \hat{r}_j + \underbrace{(\hat{r}_{p_1+p_2} + \hat{r}_{p_1+p_2+1})}_{=0} + \sum_{j=p_1+p_2+2}^{p_1+p_2+p_3-1} \hat{r}_j$$

$$+ \ldots + \underbrace{(\hat{r}_{p_1+p_2+\ldots+p_{l-1}} + \hat{r}_{p_1+p_2+\ldots+p_{l-1}+1})}_{=0} + \sum_{j=p_1+p_2+\ldots+p_{l-1}+2}^{p_1+p_2+\ldots+p_{l-1}+p_l} \hat{r}_j = 0 \tag{25}$$

Eq. (25) can be written as

$$\sum_{j=1}^{p} \hat{r}_j = 0 \tag{26}$$

This can be seen as the junction of l sets of unit vectors with the realization of a Fermat problem on each of these sets or as a Fermat problem to be realized on a set of p unit vectors.

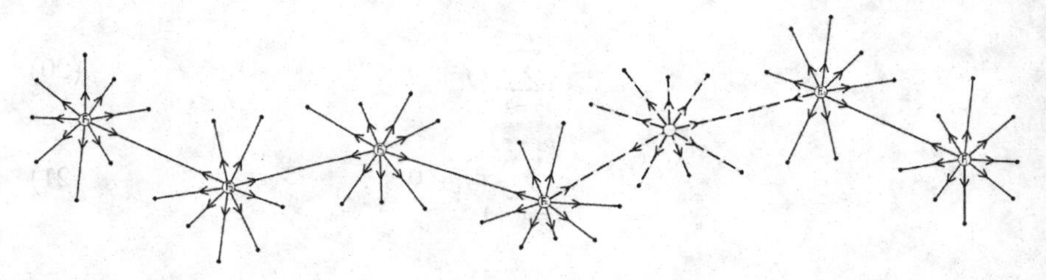

Figure 3. The junction of sets of unit vectors with a Fermat problem to be solved on each one of the partitions of a large set.

3. Application to the Formation of Protein Tertiary Structures

An elementary idea has been evolving along decades, related to the formation of structures like Fig. 1. An approach based on the generalized Fermat problem and its relation with the minimization of potential energy of those configurations[3,4,5] will clarify this idea.

Let us assume that the structure of Fig. 1 is on an unstable equilibrium. The search for an equilibrium state corresponds to the minimization of the potential energy of interaction between central atomic set (F) and the atomic sites of the vertices.

$$U_F = \sum_{j=1}^{p} \frac{c_{jF}}{R_{jF}} \tag{27}$$

where c_{jF} is a constant characterizing the interactions and

$$R_{jF} = \|\vec{r}_F - \vec{r}_j\| \tag{28}$$

is the Euclidean norm.

The minimization process will lead to

$$0 = \frac{\partial U_F}{\partial \vec{r}_F} = -\sum_{j=1}^{p} \frac{c_{jF}(\vec{r}_F - \vec{r}_j)}{(R_{jF})^3} = -\sum_{j=1}^{p} \frac{c_{jF}}{(R_{jF})^2}\,\hat{r}_{jF} \tag{29}$$

An usual assumption is that all interaction forces of the atom site F with their neighbours are equal[5]. We can then write,

$$\frac{c_{jF}}{(R_{jF})^2}\sum_{j=1}^{p} \hat{r}_{jF} = 0 \tag{30}$$

We see that the problem of minimization of potential energy is identified to the problem of finding the Fermat point as has been shown on the derivation of Eq. (5).

Let us now consider a feasible stage, subsequent to that of Fig. 1. This is to be seen as a step on the process of searching for more stable configurations. A structure like that of Fig. 3 can be then obtained by subsequent partitioning processes.

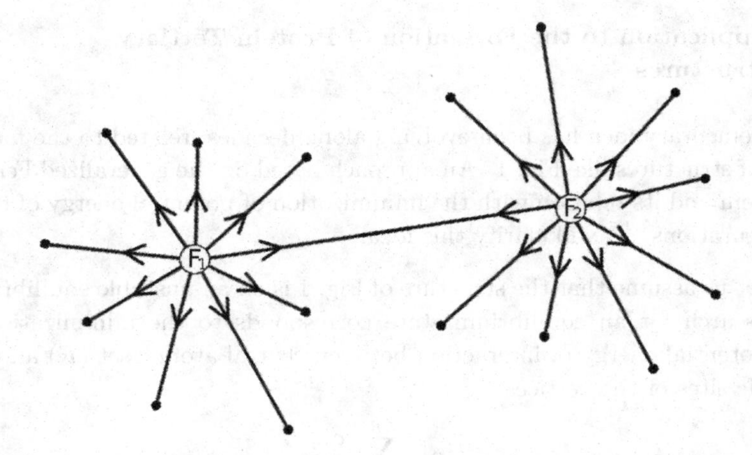

Figure 4. On a feasible searching for a more stable configuration, the configuration of Fig. 1 has been partitioned in two interating configurations with p_1 and p_2 external atom sites and $p_1 + p_2 = p$.

A remark about the need of avoiding steric hindrance should be introduced here and can be written as

$$R_{jF_1} < R_{kF_1}\,, \quad R_{kF_2} < R_{jF_2} \tag{31}$$

and

$$\min R_{jk} < R_{F_1F_2} \quad \begin{array}{l} j = 1, \ldots, p_1 \\ k = p_1 + 1, \ldots, p \end{array}$$

We should also note that on a stage like that of Fig. 4, we do not need to consider the same potential energy problem in order to obtain an analogous result of the Eq. (30) above.

As an example, let the potential energy of the configuration of Fig. 4 to be given by

$$U_{F_1F_2} = \sum_{j=1}^{p_1} \frac{c_{jF_1}}{R_{jF_1}} + \sum_{j=p_1+1}^{p} \frac{c_{jF_2}}{(R_{jF_2})^c} \tag{32}$$

where c is a positive constant.

We then have for a minimum:

$$0 = \frac{\partial U_{F_1F_2}}{\partial \overrightarrow{r}_{F_1}} = -\sum_{j=1}^{p_1} \frac{c_{jF_1}}{(R_{jF_1})^2} \hat{r}_{jF_1} \tag{33}$$

$$0 = \frac{\partial U_{F_1F_2}}{\partial \overrightarrow{r}_{F_2}} = -c \sum_{j=p_1+1}^{p} \frac{c_{jF_2}}{(R_{jF_2})^{c+1}} \hat{r}_{jF_2} \tag{34}$$

As we have done the derivation of Eq. (30), from the assumption of equality of interaction forces of the central atom sites F_1, F_2 to their respective neighbours, we can write

$$0 = \frac{c_{jF_1}}{(R_{jF_1})^2} \sum_{j=1}^{p_1} \hat{r}_{jF_1} \tag{35}$$

$$0 = \frac{cc_{jF_2}}{(R_{jF_2})^{c+1}} \sum_{j=p_1+1}^{p} \hat{r}_{jF_2} \tag{36}$$

and analogous to Eq. (25), we can write:

$$\sum_{j=1}^{p_1-1} \hat{r}_{jF_1} + \underbrace{(\hat{r}_{p_1F_1} + \hat{r}_{p_1+1\,F_2})}_{=0} + \sum_{j=p_1+2}^{p} \hat{r}_{jF_2} = 0 \tag{37}$$

or,

$$\sum_{j=1}^{p} \hat{r}_{jF_1F_2} = 0 \tag{38}$$

which is Eq. (30) as applied to the configuration depicted at Fig. 4.

4. A Class of Non-Euclidean Metrics

The formulation of section 2, can be extended by using Non-euclidean distances like

$$R_{jF}^{(b)} = \left(\sum_{k=1}^{n} (x_{kF} - x_{kj})^b \right)^{1/b} \tag{39}$$

where $b \in \mathbb{N}$, $b > 2$.

Analogously to Eq. (9), we have:

$$\sum_{k=1}^{n} \frac{(x_{kF} - x_{kj})^b}{\left(R_{jF}^{(b)}\right)^b} = 1 = \sum_{k=1}^{n} \left(\cos \alpha_{jk}^{(b)}\right)^b \tag{40}$$

where $\cos \alpha_{jk}^{(b)}$ is the direction cosine

$$\cos \alpha_{jk}^{(b)} = \frac{\vec{r}_j - \vec{r}_F}{R_{jF}^{(b)}} \cdot \hat{\imath}_k$$

of the vector $(\vec{r}_F - \vec{r}_j)$ w.r.t. the kth axis of coordinates.

It is easy to see that

$$\frac{\partial R_{jF}^{(b)}}{\partial x_{kF}} = \frac{(x_{kF} - x_{kj})^{b-1}}{\left(R_{jF}^{(b)}\right)^{b-1}} = \left(\cos \alpha_{jk}^{(b)}\right)^{b-1} \tag{41}$$

and since the Fermat problem is now written as

$$\sum_{j=1}^{p} R_{jF}^{(b)} = \text{minimum} \tag{42}$$

we can write from Eq. (41),

$$\sum_{j=1}^{p} (\cos \alpha_{kj})^{b-1} = 0\,, \quad k = 1, 2, \ldots, n \tag{43}$$

It would also be useful to announce the variation of the distance $R_{jF}^{(b)}$ with the b-exponent. We have:

$$\frac{\partial R_{jF}^{(b)}}{\partial b} = -\frac{1}{b} R_{jF}^{(b)} \log R_{jF}^{(b)} + \frac{1}{b} \frac{\sum_{k=1}^{n} (x_{kF} - x_{kj})^b \log |x_{kF} - x_{kj}|}{\left(R_{jF}^{(b)}\right)^{b-1}} \tag{44}$$

and from Eq. (41)

$$\frac{\partial \log R_{jF}^{(b)}}{\partial t} = \frac{\dot{b}}{b} \sum_{k=1}^{n} \left(\cos \alpha_{jk}^{(b)}\right)^b \log \left|\cos \alpha_{jk}^{(b)}\right| \,, \; b = b(t)\,, \; j = 1, \ldots, p. \tag{45}$$

We now introduce the probability of measuring the distance $R^{(b)}_{jF}$ at time t, $b = b(t)$,

$$P^{(b)}_{jF} = \frac{R^{(b)}_{jF}}{\sum\limits_{s=1}^{p} R^{(b)}_{sF}}$$

and the probability of finding the equilibrium point at time t_0, with $b(t_0) = 2$, corresponding to the Euclidean distance:

$$P^{(b)}_{F} = \prod_{j=1}^{b} P^{(b)}_{jF}$$

we then have

$$\frac{\partial P^{(b)}_F}{\partial t} = \frac{\dot{b}}{b}(1-p)\sum_{j=1}^{p}\sum_{k=1}^{n}\left(\cos\alpha^{(b)}_{jk}\right)^b \log\left|\cos\alpha^{(b)}_{jk}\right|$$

or

$$\frac{\partial P^{(b)}_F}{\partial t} = (p-1)\frac{\dot{b}}{b}H^{(b)}_F$$

where

$$H^{(b)}_F = -\sum_{j=1}^{p}\sum_{k=1}^{n}\left(\cos\alpha^{(b)}_{jk}\right)^b \log\left|\cos\alpha^{(b)}_{jk}\right|$$

is then proposed as a measure of "geometric entropy".

An application of the formulae above for deriving a master equation with the introduction of the probability of occurrence of each external atom site in the process of molecular formation will be published elsewhere.

Analogously to Eq. (18) we shall construct an Ansatz for deriving a class of special solutions to Eqs. (40), (43). First of all, we parameterize the angles $\alpha^{(b)}_{jk}$ by:

$$\alpha^{(b)}_{jk} = \theta_1 + \frac{2\pi k}{n} + \frac{2\pi j}{p}, \quad k = 1, 2, \ldots, n \quad j = 1, 2, \ldots, p \tag{46}$$

We then have:

$$\sum_{j=1}^{p} e^{i\alpha_{jk}^{(b)}} = e^{i\left(\theta_1+\frac{2\pi k}{n}\right)} \sum_{j=1}^{p} e^{i\frac{2\pi j}{p}} \equiv 0 \tag{47}$$

and

$$\sum_{k=1}^{n} e^{i\alpha_{jk}^{(b)}} = e^{i\left(\theta_1+\frac{2\pi j}{p}\right)} \sum_{k=1}^{n} e^{i\frac{2\pi k}{n}} \equiv 0 \tag{48}$$

From Eqs. (47), (48), we can write,

$$\sum_{j=1}^{p} e^{i(2s-1)\frac{2\pi j}{p}} = 0\,, \quad e^{i(2s-1)\frac{2\pi}{p}} \neq 1 \tag{49}$$

$$\sum_{k=1}^{n} e^{i(2s)\frac{2\pi k}{n}} = 0\,, \quad e^{i2s\frac{2\pi}{n}} \neq 1 \tag{50}$$

We then see that in order to write formulae (49), (50) as

$$\sum_{j=1}^{p} T_{2s-1}\left(\cos\alpha_{jk}^{(b)}\right) = 0\,, \quad s \in \mathbb{N} \tag{51}$$

$$\sum_{k=1}^{n} T_{2s}\left(\cos\alpha_{jk}^{(b)}\right) = 0\,, \quad s \neq 0 \tag{52}$$

where $T_{2s}\left(\cos\alpha_{jk}^{(b)}\right)$, $T_{2s-1}\left(\cos\alpha_{jk}^{(b)}\right)$ are the $(2s)$th, $(2s-1)$th Chebyshev polynomials in the variable $\cos\alpha_{jk}^{(b)}$, we should have the greatest common factor of $2s-1, p$ and $2s, n$ equal to 1 (g.c.f.$[2s-1, p] = 1$, g.c.f.$[2s, n] = 1$) if $(2s-1) \geq p$ and $2s \geq n$. We may then consider $(2s-1) < p$ and $2s < n$ in order to guarantee the equivalence of Eqs. (49), (50) to Eqs. (51), (52), respectively.

We now make an additional restriction of, as an even integer number, $b = 2m$ on the construction of an other Ansatz.

From the definition of Chebyshev polynomials[8], we can write,

$$\left(\cos\alpha_{jk}^{(2m)}\right)^{2m} = \sum_{s=0}^{m} A_{2s}^{(2m)} T_{2s}\left(\cos\alpha_{jk}^{(2m)}\right) \tag{53}$$

$$\left(\cos\alpha_{jk}^{(2m)}\right)^{2m-1} = \sum_{s=1}^{m} B_{2s-1}^{(2m-1)} T_{2s-1}\left(\cos\alpha_{jk}^{(2m)}\right) \tag{54}$$

In order to trivially satisfy Eqs. (40), (43) with $b = 2m$, we should use Eqs. (51), (52) ($b = 2m$) into Eqs. (53), (54). We have,

$$1 = \sum_{k=1}^{n}\left(\cos\alpha_{jk}^{(2m)}\right)^{2m} = \sum_{k=1}^{n}\sum_{s=0}^{m} A_{2s}^{(2m)} T_{2s}\left(\cos\alpha_{jk}^{(2m)}\right) \tag{55}$$

$$0 = \sum_{j=1}^{p}\left(\cos\alpha_{jk}^{(2m)}\right)^{2m-1} = \sum_{j=1}^{p}\sum_{s=1}^{m} B_{2s-1}^{(2m)} T_{2s-1}\left(\cos\alpha_{jk}^{(2m)}\right) \tag{56}$$

Eq. (56) is identically satisfied and Eq. (55) will be satisfied by determining A_0^{2m} from orthogonalization properties of the Chebyshev polynomials through

$$\begin{aligned}
&\int_{-1}^{+1} \frac{\left(\cos\alpha_{jk}^{(2m)}\right)^{2m} T_{2l}\left(\cos\alpha_{jk}^{(2m)}\right)}{\sin\alpha_{jk}^{(2m)}}\,\mathrm{d}\left(\cos\alpha_{jk}^{(2m)}\right) \\
&= \sum_{s=0}^{m} A_{2s}\int_0^{\pi} T_{2s}\left(\cos\alpha_{jk}^{(2m)}\right) T_{2l}\left(\cos\alpha_{jk}^{(2m)}\right)\mathrm{d}\alpha_{jk}^{(2m)} \\
&= A_{2l}^{(2m)}\frac{\pi}{2}, \quad l \neq 0 \\
&= A_0^{(2m)}\pi, \quad l = 0
\end{aligned} \tag{57}$$

We then have from Eq. (57):

$$A_0^{(2m)} = \frac{1}{\pi}\int_0^{\pi}\left(\cos\alpha_{jk}^{(2m)}\right)^{2m}\mathrm{d}\alpha_{jk}^{(2m)}, \tag{58}$$

and

$$A_0^{(0)} = 1,\ A_0^{(2)} = \frac{1}{2},\ A_0^{(4)} = \frac{3}{8},\ A_0^{(6)} = \frac{10}{32},\ \ldots$$

We are now able to introduce the following Ansatz for the $\cos\alpha_{jk}^{(2m)}$ in terms of a $p \times n$ matrix:

$$\cos\alpha^{(2m)} =$$

$$\frac{1}{\left(nA_0^{(2m)}\right)^{\frac{1}{2m}}}\begin{pmatrix} \cos\left(\theta_1+2\pi\left(\frac{1}{n}+\frac{1}{p}\right)\right) & \cos\left(\theta_1+2\pi\left(\frac{2}{n}+\frac{1}{p}\right)\right) & \dots & \cos\left(\theta_1+2\pi\left(1+\frac{1}{p}\right)\right) \\ \cos\left(\theta_1+2\pi\left(\frac{1}{n}+\frac{2}{p}\right)\right) & \cos\left(\theta_1+2\pi\left(\frac{2}{n}+\frac{2}{p}\right)\right) & \dots & \cos\left(\theta_1+2\pi\left(1+\frac{2}{p}\right)\right) \\ \vdots & \vdots & \ddots & \vdots \\ \cos\left(\theta_1+2\pi\left(\frac{1}{n}+\frac{p-1}{p}\right)\right) & \cos\left(\theta_1+2\pi\left(\frac{2}{n}+\frac{p-1}{p}\right)\right) & \dots & \cos\left(\theta_1+2\pi\left(1+\frac{p-1}{p}\right)\right) \\ \cos\left(\theta_1+2\pi\left(\frac{1}{n}+1\right)\right) & \cos\left(\theta_1+2\pi\left(\frac{2}{n}+1\right)\right) & \dots & \cos\theta_1 \end{pmatrix} \tag{59}$$

where θ_1 is an arbitrary parameter. Eqs. (40), (43) with $b = 2m$ can be written,

$$\sum_{k=1}^{n}\left(\frac{\cos\left(\theta_1+2\pi\left(\frac{k}{n}+\frac{j}{p}\right)\right)}{\left(nA_0^{(2m)}\right)^{1/2m}}\right)^{2m} = 1 \tag{60}$$

$$\sum_{j=1}^{p}\left(\frac{\cos\left(\theta_1+2\pi\left(\frac{k}{n}+\frac{j}{p}\right)\right)}{\left(nA_0^{(2m)}\right)^{1/2m}}\right)^{2m} = 0 \tag{61}$$

The value $m = 1$ do correspond to Euclidean metrics.

The Euclidean problem $m = 1$, $n = 2$, $p = 3$ — the 2-dimensional Euclidean Steiner's problem for 3 given points which is the only one to admit a geometric construction cannot be solved with the Ansatz of Eq. (59), since from Eqs. (49), (50), we have,

$$e^{i(2m-1)\frac{2\pi}{p}} = e^{i\frac{2\pi}{3}} \neq 1\,, \quad e^{i2m\frac{2\pi}{n}} = e^{i2\pi} = 1 \tag{62}$$

respectively.

We see that the 2nd equality above, violates the necessary condition for the validity of Eq. (50). This means that for the analysis of this case, we should use the Ansatz given in Eq. (18) of section 2 which is valid for Euclidean distances only.

5. "Steiner Loci" and Application to Pattern Recognition

The original Steiner problem of determining the coordinates of a point such that the sum of its distances to three other giving points is a minimum, has very well known geometrical constructions in the case of Euclidean distances. We shall use the development of the previous sections to solve the associated problem of determining the geometrical loci of the points such that the sum of their distances to two given points and a variable point is a minimum.

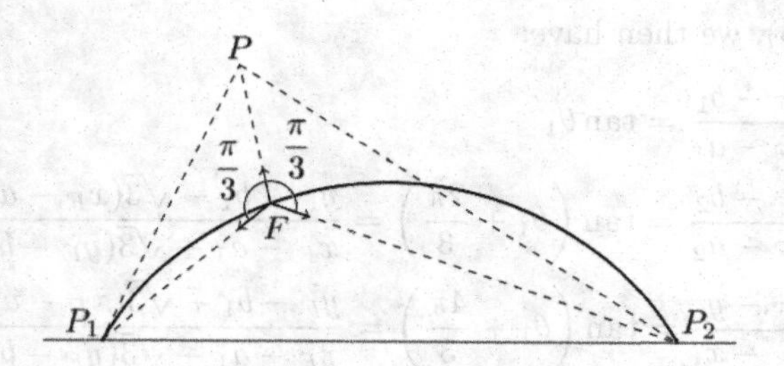

Figure 5. To each position of a variable point P and two fixed points P_1, P_2, there is a corresponding Fermat point (Steiner for $p = 3$) such that $PF + P_1F + P_2F$ is a minimum.

We then write the 2 x 3 matrix:

$$\cos\alpha^{(2)} = \begin{pmatrix} \cos\left(\theta_1 + \frac{2\pi}{3}\right) & \frac{\cos\theta_2}{\sin\theta_1}\sin\left(\theta_1 + \frac{2\pi}{3}\right) \\ \cos\left(\theta_1 + \frac{4\pi}{3}\right) & \frac{\cos\theta_2}{\sin\theta_1}\sin\left(\theta_1 + \frac{4\pi}{3}\right) \\ \cos\theta_1 & \cos\theta_2 \end{pmatrix} \tag{63}$$

There is $2\text{x}3 - (2+3) = 1$, only one arbitrary parameter

$$\cos^2\theta_1 + \cos^2\theta_2 = 1 \quad \Rightarrow \quad \theta_2 = \frac{\pi}{2} - \theta_1 \tag{64}$$

and the 2 x 3 matrix can be written:

$$\cos\alpha^{(2)} = \begin{pmatrix} \cos\left(\theta_1 + \frac{2\pi}{3}\right) & \sin\left(\theta_1 + \frac{2\pi}{3}\right) \\ \cos\left(\theta_1 + \frac{4\pi}{3}\right) & \sin\left(\theta_1 + \frac{4\pi}{3}\right) \\ \cos\theta_1 & \sin\theta_1 \end{pmatrix} \tag{65}$$

According to Eq. (40), the elements of the matrix should be written:

$$\cos\alpha_{jk}^{(2)} = \frac{x_{kF} - x_{kj}}{R_{jF}^{(2)}}$$

where

$$x_{1l} = a_l \,, \quad x_{2l} = b_l \,, \quad l = 1, 2$$
$$x_{13} = x \,, \quad x_{23} = y$$
$$x_{1F} = x_F \,, \quad x_{2F} = y_F$$

From Eq. (65), we then have:

$$\frac{y_F - b_1}{x_F - a_1} = \tan\theta_1 \tag{66}$$

$$\frac{y_F - b_2}{x_F - a_2} = \tan\left(\theta_1 + \frac{2\pi}{3}\right) = \frac{y_F - b_1 - \sqrt{3}(x_F - a_1)}{x_F - a_1 + \sqrt{3}(y_F - b_1)} \tag{67}$$

$$\frac{y_F - y}{x_F - x} = \tan\left(\theta_1 + \frac{4\pi}{3}\right) = \frac{y_F - b_1 + \sqrt{3}(x_F - a_1)}{x_F - a_1 - \sqrt{3}(y_F - b_1)} \tag{68}$$

Eqs. (66)–(68) lead to:

$$\frac{y_F - b_1}{x_F - a_1} = \frac{-\sqrt{3}\,x_F + B}{\sqrt{3}\,y_F + A} = \frac{\sqrt{3}\,x_F + D}{-\sqrt{3}\,y_F + C} \tag{69}$$

where

$$\begin{aligned} A &= a_2 - a_1 - b_2\sqrt{3} \\ B &= b_2 - b_1 + a_2\sqrt{3} \\ C &= x - a_1 + y\sqrt{3} \\ D &= y - b_1 - x\sqrt{3} \end{aligned} \tag{70}$$

From the 2nd equality on Eq. (69), we can also write:

$$\frac{y_F - b_1}{x_F - a_1} = \frac{B + D}{A + C} \tag{71}$$

The first equality on Eq. (69) will lead to:

$$\left(x_F - \frac{1}{2}\left(a_1 + \frac{B\sqrt{3}}{3}\right)\right)^2 + \left(y_F - \frac{1}{2}\left(b_1 - \frac{A\sqrt{3}}{3}\right)\right)^2$$
$$= \frac{1}{4}\left[\left(a_1 - \frac{B\sqrt{3}}{3}\right)^2 + \left(b_1 + \frac{A\sqrt{3}}{3}\right)^2\right] \tag{72}$$

This the equation of a circle. Its centre and radius does not depend of the coordinates of the third variable vertex P of the triangles P_1P_2P. Given two points P_1, P_2, the resulting circle is a geometrical locus of the Fermat (Steiner) point of all triangles P_1P_2P.

From the 2nd equality in Eq. (69) and Eq. (71) we can obtain

$$x_F = \frac{(BC-AD)(A+C)+\sqrt{3}(B+D)\Big(a_1(B+D)-b_1(A+C)\Big)}{\Big((A+C)^2+(B+D)^2\Big)\sqrt{3}} \tag{73}$$

$$y_F = \frac{(BC-AD)(B+D)-\sqrt{3}(A+C)\Big(a_1(B+D)-b_1(A+C)\Big)}{\Big((A+C)^2+(B+D)^2\Big)\sqrt{3}} \tag{74}$$

Actually, the result of Eqs. (72)–(74) corresponds to the converse of Steiner's theorem of the straigthedge Euclidean geometry[5], since a curve given by $y = y(x)$ will be represented as an arc of the circle given by Eq. (72).

We simplify the development above by choosing a coordinate system given by

$$a_1 = a\,, \quad a_2 = -a\,, \quad b_1 = 0\,, \quad b_2 = 0 \tag{75}$$

and we have

$$x_F = \frac{4a\sqrt{3}}{3}\,\frac{x(y+a\sqrt{3})}{x^2+(y+a\sqrt{3})^2} \tag{76}$$

$$y_F = \frac{a\sqrt{3}}{3}\,\frac{(y+a\sqrt{3})^2-3x^2}{x^2+(y+a\sqrt{3})^2} \tag{77}$$

$$x_F^2 + \left(y_F + \frac{a\sqrt{3}}{3}\right)^2 = \frac{4a^2}{3} \tag{78}$$

The representation of the region corresponding to the solution of the problem introduced by Eq. (65) is very convenient for a Pattern Recognition method. To every pair of points P_1, P_2, there will be a region like that of Fig. 6. In the case of a dense set of curves $y = y(x)$ v.g. level curves of a surface (obtained from a modelling process)[9], we can associate a sparse set of circle arcs to it through the transformation of Eqs. (76), (77).

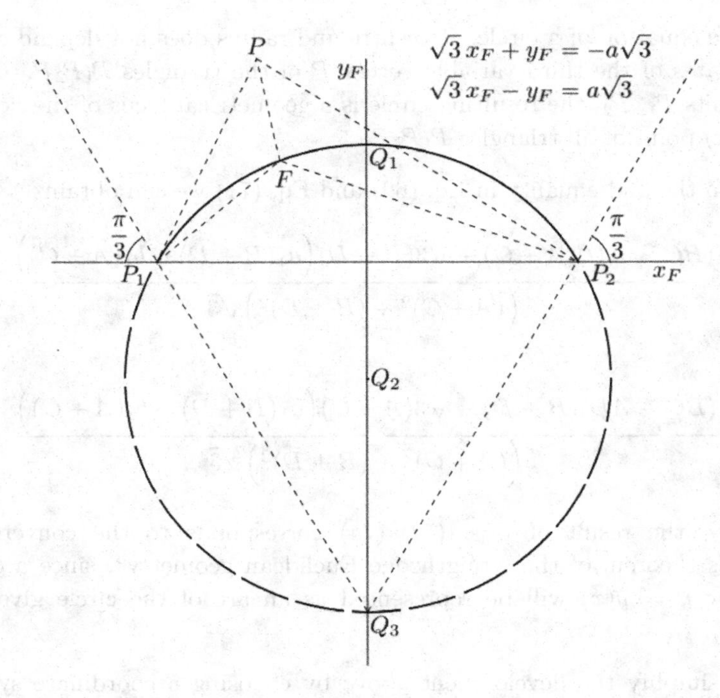

Figure 6. The solution of the Euclidean $n = 2$, $p = 3$ problem. $P_1 = (-a, 0)$, $P_2 = (a, 0)$, $Q_1 = \left(0, a\frac{\sqrt{3}}{3}\right)$, $Q_2 = \left(0, -a\frac{\sqrt{3}}{3}\right)$, $Q_3 = (0, -a\sqrt{3})$. The point P is allowed to move arbitrarily in the region which lower boundary is given by the curve Max $\left(\sqrt{3}\,x_F + y_F = a\sqrt{3},\ x_F^2 + \left(y_F + \frac{a\sqrt{3}}{3}\right)^2 = \frac{4a^2}{3},\ \sqrt{3}\,x_F - y_F = a\sqrt{3}\right)$. The points along the straigth lines $y = \pm\sqrt{3}(x - a)$ and $x < -a$, $x > a$ are mapped on the points P_1, P_2, respectively.

An example of transforming the level curves associated to the probability occurrences of amino acids of protein families will be given at the end of this section and will be presented in detail on a forthcoming publication[9].

We will now describe the properties of the transformation (76), (77). We think of a combined transformation $(x, y) \to (x_F, y_F) \to (x_{FF}, y_{FF})$. From the Jacobian matrices to be defined by

$$\begin{pmatrix} \mathrm{d}x_F \\ \mathrm{d}y_F \end{pmatrix} = J_F \begin{pmatrix} \mathrm{d}x \\ \mathrm{d}y \end{pmatrix} \tag{79}$$

$$\begin{pmatrix} \mathrm{d}x_{FF} \\ \mathrm{d}y_{FF} \end{pmatrix} = J_{FF} \begin{pmatrix} \mathrm{d}x_F \\ \mathrm{d}y_F \end{pmatrix} \tag{80}$$

where

$$\det J_F = 0\,, \quad \det J_{FF} = 0 \tag{81}$$

The Jacobian matrices will satisfy

$$J_{FF} J_F^2 = J_F \tag{82}$$

$$J_F J_{FF} J_F = J_{FF} J_F^2 \tag{83}$$

$$(J_{FF} J_F)^2 = J_{FF} J_F \tag{84}$$

Eq. (84) corresponds to the idempotency of the combined transformation $J_{FF} J_F$.

The consistency of the Eqs. (82)–(84) will follow by:

From Eq. (83) we have

$$(J_{FF} J_F)^2 = J_{FF}^2 J_F^2 \tag{85}$$

From Eqs. (85) and (84):

$$J_{FF}^2 J_F^2 = J_{FF} J_F \tag{86}$$

Eq. (86) can be also obtained from Eq. (82) q.e.d.

In order to emphasize the advantage of the modelling with the Steiner Geometrical Loci as described above we now consider an example of pattern recognition through level curves of two-dimensional surfaces which model the probability of occurrences of each amino acid of a given protein family. There are then n surfaces $p_j\big(x(a), y(a)\big)$, $j = 1, 2, \ldots, n$ corresponding to each a-amino acid, where

$$a = \mathrm{A, C, D, E, F, G, H, I, K, L, M, N, P, Q, R, S, T, V, W, Y},$$

of a previously chosen family and a level curve associated to each j-value. The alternative modelling is to consider the partitioning of the set of j-values into disjoint sets

$$(\{j\}, j = 1, 2, \ldots, n) = \{1, \ldots, j_1\} \cup \{j_1 + 1, \ldots, j_2\} \cup \ldots \cup \{j_{n-1} + 1 \ldots n\}$$

The number m of subsets will depend on the amino acid and the protein family. All j_s-values of each subset do correspond to only one intersection plane of the subset of j_s-surfaces. In other words, the j_s-values of each subset do correspond to a same intersection plane. The resulting j_s-level curves of each subset are then represented by only one arc of circle (one of the Steiner Loci curves).

We show in the figures below a comparison of the results obtained for the probability of occurrence of the A-amino acid of the protein family PF00060, of Pfam database, version 27.0. Fig. 7(a) do correspond to usual histogram representation. Fig. 7(b) is the set of n level curves in a saddle point approximation method derived from a Poisson statistical approach[6].

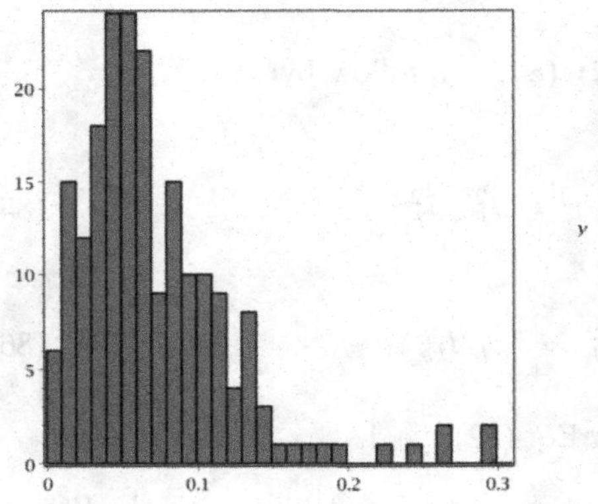

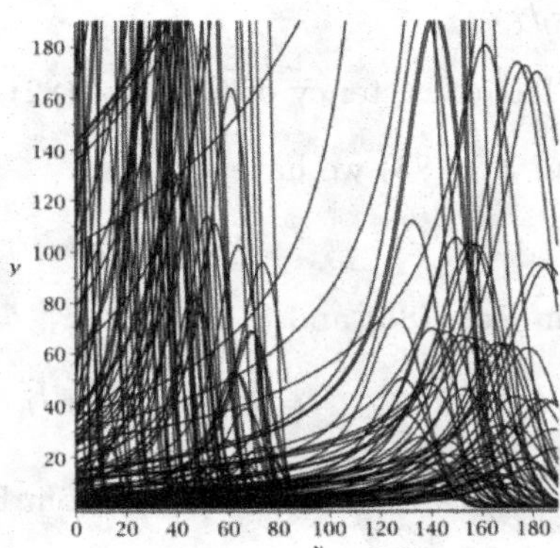

Figure 7. (a) Histogram of probability occurrence of the A-amino acid from PF00060. (b) The entire set of level curves corresponding to all feasible planes of intersection of statistical surfaces for the A-amino acid of PF00060.

Figs. 8(a), 8(b), 8(c) are the level curves for three selected subsets of intersection planes p_{j_s} from the feasible set shown at Fig. 7(b)[9]

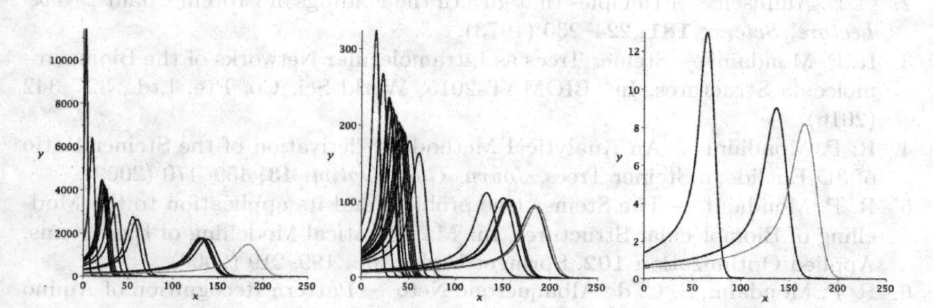

Figure 8. Level curves corresponding to subsets of j_s-values of the $p_j(x(a), y(a))$ probability of occurrence. (a) $p_{j_s} = 0.01$, (b) $p_{j_s} = 0.04$, (c) $p_{j_s} = 0.14$.

Fig. 9 is the representation of the results in terms of arcs of circles (Steiner Loci) of the modelling introduced in this section. There is m arcs of circle corresponding to the n values of p_j. This means that all curves on each of the Figs. 8 above do correspond to a unique value of p_j. The sparseness property of this representation is very convenient as compared to the representation given by Fig. 7(b) above. Other applications of the Steiner Loci representation are now in progress and will be published elsewhere.

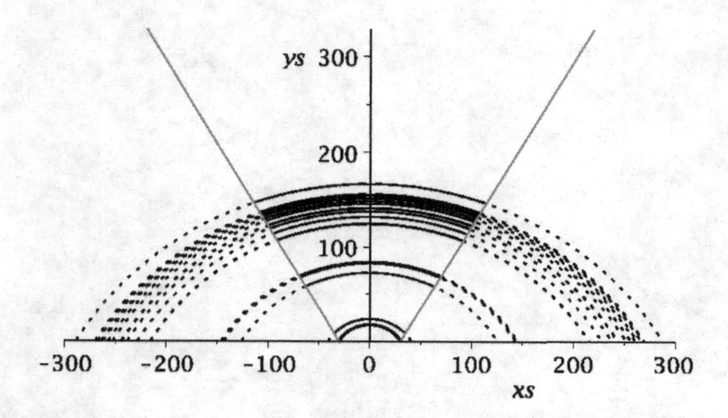

Figure 9. The Steiner Loci curves of the A-amino acid of the PF00060 protein family.

References

1. E. Schrödinger — What is Life?, Cambridge Univ. Press, Cambridge, UK (1992).
2. C. B. Anfinsen — Principles that govern the Foldings of Protein Chains, *Nobel Lecture, Science* **181**, 224–230 (1973).
3. R. P. Mondaini — Steiner Trees as Intramolecular Networks of the Biomacromolecula Structures, In: BIOMAT 2015, World Sci. Co. Pte. Ltd., 327–342 (2016).
4. R. P. Mondaini — An Analytical Method for derivation of the Steiner Ratio of 3D Euclidean Steiner Trees, *Journ. Glob. Optim.* **43**, 459–470 (2009).
5. R. P. Mondaini — The Steiner tree problem and its application to the Modelling of Biomolecular Structures, In: Mathematical Modelling of Biosystems, Applied Optimization **102**, Springer Verlag, pp. 199–219 (2008).
6. R. P. Mondaini, S. C. de Albuquerque Neto — Pattern Recognition of Amino acids by a Poisson Statistical Approach, In: 31st Int. Colloquim on Group Theoretical Methods in Physics, June 19–25, 2016, Springer Verlag, to be published (2017).
7. R. P. Mondaini, S. C. de Albuquerque Neto — The Pattern Recognition of Probability Distributions of Amino acids in protein families, pp. 29–50 in the present book.
8. W. H. Press, S. A. Teukolsky, W. T. Wettering, B. P. Flannery — Numerical Recipes - The Art of Scientific Computing, 2nd edition, Cambridge Univ. Press, Cambridge, UK (1994).
9. R. P. Mondaini, S. C. de Albuquerque Neto — The Pattern Recognition of Amino acids in protein families - The Saddle point approximation - Lecture given at the 16th BIOMAT International Symposium on Mathematical and Computational Biology, Oct. 31–Nov. 04, 2016, Chern Institute of Mathematics, Tianjin, P. R. China, to be published.

SYMMETRY AND MINIMUM PRINCIPLE AT THE BASIS OF THE GENETIC CODE

A. SCIARRINO

I.N.F.N., Sezione di Napoli
Complesso Universitario di Monte S. Angelo
Via Cinthia, I-80126 Napoli, Italy
E-mail: nino.sciarrino@gmail.com

P.SORBA

LAPTH,Laboratoire d'Annecy-le-Vieux de Physique Théorique CNRS
Université de Savoie
Chemin de Bellevue, BP 110
F-74941 Annecy-le-Vieux, France
E-mail: paul.sorba@lapth.cnrs.fr

The importance of the notion of symmetry in physics is well established: could it also be the case for the genetic code? In this spirit, a model for the Genetic Code based on continuous symmetries and entitled the "Crystal Basis Model" has been proposed a few years ago and applied to different problems, such as the elaboration and verification of sum rules for codon usage probabilities, relations between physico-chemical properties of amino-acids and some predictions. Defining in this context a "bio-spin" structure for the nucleotides and codons, the interaction between a couple of codon-anticodon can simply be represented by a (bio) spin-spin potential. Then, imposing the minimum energy principle, an analysis of the evolution of the genetic code can be performed with good agreement with the generally accepted scheme. A more precise study of this interaction model provides informations on codon bias, consistent with data.

1. Introduction

The sciences of life offer an important domain of investigations for the physicist. Already about seventy years ago, Erwin Schrödinger provided in his book "*What is life?*"[1] some ideas about the possible role of a "new physics" in this domain, imagining for example mutations to be directly linked to quantum leads. As can be read there:

" living matter, while not eluding the "laws of physics" as established up to date, is likely to involve "other laws of physics" hitherto unknown, which

however, once they have been revealed, will form just as integral a part of science as the former".

Among the mathematical tools which played in the second part of the twentieth century and are still playing an essential role in theoretical physics, and in particular in particle physics, is the one of Group theory, this concept being usually called in physics *Symmetry*, or *Invariance.* It is this notion which is at the basis of our model for describing the genetic code and developing a theoretical approach of its biological properties.

The idea of symmetry, or invariance, can be used in different ways, but to illustrate the one we need today, let us take an example. Consider an electron e^-. An important physical quantity attached to it is its spin. And actually, as you know, there are two states for the spin of the electron, called up and down, or + and − (or $+1/2$ and $-1/2$ following the notation you choose, and we use to say that the spin of e^- is $1/2$). Mathematically, these two states can be seen as orthogonal vectors of the 2-dim complex Euclidean space,[a] on which acts the group of 2 by 2 unitary matrices, called $\mathcal{SU}(2)$, transforming one state into another one. It is a Lie group and considering its Lie algebra, there exists an element on it, a 2×2 matrix with eigenvalues $+1/2$ and $-1/2$ associated to the eigenvectors which are the states up and down. If you consider a vector boson, (e.g. the W boson which mediates the weak interaction) there are three states of spin, denoted $+1$, 0, -1 and we can represent the elements of the group $\mathcal{SU}(2)$ by 3×3 matrices acting on a 3-dim Hilbert space. It is this notion that we will use to construct our model describing the genetic code.

At this point, let us mention two essential aspects of genetics on which we propose to use our model: the DNA structure on one hand and the mechanism of polypeptide fixation from codons on the other hand. But, it might be good to start by reminding some essential features on the genetic code. First, as well known, the DNA macromolecule is constituted by two chains of nucleotides wrapped in a double helix shape. There are four different nucleotides, characterized by their bases: adenine (A) and guanine (G) deriving from purine, and cytosine (C) and thymine (T) coming from pyrimidine. Note also the A (reps. T) base in one strand is connected with two hydrogen bonds to a T (resp. A) base in the other strand, while a C (resp. G) base is related to a G (reps. C) base with three hydrogen bonds. The genetic information is transmitted to the cytoplasm via the messenger ribonucleic acid (mRNA). During this operation, called transcription, the

[a]Such a space is denoted as a 2-dim Hilbert space.

A, G, C, T bases in the DNA are associated respectively to the U, C, G, A bases, U denoting the uracile base. Then it will be through a ribosome that a triplet of nucleotides or codon will be related to an amino acid (a.a.). More precisely, a codon is defined as an ordered sequence of three nucleotides, e.g. AAG, ACG, etc., and one enumerates in this way $4 \times 4 \times 4 = 64$ different codons. In the universal eukariotic code (see Table 1), 61 of such triplets can be connected in an unambiguous way to the amino-acids, except the three following triplets UAA, UAG and UGA, which are called non-sense or stop codons, the role of which is to stop the biosynthesis. Indeed the genetic code is the association between codons and amino-acids. But since one distinguishes only 20 amino-acids[b] related to the 61 codons, it follows that the genetic code is degenerate. Still considering the standard eukariotic code, one observes sextets, quadruplets, triplet, doublets and singlet of codons, each multiplet corresponding to a specific amino-acid. In the

Table 1. The eukariotic code.

codon	amino acid	codon	amino acid	codon	amino acid	codon	amino acid
CCC	Pro P	UCC	Ser S	GCC	Ala A	ACC	Thr T
CCU	Pro P	UCU	Ser S	GCU	Ala A	ACU	Thr T
CCG	Pro P	UCG	Ser S	GCG	Ala A	ACG	Thr T
CCA	Pro P	UCA	Ser S	GCA	Ala A	ACA	Thr T
CUC	Leu L	UUC	Phe F	GUC	Val V	AUC	Ile I
CUU	Leu L	UUU	Phe F	GUU	Val V	AUU	Ile I
CUG	Leu L	UUG	Leu L	GUG	Val V	AUG	Met M
CUA	Leu L	UUA	Leu L	GUA	Val V	AUA	Ile I
CGC	Arg R	UGC	Cys C	GGC	Gly G	AGC	Ser S
CGU	Arg R	UGU	Cys C	GGU	Gly G	AGU	Ser S
CGG	Arg R	UGG	Trp W	GGG	Gly G	AGG	Arg R
CGA	Arg R	UGA	Stop	GGA	Gly G	AGA	Arg R
CAC	His H	UAC	Tyr Y	GAC	Asp D	AAC	Asn N
CAU	His H	UAU	Tyr Y	GAU	Asp D	AAU	Asn N
CAG	Gln Q	UAG	Stop	GAG	Glu E	AAG	Lys K
CAA	Gln Q	UAA	Stop	GAA	Glu E	AAA	Lys K

mathematical framework we have proposed[2], the codons appear as composite states of nucleotides. More precisely, the codons are obtained as tensor products of nucleotides, the four nucleotides being assigned to the fundamental representation of the quantum group $\mathcal{U}_q(sl(2) \oplus sl(2))$ in the limit of the deformation parameter $q \to 0$. The use of a quantum group in the limit $q \to 0$ is essential to take into account the nucleotide ordering (see Table 1). Of course, the reader who is not interested in the mathematical aspects can jump over them and focus his attention on the biophysical

[b]Alanine (Ala), Arginine (Arg), Asparagine (Asn), Aspartic acid (Asp), Cysteine (Cys), Glutamine (Gln), Glutamic acid (Glu), Glycine (Gly), Histidine (His), Isoleucine (Ile), Leucine (Leu), Lysine (Lys), Methionine (Met), Phenylalanine (Phe), Proline (Pro), Serine (Ser), Threonine (Thr), Tryptophane (Trp), Tyrosine (Tyr), Valine (Val).

results which are presented hereafter. However, for the reader who wishes to better understand our approach, we have devoted a rather developed "tutorial" on group theory at the end of this review. The first part of this appendix deals with general notions and properties of Lie groups while the second part shows explicitly how are constructed the codons as representation states of the quantum group above mentioned.

We have distinguished two parts in this review.

The first one starts with a rapid recalling of the main aspects of our model that we called "Crystal Basis Model". It is followed by two examples of applications. The first one concerns the setting of sum rules for codon usage probabilities[3]: it is deduced that the sum of usage probabilities of codons with C and A in the third position for the quartets and/or sextets is independent of the biological species for vertebrates.

The second application deals with the physical-chemical properties of amino-acids for which a set of relations have been derived and compared with the experimental data[4]. A prediction for the not yet measured thermodynamical parameters of three amino-acids is also proposed.

Another important notion in physics is the principle of minimal action, or use of minimum of energy. This is the second main idea that we will keep with us in the second part of this review in which a codon-anticodon interaction potential is proposed[5], still in the framework of the "Crystal Basis Model". Such a study will first allow to determine the structure of the minimum set of 22 anticodons allowing the translational-transcription for animal mitochondrial code. The results are in very good agreement with the observed anticodons. Then, the evolution of the genetic code is considered, with 20 amino-acids encoded from the beginning, from the viewpoint of codon-anticodon interaction. Following the same spirit as above, a determination of the structure of the anticodons in the Ancient, Archetypal and Early Genetic codes is obtained[6]. Most of our results agree with the generally accepted scheme.

Finally, keeping still at hand the minimization of our codon-anticodon interaction potential, codon bias are discussed, providing inequalities between codon usage probabilities for quartets of codons[7]. Performing this study separately for the Early and for the Eukariotic genetic code, we observe a consistency with the obtained results as well as good agreement with the available data. Last but not least, an analysis of the coherent change of sign, in the evolution from the Early to the Eukaryotic code, of the two parameters regulating our interaction potential is performed.

Some general remarks are gathered in the conclusion, while, as already mentioned, a large appendix is devoted to the mathematical aspect of symmetry.

As this paper is essentially a review of the Crystal Basis Model, we have limited the references and only provided those directly connected to our approach. The interested reader can find in each quoted paper the relative biography.

2. PART 1: Crystal Basis Model and Application

2.1. *A group theoretical model of the genetic code*

We consider the four nucleotides as basic states of the $(\frac{1}{2}, \frac{1}{2})$ representation of the $\mathcal{U}_q(sl(2) \oplus sl(2))$ quantum enveloping algebra in the limit $q \to 0$.[2] A triplet of nucleotides will then be obtained by constructing the tensor product of three such four-dimensional representations. Actually, this approach mimicks the group theoretical classification of baryons made out from three quarks in elementary particles physics, the building blocks being here the A, C, G, T/U nucleotides. The main and essential difference stands in the property of a codon to be an *ordered* set of three nucleotides, which is not the case for a baryon.

Constructing such pure states is made possible in the framework of any algebra $\mathcal{U}_{q\to 0}(\mathcal{G})$ with $\mathcal{G}$ being any (semi)-simple classical Lie algebra owing to the existence of a special basis, called crystal basis, in any (finite dimensional) representation of $\mathcal{G}$. The algebra $\mathcal{G} = sl(2) \oplus sl(2)$ appears the most natural for our purpose. The complementary rule in the DNA–mRNA transcription may suggest to assign a *quantum number* with opposite values to the couples (A,T/U) and (C,G). The distinction between the purine bases (A,G) and the pyrimidine ones (C,T/U) can be algebraically represented in an analogous way. Thus considering the fundamental representation $(\frac{1}{2}, \frac{1}{2})$ of $sl(2) \oplus sl(2)$ and denoting $\pm$ the basis vector corresponding to the eigenvalues $\pm\frac{1}{2}$ of the J_3 generator in any of the two $sl(2)$ corresponding algebras, we will assume the following "biological" spin structure:

$$\begin{array}{ccc} & sl(2)_H & \\ C \equiv (+,+) & \longleftrightarrow & U \equiv (-,+) \\ sl(2)_V \updownarrow & & \updownarrow sl(2)_V \\ G \equiv (+,-) & \longleftrightarrow & A \equiv (-,-) \\ & sl(2)_H & \end{array} \qquad (1)$$

the subscripts H (:= horizontal) and V (:= vertical) being just added to specify the algebra.

Now, we consider the representations of $\mathcal{U}_q(sl(2))$ and more specifically the crystal bases obtained when $q \to 0$. Introducing in $\mathcal{U}_{q\to 0}(sl(2))$ the operators J_+ and J_- after modification of the corresponding simple root vectors of $\mathcal{U}_q(sl(2))$, a particular kind of basis in a $\mathcal{U}_q(sl(2))$-module can be defined. Such a basis is called a crystal basis and carries the property to undergo in a specially simple way the action of the J_+ and J_- operators: as an example, for any couple of vectors u, v in the crystal basis $\mathbf{B}$, one gets $u = J_+ v$ if and only if $v = J_- u$. More interesting for our purpose is the crystal basis in the tensorial product of two representations. Then the following theorem holds[8] (written here in the case of $sl(2)$):

Theorem 2.1. *Let* $\mathbf{B}_1$ *and* $\mathbf{B}_2$ *be the crystal bases of the* M_1 *and* M_2 $\mathcal{U}_{q\to 0}(sl(2))$*-modules respectively. Then for* $u \in \mathbf{B}_1$ *and* $v \in \mathbf{B}_2$*, we have:*

$$J_-(u \otimes v) = \begin{cases} J_- u \otimes v \; \exists\, n \geq 1 \text{ such that } J_-^n u \neq 0 \text{ and } J_+^n v = 0 \\ u \otimes J_- v \text{ otherwise} \end{cases} \qquad (2)$$

$$J_+(u \otimes v) = \begin{cases} u \otimes J_+ v \; \exists\, n \geq 1 \text{ such that } J_+^n v \neq 0 \text{ and } J_-^n u = 0 \\ J_+ u \otimes v \text{ otherwise} \end{cases} \qquad (3)$$

Note that the tensor product of two representations in the crystal basis is not commutative. In the case of our model, we only need to construct the n-fold tensor product of the fundamental representation $(\frac{1}{2}, \frac{1}{2})$ of $\mathcal{U}_{q\to 0}(sl(2) \oplus sl(2))$ by itself.

In Table 2 we report the assignments of the codons of the eukariotic code (the upper label denotes different irreducible representations) and, respectively the amino-acid content of the $\otimes^3(\frac{1}{2}, \frac{1}{2})$ representations. The codon content in each of the obtained irreducible representations is also expressed at the end of this subsection.

Let us insist on the choice of the crystal basis, which exists only in the limit $q \to 0$. In a codon the order of the nucleotides is of fundamental importance (e.g. CCU $\to$ Pro, CUC $\to$ Leu, UCC $\to$ Ser). If we want to consider the codons as composite states of the (elementary) nucleotides, this surely cannot be done in the framework of Lie (super)algebras. Indeed in the Lie theory, the composite states are obtained by performing tensor products of the fundamental irreducible representations. They appear as linear combinations of the elementary states, with symmetry properties determined from the tensor product (i.e. for $sl(n)$, by the structure of the corresponding Young tableaux).

On the contrary the crystal basis provides us with the mathematical structure to build composite states as *pure* states, characterised by the order of the constituents. In order to dispose of such a basis, we need to consider the limit $q \to 0$. Note that in this limit we do not deal anymore either with a Lie algebra or with an universal deformed enveloping algebra.

To represent a codon, we have to perform the tensor product of three $(\frac{1}{2},\frac{1}{2})$ representations of $\mathcal{U}_{q\to 0}(sl(2)\oplus sl(2))$. However, it is well-known (see Table 1) that in a multiplet of codons relative to a specific amino-acid, the two first bases constituent of a codon are "relatively stable", the degeneracy being mainly generated by the third nucleotide. We consider first the tensor product:

$$(\tfrac{1}{2},\tfrac{1}{2}) \otimes (\tfrac{1}{2},\tfrac{1}{2}) = (1,1) \oplus (1,0) \oplus (0,1) \oplus (0,0) \tag{4}$$

where inside the parenthesis, $j = 0, \frac{1}{2}, 1$ is put in place of the $2j+1 = 1,2,3$ respectively dimensional $sl(2)$ representation. We get, using Theorem 2.1, the following tableau:

$$\begin{array}{llll} \to su(2)_H & (0,0) \;\; (\mathrm{CA}) & & (1,0) \;\; (\mathrm{CG\;UG\;UA}) \\ \downarrow & & & \\ su(2)_V & (0,1) \begin{pmatrix} \mathrm{CU} \\ \mathrm{GU} \\ \mathrm{GA} \end{pmatrix} & & (1,1) \begin{pmatrix} \mathrm{CC\;UC\;UU} \\ \mathrm{GC\;AC\;AU} \\ \mathrm{GG\;AG\;AA} \end{pmatrix} \end{array} \tag{5}$$

From Table 2, the dinucleotide states formed by the first two nucleotides in a codon can be put in correspondence with quadruplets, doublets or singlets of codons relative to an amino-acid. Note that the sextets (resp. triplets) are viewed as the sum of a quadruplet and a doublet (resp. a doublet and

Table 2. Assignments of the codons of the eukariotic code in the crystal basis model. The upper label denotes different irreducible representations.

codon	amino acid	J_H	J_V	$J_{H,3}$	$J_{V,3}$	codon	amino acid	J_H	J_V	$J_{H,3}$	$J_{V,3}$
CCC	Pro P	3/2	3/2	3/2	3/2	UCC	Ser S	3/2	3/2	1/2	3/2
CCU	Pro P	(1/2	3/2)^{1}	1/2	3/2	UCU	Ser S	(1/2	3/2)^{1}	−1/2	3/2
CCG	Pro P	(3/2	1/2)^{1}	3/2	1/2	UCG	Ser S	(3/2	1/2)^{1}	1/2	1/2
CCA	Pro P	(1/2	1/2)^{1}	1/2	1/2	UCA	Ser S	(1/2	1/2)^{1}	−1/2	1/2
CUC	Leu L	(1/2	3/2)^{2}	1/2	3/2	UUC	Phe F	3/2	3/2	−1/2	3/2
CUU	Leu L	(1/2	3/2)^{2}	−1/2	3/2	UUU	Phe F	3/2	3/2	−3/2	3/2
CUG	Leu L	(1/2	1/2)^{3}	1/2	1/2	UUG	Leu L	(3/2	1/2)^{1}	−1/2	1/2
CUA	Leu L	(1/2	1/2)^{3}	−1/2	1/2	UUA	Leu L	(3/2	1/2)^{1}	−3/2	1/2
CGC	Arg R	(3/2	1/2)^{2}	3/2	1/2	UGC	Cys C	(3/2	1/2)^{2}	1/2	1/2
CGU	Arg R	(1/2	1/2)^{2}	1/2	1/2	UGU	Cys C	(1/2	1/2)^{2}	−1/2	1/2
CGG	Arg R	(3/2	1/2)^{2}	3/2	−1/2	UGG	Trp W	(3/2	1/2)^{2}	1/2	−1/2
CGA	Arg R	(1/2	1/2)^{2}	1/2	−1/2	UGA	Ter	(1/2	1/2)^{2}	−1/2	−1/2
CAC	His H	(1/2	1/2)^{4}	1/2	1/2	UAC	Tyr Y	(3/2	1/2)^{2}	−1/2	1/2
CAU	His H	(1/2	1/2)^{4}	−1/2	1/2	UAU	Tyr Y	(3/2	1/2)^{2}	−3/2	1/2
CAG	Gln Q	(1/2	1/2)^{4}	1/2	−1/2	UAG	Ter	(3/2	1/2)^{2}	−1/2	−1/2
CAA	Gln Q	(1/2	1/2)^{4}	−1/2	−1/2	UAA	Ter	(3/2	1/2)^{2}	−3/2	−1/2
GCC	Ala A	3/2	3/2	3/2	1/2	ACC	Thr T	3/2	3/2	1/2	1/2
GCU	Ala A	(1/2	3/2)^{1}	1/2	1/2	ACU	Thr T	(1/2	3/2)^{1}	−1/2	1/2
GCG	Ala A	(3/2	1/2)^{1}	3/2	−1/2	ACG	Thr T	(3/2	1/2)^{1}	1/2	−1/2
GCA	Ala A	(1/2	1/2)^{1}	1/2	−1/2	ACA	Thr T	(1/2	1/2)^{1}	−1/2	−1/2
GUC	Val V	(1/2	3/2)^{2}	1/2	1/2	AUC	Ile I	3/2	3/2	−1/2	1/2
GUU	Val V	(1/2	3/2)^{2}	−1/2	1/2	AUU	Ile I	3/2	3/2	−3/2	1/2
GUG	Val V	(1/2	1/2)^{3}	1/2	−1/2	AUG	Met M	(3/2	1/2)^{1}	−1/2	−1/2
GUA	Val V	(1/2	1/2)^{3}	−1/2	−1/2	AUA	Ile I	(3/2	1/2)^{1}	−3/2	−1/2
GGC	Gly G	3/2	3/2	3/2	−1/2	AGC	Ser S	3/2	3/2	1/2	−1/2
GGU	Gly G	(1/2	3/2)^{1}	1/2	−1/2	AGU	Ser S	(1/2	3/2)^{1}	−1/2	−1/2
GGG	Gly G	3/2	3/2	3/2	−3/2	AGG	Arg R	3/2	3/2	1/2	−3/2
GGA	Gly G	(1/2	3/2)^{1}	1/2	−3/2	AGA	Arg R	(1/2	3/2)^{1}	−1/2	−3/2
GAC	Asp D	(1/2	3/2)^{2}	1/2	−1/2	AAC	Asn N	3/2	3/2	−1/2	−1/2
GAU	Asp D	(1/2	3/2)^{2}	−1/2	−1/2	AAU	Asn N	3/2	3/2	−3/2	−1/2
GAG	Glu E	(1/2	3/2)^{2}	1/2	−3/2	AAG	Lys K	3/2	3/2	−1/2	−3/2
GAA	Glu E	(1/2	3/2)^{2}	−1/2	−3/2	AAA	Lys K	3/2	3/2	−3/2	−3/2

a singlet). Let us define the "charge" Q of a dinucleotide state by

$$Q = J_{H,3} + \frac{1}{4} C_V (J_{V,3} + 1) - \frac{1}{4} \tag{6}$$

$J_{\alpha,3}$ ($\alpha = H, V$) stands for the diagonalised $sl(2)_\alpha$ generator. The operator C_α is a Casimir operator of $\mathcal{U}_{q\to 0}(sl(2)_\alpha)$ in the crystal basis. It commutes with $J_{\alpha\pm}$ and $J_{\alpha,3}$ and its eigenvalues on any vector basis of an irreducible representation of highest weight j is $j(j+1)$, that is the same as the undeformed standard second degree Casimir operator of $sl(2)$. Its explicit expression is

$$C_\alpha = (J_{\alpha,3})^2 + \tfrac{1}{2} \sum_{n\in\mathbb{Z}_+} \sum_{k=0}^{n} (J_{\alpha-})^{n-k}(J_{\alpha+})^n(J_{\alpha-})^k \tag{7}$$

Note that for $sl(2)_{q\to 0}$ the Casimir operator is an infinite series of powers of $J_{\alpha\pm}$. However in any finite irreducible representation only a finite number of terms gives a non-vanishing contribution.

The dinucleotide states are then split into two octets with respect to the charge Q: the eight *strong* dinucleotides associated to the quadruplets (as well as those included in the sextets) of codons satisfy $Q > 0$, while the eight *weak* dinucleotides associated to the doublets (as well as those included in the triplets) and eventually to the singlets of codons satisfy $Q < 0$. Let us remark that by the change $C \leftrightarrow A$ and $U \leftrightarrow G$, which is equivalent to the change of the sign of $J_{\alpha,3}$ or to reflexion with respect to the diagonals of the Eq. (1), the 8 strong dinucleotides are transformed into weak ones and vice-versa.

If we consider the three-fold tensor product, the content into irreducible representations of $\mathcal{U}_{q\to 0}(sl(2) \oplus sl(2))$ is given by:

$$(\tfrac{1}{2},\tfrac{1}{2}) \otimes (\tfrac{1}{2},\tfrac{1}{2}) \otimes (\tfrac{1}{2},\tfrac{1}{2}) = (\tfrac{3}{2},\tfrac{3}{2}) \oplus 2\,(\tfrac{3}{2},\tfrac{1}{2}) \oplus 2\,(\tfrac{1}{2},\tfrac{3}{2}) \oplus 4\,(\tfrac{1}{2},\tfrac{1}{2}) \tag{8}$$

The structure of the irreducible representations of the r.h.s. of Eq. (8) is (the upper labels denote different irreducible representations):

$$(\tfrac{3}{2},\tfrac{3}{2}) \equiv \begin{pmatrix} \text{CCC UCC UUC UUU} \\ \text{GCC ACC AUC AUU} \\ \text{GGC AGC AAC AAU} \\ \text{GGG AGG AAG AAA} \end{pmatrix}$$

$$(\tfrac{3}{2},\tfrac{1}{2})^1 \equiv \begin{pmatrix} \text{CCG UCG UUG UUA} \\ \text{GCG ACG AUG AUA} \end{pmatrix}$$

$$(\tfrac{3}{2},\tfrac{1}{2})^2 \equiv \begin{pmatrix} \text{CGC UGC UAC UAU} \\ \text{CGG UGG UAG UAA} \end{pmatrix}$$

$$(\tfrac{1}{2},\tfrac{3}{2})^1 \equiv \begin{pmatrix} \text{CCU UCU} \\ \text{GCU ACU} \\ \text{GGU AGU} \\ \text{GGA AGA} \end{pmatrix} \qquad (\tfrac{1}{2},\tfrac{3}{2})^2 \equiv \begin{pmatrix} \text{CUC CUU} \\ \text{GUC GUU} \\ \text{GAC GAU} \\ \text{GAG GAA} \end{pmatrix}$$

$$(\tfrac{1}{2},\tfrac{1}{2})^1 \equiv \begin{pmatrix} \text{CCA UCA} \\ \text{GCA ACA} \end{pmatrix} \qquad (\tfrac{1}{2},\tfrac{1}{2})^2 \equiv \begin{pmatrix} \text{CGU UGU} \\ \text{CGA UGA} \end{pmatrix}$$

$$(\tfrac{1}{2},\tfrac{1}{2})^3 \equiv \begin{pmatrix} \text{CUG CUA} \\ \text{GUG GUA} \end{pmatrix} \qquad (\tfrac{1}{2},\tfrac{1}{2})^4 \equiv \begin{pmatrix} \text{CAC CAU} \\ \text{CAG CAA} \end{pmatrix}$$

2.2. *Applications*

2.2.1. *Sum rules of codon usage probabilities*

Let XZN be a codon in a multiplet encoding an amino acid, where the labels X, Z, N stands for any of the four bases $A, C, G, U/T$. We define the relative frequency of usage of the codon XZN as the ratio between the number of times n_{XZN} the codon XZN is used in the biosynthesis of the amino acid, and the total number n_{tot} of synthesised amino acid,. Then, the frequency of usage of a codon in a multiplet is connected, in the limit of *very large* n_{tot}, to its probability of usage $P(XZN)$:

$$P(XZN) = \lim_{n_{tot}\to\infty} \frac{n_{XZN}}{n_{tot}} \tag{9}$$

with the normalization

$$P(XZA) + P(XZC) + P(XZG) + P(XZU) = 1 \tag{10}$$

The pattern of codon usage varies between species and even among tissues within a species. Most of the analyses of the codon usage frequencies have adressed to analyze the relative abundance of specified codons in different genes of the same biological species or in the comparison of the relative abundance in the same gene for different biological species. No attention, at our knowledge, has been paid to analyse codon usage frequency summed over the whole available sequences to infer global correlations between different biological species.

The aim of the paper[3] was to investigate this aspect and to predict a general law which should be satisfied by all the biological species belonging to vertebrates.

From the definition of the usage probability for a codon XZN, see Eq. (9), it follows that our analysis and predictions hold for biological species with large enough statistics of codons. In the crystal basis model of the genetic code, each codon XZN is described by a state belonging to an irreducible representation denoted $(J_H, J_V)^\xi$ (ξ specifying the representation) of the algebra $\mathcal{U}_q(sl(2)_H \oplus sl(2)_V)$ in the limit $q \to 0$. It is natural in this model to write the usage probability as a function of the biological species (b.s.), of the particular amino-acid and of the labels J_H, J_V, $J_{H,3}$, $J_{V,3}$ describing the state XZN.

Assuming the dependence of the amino-acid to be completely determined by the set of labels Js, we write

$$P(XZN) = P(b.s.; J_H, J_V, J_{H,3}, J_{V,3}) \tag{11}$$

Let us now make the hypothesis that we can write the r.h.s. of Eq. (11) as the sum of two contributions: a universal function ρ independent on the biological species at least for vertebrates and a b.s. depending function f_{bs}, i.e.

$$P(XZN) = \rho^{XZ}(J_H, J_V, J_{H,3}, J_{V,3}) \; + \; f_{bs}^{XZ}(J_H, J_V, J_{H,3}, J_{V,3}) \tag{12}$$

From the analysis of the available data, we assume that the contribution of f_{bs} is not negligible but could be smaller than the one due to ρ. As each state describing a codon is labelled by the *quantum* labels of two commuting $sl(2)$, it is reasonable, at first approximation, to assume

$$f_{bs}^{XZ}(J_H, J_V, J_{H,3}, J_{V,3}) \approx F_{bs}^{XZ}(J_H; J_{H,3}) \; + \; G_{bs}^{XZ}(J_V; J_{V,3}) \tag{13}$$

Now, let us analyse in the light of the above considerations the usage probability for the quartets Ala, Gly, Pro, Thr and Val and for the quartet sub-part of the sextets Arg (i.e. the codons of the form CGN), Leu (i.e. CUN) and Ser (i.e. UCN).

For Thr, Pro, Ala and Ser we can write, using Table 2 and Eqs. (11)–(13), with $N = A, C, G, U$,

$$P(NCC) + P(NCA) = \rho_{C+A}^{NC} + F_{bs}^{NC}(\tfrac{3}{2}; x) + G_{bs}^{NC}(\tfrac{3}{2}; y) + F_{bs}^{NC}(\tfrac{1}{2}; x') + G_{bs}^{NC}(\tfrac{1}{2}; y') \quad (14)$$

where we have denoted by ρ_{C+A}^{NC} the sum of the contribution of the universal function (i.e. not depending on the biological species) ρ relative to NCC and NCA, while the labels x, y, x', y' depend on the nature of the first two nucleotides NC, see Table 2. For the same amino acid we can also write

$$P(NCG) + P(NCU) = \rho_{G+U}^{NC} + F_{bs}^{NC}(\tfrac{3}{2}; x) + G_{bs}^{NC}(\tfrac{3}{2}; y) + F_{bs}^{NC}(\tfrac{1}{2}; x') + G_{bs}^{NC}(\tfrac{1}{2}; y') \quad (15)$$

Using the results of Table 2, we can remark that the difference between Eq. (14) and Eq. (15) is a quantity independent of the biological species,

$$P(NCC) + P(NCA) - P(NCG) - P(NCU) = \rho_{C+A}^{NC} - \rho_{G+U}^{NC} = \text{Const.} \quad (16)$$

In the same way, considering the cases of Leu, Val, Arg and Gly, we obtain with $W = C, G$

$$\begin{aligned} P(WUC) + P(WUA) - P(WUG) - P(WUU) &= \rho_{C+A}^{WU} - \rho_{G+U}^{WU} = \text{Const.} \\ P(CGC) + P(CGA) - P(CGG) - P(CGU) &= \rho_{C+A}^{CG} - \rho_{G+U}^{CG} = \text{Const.} \\ P(GGC) + P(GGA) - P(GGG) - P(GGU) &= \rho_{C+A}^{GG} - \rho_{G+U}^{GG} = \text{Const.} \end{aligned} \quad (17)$$

Since the probabilities for one quadruplet are normalised to one, from Eqs. (15)–(17) we deduce that for all the eight amino acids the sum of probabilities of codon usage for codons with last A and C (or U and G) nucleotide is independent of the biological species, i.e.

$$P(XZC) + P(XZA) = \text{Const.} \qquad (XZ = NC, CU, GU, CG, GG) \quad (18)$$

Moreover, assuming that for sextets the functions F and G depend really on the nature of the encoded amino acid rather than on the dinucleotide, we derive in a completely analogous way as above that for the amino acid Ser the sum $P'_{C+A}(S) = P(UCA) + P(AGC)$ is independent of the biological species. Note the that we normalize to 1 the probabilities of a quartet in a sextet.

A statistical discussion of the sum rules, in the more general context of correlations between the probabilities $P(XZN)$, can be found in[10].

An analysis with more recent data for more biological species can be found in[11].

2.3. *Physico-chemical properties of amino-acids: Relations and predictions*

It is a known observation that a relationship exists between the codons and the physical-chemical properties of the coded amino acids. The observed pattern is read either as a relic of some kind of interaction between the amino acids and the nucleotides at an early stage of evolution or as the existence of a mechanism relating the properties of codons with those of amino acids.

It is also observed that the relationship depends essentially on the nature of the *second* nucleotide in the codons and it holds when the second nucleotide is A, U, C, not when it is G. To our knowledge neither the anomalous behaviour of G nor the existence of a closest relationship between some of the amino acids is understood. In[4] we provided an explanation of both these facts in the framework of the crystal basis model of the genetic code.

2.3.1. *Relationship between the physical-chemical properties of amino acids*

We assume that some physical-chemical property of a given amino acid are related to the nature of the codons, in particular they depend on the following mathematical features, written in hierarchical order:

(1) the irreducible representation of the dinucleotide formed by the first two nucleotides;
(2) the sign of the charge Q Eq. (6) on the dinucleotide state;
(3) the value of the third component of $J_{V,3}$ inside a fixed irreducible representations for the dinucleotides;
(4) the upper label(s) of the codon irreducible representation(s);

Not all the physical-chemical properties are supposed to follow the scheme above; some of them are essentially given by the specific chemical structure of the amino acid itself. In the following, we analyse the physical-chemical properties of the amino acids in the light of the dinucleotide content of the irreducible representations of Eq. (5).

– Representation (0,0): the codons of the form CAN (N = C, U, G, A) all belong to the irreducible representation $(\frac{1}{2}, \frac{1}{2})^4$ and code for His and Gln, both being coded by doublets and differing by the value of $J_{V,3}$. Then we expect that the physical-chemical properties of His and Gln are very close.

– Representation (1,0): we analyse the codons CG ($Q > 0$), UG, UA (both $Q < 0$). The codons CGS (S = C, G), resp. CGW (W = U, A), belonging to irreducible representation $(3/2, 1/2)^2$, resp. $(1/2, 1/2)^2$, all code for Arg, so we do not have any relation. The codons UGS, resp. UGW, belonging to irreducible representation $(3/2, 1/2)^2$, resp. $(1/2, 1/2)^2$, code for Cys and Trp, resp. the other Cys and Ter. So we expect some affinity between the physical-chemical properties of Cys and Trp, not very strong indeed as the former is encoded by a doublet and the latter by a singlet. The codons UAN, belonging to the irreducible representation $(3/2, 1/2)^2$, code for the Tyr and Ter. So we expect some affinity between the amino acids coded by UGN and UAN, in particular between Cys and Tyr both being coded by doublets.

– Representation (0,1): we analyse the codons CU, GU (both $Q > 0$) and GA ($Q < 0$). The codons CUY and GUY (Y = C, U), resp. CUR and GUR (R = G, A), belonging to irreducible representation $(1/2, 3/2)^2$, resp. $(1/2, 1/2)^3$, code for Leu and Val. Therefore we do not have any relation between amino acids coded by the same dinucleotide, but we expect that the physical-chemical properties of Leu and Val are close since CU and GU both belong to the same irreducible representation and are both strong. The codons GAN belong to the irreducible representation $(1/2, 3/2)^2$ and they code Asp and Glu (both doublets). Then we expect the physical-chemical properties of Asp and Glu to be very close.

– Representation (1,1): the dinucleotide irreducible representation $(1, 1)$ contains five states with $Q > 0$ (CC, UC, GC, AC, GG). The codons CCN and UCN (resp. GCN and ACN) belong to four different irreducible representations and code for Pro and Ser (resp. Ala and Thr). We expect a strong affinity between the physical-chemical properties of Pro and Ser on the one hand and between the physical-chemical properties of Ala and Thr on the other hand. The codons GGN belong to two different irreducible representations and code for Gly, so we expect an affinity of physical-chemical properties of Gly with those of Pro, Ser, Ala, Thr. Now let us look at the four states with $Q < 0$ (UU, AU, AG, AA). The codons UUN belong to two different irreducible representations and code for Leu, the doublet subpart of the sextet, and for Phe (doublet). An affinity is expected between the physical-chemical properties of these two amino acids. The codons AUN belong to two different irreducible representations and code Ile (triplet) and Met (singlet) and, in fact, the values of physical-chemical properties of these two amino acids are not very different. The codons AGN belong to two different irreducible representations and code for Ser and Arg, the doublet subpart of the sextet, so an affinity between the physical-chemical properties of these codons is expected. The codons AAN belong to the same irreducible representation $(3/2, 3/2)$ and code for Asn and Lys, so the values of the physical-chemical properties of these amino acids should be close.

Note that for the three sextets (Arg, Leu, Ser) the quartet (doublet) subpart is coded by a codon with a strong (weak) dinucleotide.

2.3.2. *Discussion*

We have compared our theoretical predictions with 10 physical-chemical properties:

- the Chou-Fasman conformational parameters P_α, P_β and P_τ which gives a measure of the probability of the amino acids to form respectively a helix, a sheet and a turn. The sum $P_\alpha + P_\beta$ appears more appropriate to characterise the generic structure forming potential and the difference $P_\alpha - P_\beta$ the helix forming potential, this quantity depending more on the particular amino acid. So we compare with $P_\alpha + P_\beta$ and P_τ;
- the Grantham polarity P_G;
- the relative hydrophilicity R_f;
- the thermodynamic activation parameters at 298 K: ΔH (enthalpy, in kJ/mol), ΔG (free energy, in kJ/mol) and ΔS (entropy, in J/mole/K);
- the negative of the logarithm of the dissociation constants at 298 K: pK_a for the α-COOH group and pK_b for the α-NH_3^+ group;
- the isoelectronic point pI, i.e. the pH value at which no electrophoresis occurs.

The comparison between the theoretical relations and the experimental values shows:

($\cong$ means strong affinity, $\approx$ affinity, $\sim$ weak affinity):

- His $\cong$ Gln – The agreement, except for pI, is very good.
- Asp $\cong$ Glu – The agreement, except for P_τ, is very good.
- Asn $\cong$ Lys $\sim$ Arg, Ser – The agreement, except for pI and P_τ is very good. The comparison with the values of physical-chemical properties of Ser and Arg is satisfactory.
- Cys $\cong$ Tyr $\approx$ Trp – Except for R_f, the agreement between the first two amino acids is very good, while with Trp is satisfactory.

- Leu $\cong$ Val – The agreement is very good.
- Pro $\cong$ Ser $\approx$ Gly – The agreement is very good, except for $P_\alpha + P_\beta$ and ΔH, and with Gly more than satisfactory.
- Ala $\cong$ Thr $\approx$ Gly, Pro, Ser – The agreement is very good between the first two amino acids except for P_τ and satisfactory with the others except for the conformational parameters.
- Ile $\cong$ Met $\approx$ Phe – The agreement is very good between the first two amino acids and satisfactory with Phe.

So we predict that for Asp and Glu, one should find $\Delta H \approx 60$ kJ/mol, $-\Delta S \approx 135$ kJ/mol/K and $\Delta G \approx 100$ kJ/mol.

In conclusion, the values of physical-chemical properties show, with a few exceptions, a pattern of correlations which is expected from the assumptions of the crystal basis model. The remarked property that the amino acids coded by codons whose second nucleotide is G do not share similarity in the physical-chemical properties with other amino acids does find an explication in the model, as it is immediate to verify that there are no two states with G in second position which share simultaneously the properties of belonging to the same irreducible representation and being characterised by the same value of Q.

More details and illustrative Tables can be found in[4].

3. PART 2: A "Minimum" Principle in the Genetic Code

3.1. *A "minimum" principle in the mRNA editing*

The "minimum" principles, in their different formulations, have played and play a very relevant role in any mathematically formulated scientific theory. The key point of a "minimum" principle is to state that an event happens along the path that minimizes a suitable function. The mathematical formulation of a sequence in RNA or DNA in the crystal basis model allows to investigate if some "minimum" principle can be applied to the genetic code.

In[9], we have investigated the possibility to explain the position of a nucleotide insertion in mRNA, the so called mRNA editing. The deep mechanism which causes RNA editing is still unknown. The understanding of the event is complicated: from a thermodynamics point of view a change, i.e. C $\to$ U, takes place if it is favored in the change of entalpy or entropy, but should this be the case, the change should appear in all the organisms. Moreover from a microscopic (quantum mechanical) point of view, the change should occur in both directions, i.e. C $\leftrightarrow$ U. It seems that the primary aim of mRNA editing is the evolution and conservation of protein structures, creating a meaningful coding sequence specific for a particular amino acid sequence.

The purpose of the paper[9] was to propose an effective model to describe the RNA editing. Our model does not explain why, where and in which organisms editing happens, but it gives a framework to understand some specific features of the phenomenon.

A consequence of the crystal basis model is that any nucleotide sequence is characterized as an element of a vector space. Therefore, functions can be defined on this space and can be computed on the sequence of codons. In particular any codon is identified by a set of four half-integer labels and functions can be defined on the codons. We make the assumption that the location sites for the insertion of a nucleotide should minimize the following function for the mRNA or cDNA

$$\mathcal{A}_0 = \exp\left[-\sum_k 4\alpha_c C_H^k + 4\beta_c C_V^k + 2\gamma_c J_{3,H}^k\right] \tag{19}$$

where the sum in k is over all the codons in the edited sequence, C_H^k (C_V^k) and $J_{3,H}^k$ ($J_{3,V}^k$), are the values of the *Casimir* operator, see Eq. (7) and of the third component of the generator of the $sl(2)_H$ ($sl(2)_V$), in the irreducible representation to which the kth codon belongs, see Table 2. In (19) the simplified assumption that the dependence of $\mathcal{A}_0$ on the irreducible representation to which the codon belongs is given only by the values of the Casimir operators has been made. The parameters $\alpha_c, \beta_c, \gamma_c$ are constants, depending on the biological species.

The minimum of $\mathcal{A}_0$ has to be computed in the whole set of configurations satisfying to the constraints: (i) the starting point should be the mtDNA and (ii) the final peptide chain should not be modified. It is obvious that the global minimization of expression Eq. (19) is ensured if $\mathcal{A}_0$ takes the smallest value locally, i.e. in the neighborhood of each insertion site. The form of the function $\mathcal{A}_0$ is rather arbitrary; one of the reasons of this choice is that the chosen expression is computationally quite easily tractable. If the parameters $\alpha_c, \beta_c, \gamma_c$ are strictly positive with $\gamma_c/6 > \beta_c > \alpha_c$, the minimization of Eq. (19) explains the observed configurations in almost all the considered cases, for more details see[9].

3.2. *A "minimum" principle in the interaction codon-anticodon*

Given a codon[c] XYZ ($X, Y, Z \in \{C, A, G, U\}$) we conjecture that an anticodon $X^a Y^a Z^a$, where $Y^a Z^a = Y_c X_c$, N_c denoting the nucleotide complementary to the nucleotide N according to the Watson-Crick pairing rule,[d] pairs to the codon XYZ, i.e. it is most used to "read" the codon XYZ if it minimizes the operator $\mathcal{T}$, explicitly written in Eq. (20) and computed between the "states", which can be read from Table 3, describing the codon and anticodon in the "crystal basis model". We write both codons (c) and anticodons (a) in 5" → 3" direction. As an anticodon is antiparallel to codon, the 1st nucleotide (respectively the 3rd nucleotide) of the anticodon is paired to the 3rd (respectively the 1st) nucleotide of the codon.

$$\mathcal{T} = 8c_H\, \vec{J^c_H} \cdot \vec{J^a_H} + 8c_V\, \vec{J^c_V} \cdot \vec{J^a_V} \tag{20}$$

where:

- $c_H.c_V$ are constants depending on the "biological species" and weakly depending on the encoded a.a., as we will later specify.
- J^c_H, J^c_V (resp. J^a_H, J^a_V) are the labels of $\mathcal{U}_{q\to 0}(su(2)_H \oplus su(2)_V)$ specifying the state
  describing the codon XYZ (resp. the anticodon NY_cX_c pairing the codon XYZ).
- $\vec{J^c_\alpha} \cdot \vec{J^a_\alpha}$ $(\alpha = H, V)$ should be read as

  $$\vec{J^c_\alpha} \cdot \vec{J^a_\alpha} = \frac{1}{2}\left\{\left(\vec{J^c_\alpha} \oplus \vec{J_\alpha}^a\right)^2 - (\vec{J^c_\alpha})^2 - (\vec{J^a_\alpha})^2\right\} \tag{21}$$

  and $\vec{J^c_\alpha} \oplus \vec{J^a_\alpha} \equiv \vec{J^T_\alpha}$ stands for the irreducible representation which the codon-anticodon state under consideration belongs to, the tensor product of $\vec{J^c_\alpha}$ and $\vec{J^a_\alpha}$ being performed according to the rule of[8], choosing the codon as first vector and the anticodon as second vector. Note that $\vec{J_\alpha}^2$ should be read as the Casimir operator whose eigenvalues are given by $J_\alpha(J_\alpha + 1)$.

As we are interested in finding the composition of the 22 anticodons, minimun number to ensure a faihful translation, we shall assume that the

[c] In the paper we use the notation $N = C, A, G, U.$; $R = G, A.$ (*purine*); $Y = C, U.$ (*pyrimidine*).

[d] This property is observed to be verified in most, but not in all, the observed cases. To simplify we shall assume it.

used anticodon for each quartet and each doublet is the one which minimizes the averaged value of the operator given in Eq. (20), the average being performed over the 4 (2) codons for quadruplets (doublets), see next section. We have found that the anticodons minimizing the conjectured operator $\mathcal{T}$ given in Eq. (20), averaged over the concerned multiplets, are in very good agreement, the results depending only on the signs of the two coupling constants, with the observed ones, even if we have made comparison with a limited database.

The fact that the crystal basis model is able to explain, in a relatively simple way, the observed anticodon-codon pairing which has its roots on the stereochemical properties of nucleotides strongly suggests that our modelisation is able to incorporate some crucial features of the complex physico-chemical structure of the genetic code. Incidentally let us remark that the model explains the symmetry codon anticodon remarked. Let us stress that our modelisation has a very peculiar feature which makes it very different from the standard 4-letter alphabet, used to identify the nucleotides, as well as with the usual modelisation of nucleotide chain as spin chain. Indeed the identification of the nucleotides with the fundamental irrep. of $\mathcal{U}_q(su(2)_H \oplus su(2)_V)$ introduces a sort of double "bio-spin", which allows the description of any ordered sequence of n nucleotides as as state of an irrep. and allows to describe interactions using the standard powerful mathematical language used in physical spin models.

In the paper[5] we have faced the problem to find the structure of the mimimum set of anticodons and, then, we have used a very simple form for the operator $\mathcal{T}$. We have not at all discussed the possible appearance of any other anticodon, which should require a more quantitative discussion. For such analysis, as well as for the eukaryotic code, the situation may be different and more than an anticodon may pair to a quartet.

The pattern, which in the general case may show up, is undoubtedly more complicated, depending on the biological species and on the concerned biosynthesis process, but it is natural to argue that the usage of anticodons exhibits the general feature to assure an "efficient" translation process by a number of anticodons, minimum with respect to the involved constraints. A more refined and quantitative analysis, which should require more data, depends on the value of these constants.

However our analysis strongly suggests that the minimum number of anticodons should be 32 (3 for the sextets, 2 for quadruplets and triplet and 1 for doublets and singlets).

In conclusion, we have found that the anticodons minimizing the conjectured operator $\mathcal{T}$ given in Eq. (20), averaged over the concerned multiplets, are in very good agreement, the results depending only on the signs of the two coupling constants, with the observed ones, even if we have made comparison with a limited database.

The fact that the crystal basis model is able to explain, in a relatively simple way, the observed anticodon-codon pairing which has its roots on the stereochemical properties of nucleotides strongly suggests that our modelisation is able to incorporate some crucial features of the complex physico-chemical structure of the genetic code. Incidentally let us remark that the model explains the symmetry codon anticodon remarked. Let us stress that our modelisation has a very peculiar feature which makes it very different from the standard 4-letter alphabet, used to identify the nucleotides, as well as with the usual modelisation of nucleotide chain as spin chain. Indeed the identification of the nucleotides with the fundamental irrep. of $\mathcal{U}_q(su(2)_H \oplus su(2)_V)$ introduces a sort of double "bio-spin", which allows the description of any ordered sequence of n nucleotides as as state of an irrep. and allows to describe interactions using the standard powerful mathematical language used in physical spin models.

3.3. *The "minimum" principle in the evolution of genetic code*

Using the minimum principle stated in Subsection 3.2 in[6] we have analyzed and mathematically modellised the evolution of the genetic code in the framework on the so called "codon capture theory".
Let us briefly summarize and comment our results.

We determine the structure of the anticodons in the Ancient, Archetypal and Early Genetic codes, that are all reconciled in a unique frame. Most of our results agree with the generally accepted scheme. Moreover the pattern of the model is surprisingly coherent.

The pattern of the Ancient Code can be summarized by saying that in this primordial code the a.a., which would be encoded by a doublet of the type XZY are encoded by a codon ending with a C. Similarly the a.a. which would be encoded by a doublet of the type XZR are encoded by a codon ending with a G.

Indeed, in the Ancient Genetic Code, the sign of c_V for the weak dinucleotides is undetermined, i.e. the minimization does not depend on the sign of c_V. In our model, this means that there is no distinction between

Table 3. The vertebral mitochondrial code. In bold (italic) the anticodons reading quadruplets (resp. doublets).

codon	a.a.	anticodon	codon	a.a.	anticodon
CCC	P		UCC	S	
CCU	P		UCU	S	
CCG	P	**UGG**	UCG	S	**UGA**
CCA	P		UCA	S	
CUC	L		UUC	F	
CUU	L		UUU	F	*GAA*
CUG	L	**UAG**	UUG	L	
CUA	L		UUA	L	*UAA*
CGC	R		UGC	C	
CGU	R		UGU	C	*GCA*
CGG	R	**UCG**	UGG	W	
CGA	R		UGA	W	*UCA*
CAC	H		UAC	Y	
CAU	H	*GUG*	UAU	Y	*GUA*
CAG	Q		UAG	Te	—
CAA	Q	*UUG*	UAA	Ter	—
GCC	A		ACC	T	
GCU	A		ACU	T	
GCG	A	**UGC**	ACG	T	**UGU**
GCA	A		ACA	T	
GUC	V		AUC	I	
GUU	V		AUU	I	*GAU*
GUG	V	**UAC**	AUG	M	
GUA	V		AUA	M	*CAU*
GGC	G		AGC	S	
GGU	G		AGU	S	*GCU*
GGG	G	**UCC**	AGG	Ter	—
GGA	G		AGA	Ter	—
GAC	D		AAC	N	
GAU	D	*GUC*	AAU	N	*GUU*
GAG	E		AAG	K	
GAA	E	*UUC*	AAA	K	*UUU*

C (U) and G (A). This is coherent since at this stage there is not yet a distinction between the doublet XZR and XZY. On the contrary for strong dinucleotides for which the role of XZR and XZY is the same up to the Standard Genetic Code, the sign is fixed and it does not change during the evolution. For strong dinucleotides and almost half of the weak ones[e] there is a change in c_H just when the codon degeneracy appears, that is going from the Ancient to the Archetypal code, and the "wobble mechanism" is called in. For all weak dinucleotides, the sign of c_V is now determined and there is a further change in the sign of c_H and of c_V when the correspondence between doublets and a.a. is fixed.

Let us remark that:

- for each codon there are at least two anticodons with the same value of $\mathcal{T}$ and viceversa. This degeneracy can be removed by further terms of the interaction, not yet taken into account, but it can be also read as the "codon disappearance" before a new readjustment of the code.
- we remark that the presently less used (in the average) codons, for the a.a. encoded by doublets, are those with last nucleotide G or U, while the most used are those with last nucleotide A or C. So it is natural to ask the question: why most of the ancestral codons encoding a.a. in the Ancient Code are now repressed? Naively, one should expect that the ancestral codon should be the most used one.
- in our model the sign of c_H for a.a. encoded by XZY in the Early Code is the same than the one in the Ancient Code, while for a.a. encoded by XZR is the opposite. Cys is an exception, but in this case the anticodon is different in the two codes. This kind of argument cannot be immediately applied to a.a. encoded by quartets because in most cases the anticodon in the Archetypal, Early or Mitochondrial Code is not the same as the one appearing in the Ancient Code and, moreover, there is an important effect due to the averaging over four codons.

Analogous analysis of the codon usage frequencies for species following the Standard Code confirms generally such a pattern, but the presence of anticodons in the Standard Code is more complicated, so we do not want to refer to these data.

Moreover, in our model naturally the anticodon with first nucleotide A does never appear, in good agreement with the observed data.

In our model we can express the evolution of the genetic code through

---

[e] Let us remark that the sign of c_H does not change for the weak dinucleotides which has the value of $J_{3,H} = 0$.

the following pattern of the codon-anticodon interaction as

$$
\begin{aligned}
&< XZN|\mathcal{T}|N^a Z^a_c X^a_c >= \\
&< XZN| \left(8c_H \vec{J^c_H} \cdot \vec{J^a_H} + 8c_V \vec{J^c_V} \cdot \vec{J^a_V} \right) \delta_{M^a, N^a_c} |M^a Z^a_c X^a_c > \\
&\Longrightarrow_{Evolution} \ < XZN|8c_H \vec{J^c_H} \cdot \vec{J^a_H} + 8c_V \vec{J^c_V} \cdot \vec{J^a_V}|M^a Z^a_c X^a_c > \quad (22)
\end{aligned}
$$

In the first row of Eq. (22), the presence of the Kronecher delta δ_{M^a, N^a_c} enforces the Watson-Crick coupling mechanism implying $M^a = N^a_c$, while in the second row M^a can be any nucleotide and the selection is implemented by the value of the operator $\mathcal{T}$, computed between the concerned states and, eventually, averaged over the multiplet taking into account the codon usage probabilities. As example of typical behavior of the constant c_H for weak dinucleotides, we consider the case of the AA dinucleotide:

$$
c_H^{AAN} > 0 \Longrightarrow c_H^{AAN} < 0 \Longrightarrow \begin{cases} c_H^{AAY} > 0 \\ \\ c_H^{AAR} < 0 \end{cases} \quad (23)
$$

The change of the sign in the coupling constants is a mathematical description to frame the modification of the interaction codon-anticodon due to the change of the molecular structure of the nucleotides in the anticodons and of the (non local) structure of the tRNA.

Of course we have to assume that the constants c_H and c_V depend on the "time" even if, at this stage, only the change of the sign in the coupling constants has been considered.

Presumably the genetic code has not evolved along one path, It could be that multiple branching points showed up in the course of the evolution with the advent of different genetic codes, and then the standard genetic code would have emerged as the one exhibiting selective advantages. For example, one can imagine that not all the changes of the signs of c_H and c_V would have occurred at the same time, and, therefore, that several intermediate codes would have arisen between, say, the Ancient and the Archetypal Code. As a consequence, we believe it is more reasonable not to write a time-dependent evolution equation, but to verify that the existing genetic code, that is the branching point which has survived, satisfies the required optimality conditions.

3.4. *The "minimum" principle to explain the codon bias*

As already stated the genetic code is degenerate in the sense that a multiplet is used to encode most of the amino-acids. Some codons in the multiplets are used much more frequently than others to encode a particular amino-acid, i.e. there is a "codon usage bias". The non-uniform usage of synonymous codons is a widespread phenomenon and it is experimentally observed that the pattern of codon usage varies between species.

The main reasons for the codon usage biases are believed to be: the genetic coding error minimization, the CG content, the abundance of specific anticodons in the tRNA. No clear indication comes out for the existence of one or more factors which universally engender the codon bias, on the contrary the role of some factors is controversial.

In paper[7] we have analyzed possible effects of the codon-anticodon interactions defined by the operator given in Eq. (20) on the codon bias, according to the approach introduced in[5], and to propose semi-quantitative predictions of the codon bias. Moreover we briefly analyzed the codon usage bias variation along the evolution of the genetic code on the basis of the model developed in[6]. In the following, we will be concerned about amino acids encoded by quartets. For the ones encoded by a sextet, that we consider as the sum of a quartet and a doublet, only the quartet will be considered. The method we developed is essentially based on the determination of the minimum values of an operator which can be seen as an interaction potential between a codon and its corresponding anticodon. A possible general pattern of the bias is searched by deriving inequalities for the codon usage probabilities.

With reference to Subsection 3.2 we have to minimize an expression of the type:

$$\begin{aligned} \mathcal{T}_{av}(N'Y''X'', XYN) &= \sum_N P_N < N'Y''X''|\mathcal{T}|XYN > \\ &= P_C < N'Y''X''|\mathcal{T}|XYC > + P_U < N'Y''X''|\mathcal{T}|XYU > \\ &+ P_G < N'Y''X''|\mathcal{T}|XYG > + P_A < N'Y''X''|\mathcal{T}|XYA > \end{aligned} \tag{24}$$

Let us recall that the expression $< N'Y''X''|\mathcal{T}|XYN >$ has to be read as

$$\begin{aligned} &< N'Y''X''|\mathcal{T}|XYN > \equiv \mathcal{T}\ (|XYN > \otimes |N'Y''X'' >) \\ &= \mathcal{T}\ (|J_H^c, J_V^c; J_{H,3}^c, J_{V,3}^c > \otimes |J_H^a, J_V^a; J_{H,3}^a, J_{V,3}^a >) \\ &= \lambda\ (|J_H^c, J_V^c; J_{H,3}^c, J_{V,3}^c > \otimes |J_H^a, J_V^a; J_{H,3}^a, J_{V,3}^a >) \end{aligned} \tag{25}$$

where we have used the correspondence

$$\begin{aligned} |XYN > \quad &\rightarrow \quad |J_H^c, J_V^c; J_{H,3}^c, J_{V,3}^c > \\ |N'Y''X'' > \&; &\rightarrow |J_H^a, J_V^a; J_{H,3}^a, J_{V,3}^a > \end{aligned} \tag{26}$$

and λ is the eigenvalue of $\mathcal{T}$ on the state $|J_H^c, J_V^c; J_{H,3}^c, J_{V,3}^c> \otimes |J_H^a, J_V^a; J_{H,3}^a, J_{V,3}^a>$, see[5] for more details. As the P_{XYN} have to satisfy, in addition to Eq. (10), a set of unknown constraints, we cannot impose the minimization condition in a rigorous manner, so we proceed by a heuristic method. Using the results of Subsection 2.2.1, we are left with only two probabilities in Eq. (24) and we try to argue which from the two present P_{XYN} is enhanced respect to the other one. For this aim we compare the two probabilities which appear after the substitution of the other two, using Eq. (18), that have the greatest coefficient. In this way we will get in our expression a constant terms, depending on c_H, c_V and, generally, on K (K being the constant which appears in the r.h.s. of Eq. (18), which has the highest possible value, without, possibly, any specific assumption on the value of the parameters, except for the assumed sign. Then, in order to minimize the expression, it is reasonable to require that the probability with the lowest coefficient has a higher value than the other one. Nextly, in some cases, we can derive another inequality for the complementary probabilities, according to $K > 0,5$ or $K < 0,5$.

For a more detailed discussion of the difference between the minimization procedure for the Early Genetic Code and the Eukaryotic Genetic Code, as well as on the assumed behavior of the coefficient c_H and c_V we refer to[7].

The outcomes derived, which we summarize in Table 4, are in an amazing agreement with the observed data, nevertheless the over-simplifying assumptions of our theoretical scheme and despite that in the real world the number of operating anticodons is greater that the minimum number 31, which implies that the matching of an a.a. encoded by a quartet is done by more than two anticodons. Moreover let us remark that the results found in the Early Genetic Code survive in the Eukaryotic Genetic Code, suggesting that we have caught some feature of a very relevant mechanism. So we argue that codon-anticodon interaction plays a relevant role in the codon usage bias. Moreover it seems that, in despite of its apparently fragmented behavior, the codon bias exhibits a sort of universal feature that our approach and the Crystal Basis Model is able to take into account. Let us remark that, in general, for plants and bacteria and, for some extent for invertebrates, the agreement is less satisfactory. Likely this reflects the fact that the choice of the specimen for these species is too rough. The experimental data should be taken from smaller, suitably chosen, subsets of the species.

Our model seems to support the idea that the codon usage bias reflects two aspects of the tRNA population: firstly, where there are multiple species of tRNAs with different anticodons, the codons translated by the most abundant tRNA species are preferred; secondly, when a tRNA can translate more than one codon, the codon best recognized by the anticodon is preferred. In our language the most abundant tRNA and the best recognized codon are the ones which minimize the $\mathcal{T}$ operator. The good agreement of our results with data suggests that it may be interesting to perform a more detailed analysis of the two parameters controlling our codon-anticodon interaction potential, in a general context taking into account the complexity and the evolution of species. In particular, what does it mean when similar parameters c_H and c_V conditions allow a good prediction of codon usage for different biological sets? For example, for Pro and Ser, it seems that, on one hand, vertebrates and invertebrates share similar on conditions c_H and c_V and, on the other hand, plants (and bacteria for Pro) do the same. Does the species dependent codon bias depend on the time appearance of the amino acids?

Finally there is much debate on the exact reasons for the selection of translationally optimal codons: to increase the translational efficiency or the accuracy of the translation? At the moment our model does not give at all any indication for the favoured mechanism. Possibly some hints can be obtained by a study of the mutation bias along with the minimization of the codon-anticodon interaction.

4. Conclusions

The above presented model for the genetic code can be seen as a attempt towards a theoretical approach in the complex domain of the sciences of life. Two main notions, already contained in the title of this review, have been used: Symmetry and Minimum Energy Principle. The first one had a particularly spectacular development during the twentieth century, as well as in mathematics under the general label of "Group Theory" as in several domains of fundamental physics such as Relativity, Quantum Physics and High Energy Physics. The second one, generally called the "principle of least action", appeared earlier with the works of Leibniz, Fermat, Euler and is generally attributed to Maupertuis, who, during the *Siècle des Lumières*, felt that *"Nature is thrifty in all its actions"*.

As developed in this contribution, the Group Theory approach we proposed seems well adapted to represent the constituent actors in the genetic code and to describe some of their effects: in other words, the Crystal Basis

Table 4. Inequalities derived in the Early and in the Eukaryotic Genetic Code. The value of the parameters c_H and c_V is different in the two codes.

	Early Code		**Eukaryotic Code**	
a.a.	Parameters	Inequalities	Parameters	Inequalities
Thr	—	$P_C > P_G$	$\|c_H\| < 3\|c_V\|$ $\|c_H\| > 3\|c_V\|$	$P_C > P_U$ $P_C > P_G$
Arg	—	—	$\|c_V\| < \|c_H\| < 2\|c_V\|$ $\|c_H\| < \|c_V\|$	$P_C > P_G$ $P_C > P_U$
Pro	$\|c_V\| < c_H$ $\|c_V\| > c_H$	$P_A > P_U$ $P_A > P_G$	$\|c_V\| < 1/4\|c_H\|$ $\|c_V\| > 1/4\|c_H\|$	$P_C > P_U$ $P_U > P_C, P_A > P_G$
Leu	—	$P_G > P_C$	$\|c_H\| < 2/3\|c_V\|$, $\|c_H\| > 2/3\|c_V\|$,	$P_U > P_C$ $P_C > P_U$, $P_G > P_A$
Ala	$\|c_V\| < c_H$ $\|c_V\| > c_H$	$P_U > P_A$, $P_C > P_G$ $P_U > P_C$, $P_A > P_G$	—	$P_C > P_U$
Gly	—	—	$\|c_H\| < \|c_V\|$ $\|c_H\| > \|c_V\|$	$P_C > P_G$ $P_G > P_C$, $P_A > P_U$
Val	$\|c_V\| < 4c_H$ $\|c_V\| > 4c_H$	$P_C > P_G, P_U > P_A$ $P_C > P_U$, $P_G > P_A$	$\|c_H\| < 3\|c_V\|$ $\|c_H\| > 3\|c_V\|$	$P_C > P_U, P_G > P_A$ $P_C > P_G$, $P_U > P_A$
Ser	$\|c_V\| < 3c_H$ $\|c_V\| > 3c_H$	$P_G > P_C, P_A > P_U$ $P_G > P_A, P_C > P_U$	$\|c_V\| < \frac{5}{4}\|c_H\|$ $\|c_V\| > \frac{5}{4}\|c_H\|$	$P_C > P_U$ $P_U > P_C$, $P_A > P_G$

Model allows a rather powerful "parametrization" of our problem. On the other hand, the least action principle is a more accepted and used notion in the different domains of physics, may be owing to its naturalness. In this spirit, the spin-spin like potential we have built to describe the codon-anticodon interaction can also appear appealing owing to its conceptual simplicity.

For these reasons, it seems to us worthwhile to pursue investigations in the framework of our model. Such developments could be carried out as well as in the mathematical side as in the phenomenological one. Let us for example note the construction of a distance between two sequences of DNA or RNA[13] still in the framework of our model: this work deserves to be developed and applied. In the context of evolution of the Genetic Code, it looks worthwhile to study in more details the behavior of the parameters CH and CV the role of which is determinant. More generally, a refinement of the interaction potential deserves to be considered.

Finally, let us end this contribution by emphasizing that, among the important questions which deserve to be considered, the adaptability of the code with the increasing complexity of the organisms is a crucial one. It will be worthwhile to see to what extend our methods can be used for such a problem. In[14] a mathematical model, always in our framework, has been presented in which the main features (numbers of encoded a.a., dimensions and structure of synonymous codon multiplet) are obtained, requiring stability of the genetic code against mutations modeled by suitable operators.

Thus, our scheme appears rather well adapted to reproduce the features of the codon capture theory as well as to provide a mathematical framework for a more quantitative and detailed description of the theory.

5. Appendix: A Short Tutorial on Group Theory

The aim of this short lesson on symmetry is to provide the necessary informations for the reader who desires to follow the mathematical aspects of group theory used in the crystal basis model developed in this review. The first part deals with general notions of a Lie group, while in the second part the notions of quantum group and so-called crystal basis are presented.

5.1. *A I: Symmetry and group theory*

Definition 1: A group $\mathcal{G}$ is a set of elements together with a composition law — we denote it by " . " — such that:

(1) $\forall\, x, y \in \mathcal{G} \qquad x.y \in \mathcal{G} \qquad (internal\, law)$
(2) $\forall\, x, y, z \in \mathcal{G} \qquad (x.y).z = x.(y.z) \qquad (associativity)$
(3) $\exists\, e \in \mathcal{G},\ |\ \forall\, x \in \mathcal{G} \qquad x.e = e.x = x \qquad (e \equiv identity)$
(4) $\forall\, x \in \mathcal{G},\ |\ \exists\, x^{-1} \in \mathcal{G} \qquad x.x^{-1} = x^{-1}.x = e \qquad (x^{-1} inverse\, of\, x)$

Examples: $Z = \{n\}$, n integer or $\mathbf{R} = \{real\, numbers\}$ with " + " as internal law. $\mathbf{P_n}$: group of permutations of n objects. Set of rotations in the plane around an origin.

Actually, in Physics, groups are never considered abstractly, but as action on some set $\mathbf{S}$: we will talk about group of transformations. More precisely:

Definition 2: Let $\mathcal{G}$ be a group, $\mathbf{S}$ a set. An action of $\mathcal{G}$ on $\mathbf{S}$ is an application: $\mathcal{G} \times \mathbf{S} \to \mathbf{S}$ that is: $(g, s) \to g(s)$ if $g \in \mathcal{G}$ and $s \in \mathbf{S}$ such that: $\forall\, g, g\prime \in \mathcal{G}$ and $\forall\, s \in \mathbf{S} : g(g\prime(s)) = (g.g\prime)(s)$ and $e(s) = s$

A (transformation) group needs to be "represented".

Definition 3: A linear representation of a group $\mathcal{G}$ in a vector space $\mathbf{V}$ (itself defined on $\mathbf{R}$ or on $\mathbf{C}$) is an homomorphism D of $\mathcal{G}$ on the group of linear and invertible operators on $\mathbf{V}$, that is: $\forall\, g \in \mathcal{G} \to D(g)$ such that: $\forall g, g\prime \in \mathcal{G} : D(g).D(g\prime) = D(g.g\prime)$.

We note that $D(g)$ is a $n \times n$ matrix if $\mathbf{V}$ is of (finite) n-dimension.

We like to say that "*Nature is full of symmetries ... and of symmetry breaking*". For instance, let us imagine the three dimensional real space $\mathbf{R_3}$ as an homogeneous isotropic space and let us put (in an idealistic way) an electron e^- at the point O. Then the interaction of e^- with a second charged particle f will only depend of the distance between e^- and f: in other words, the physics will be invariant under the three dimensional group of rotations (we denote it $\mathcal{SO}(3)$, see below) around O. Now, let us introduce a magnetic field $\vec{B}$ going in a certain direction z: the interaction of e^- with $\vec{B}$ will no more be invariant under the whole $\mathcal{SO}(3)$ group, but only under a "subgroup" of it consisting of rotations in a plane perpendicular to the z axis. So, we have performed a "breaking" of the original symmetry, and only a part of the previous set of symmetry transformations will remain as good symmetries. This phenomenon is general and it is worthwhile to know the remaining symmetries. Whence the importance of the following — and natural — definition:

Definition 4: A subgroup $\mathcal{H}$ of a group $\mathcal{G}$ is a (non-empty) part $\mathcal{H} \subset \mathcal{G}$, which is a group with the composition law induced by $\mathcal{G}$. $\mathcal{H}$ is a proper subgroup of $\mathcal{G}$ if $\mathcal{H} \neq \mathcal{H}$ and $\mathcal{H} \neq \{e\}$.

Types of (symmetry) groups in physics:

- with a finite number of elements: case of crystallographic groups;
- with an infinite number of elements:
  - discrete groups (number of elements in one to one correspondence with $\mathbf{Z}$ (i.e. the set of integers);
  - continuous groups: Lie groups for gauge theories, classification of particles, spin group, group of symmetry of space-time (Poincaré group,..)

An example of Lie group:

Group of rotations in the real plane around an origin O. It is then defined by 2×2 orthogonal matrices (with real entries):

$$R(\theta) = \begin{pmatrix} \cos\theta - \sin\theta \\ \sin\theta \quad \cos\theta \end{pmatrix} \qquad 0 \leq \theta \leq 2\pi$$

such a matrix transforming the 2 dim. real vector $\vec{X} = (x, y)$ into $R(\theta)\vec{X} = \vec{X\prime}$ with components:

$$\begin{aligned} x\prime &= x\cos\theta - y\sin\theta \\ y\prime &= x\sin\theta + y\cos\theta \end{aligned} \qquad (27)$$

such that:

$$\vec{X} \cdot \vec{X\prime} = \vec{X\prime} \cdot \vec{X} \quad \vec{X} \cdot \vec{X\prime} = xx\prime + yy\prime \quad \text{(norm conservation)} \tag{28}$$

Let us recall that:

$$\cos\theta = 1 - \frac{\theta^2}{2!} + \ldots; \qquad \sin\theta = \theta - \frac{\theta^3}{3!} + \ldots \tag{29}$$

leading, after a simple computation, to:

$$R(\theta) = \mathbf{1} + \theta\mathbf{M} + \ldots = \exp(\theta\mathbf{M}) \tag{30}$$

where

$$\mathbf{M} = \begin{pmatrix} 0 & -1 \\ 1 & 0 \end{pmatrix}$$

$\mathbf{M}$ is called the infinitesimal generator of the rotation group in two dimension, itself usually denoted by $\mathcal{O}(2)$. θ is called the parameter of $\mathcal{O}(2)$. Actually, we have just considered one of the simplest Lie group: it has one and only one continuous parameter (θ determines completely the angle of rotation around the origin for the group element $R(\theta)$).

Let us make things more complicated, and consider now the group of two by two unitary matrices, that is the group of 2×2 complex matrices $\mathcal{U}$ satisfying

$$(\mathcal{U}\vec{X}, \mathcal{U}\vec{X}) = (\vec{X}, \vec{X})$$

but with the components of $\vec{X}$ being complex numbers: $x = a+ib, y = c+id$ and the scalar product $(\vec{X}, \vec{X})$ being defined by:

$$(\vec{X}, \vec{X}) = (a - ib) \times (a + ib) + (c - id) \times (c + id)$$

Then, the condition $(\mathcal{U}\vec{X}, \mathcal{U}\vec{X}) = (\vec{X}, \vec{X})$ will be satisfied, for any $\vec{X}$, if and only if: $\mathcal{U}^\dagger = \mathcal{U}^{-1}$ with $\mathcal{U}^\dagger$ obtained by replacing each entry of $\mathcal{U}$ by its complex conjugate and transposing the matrix (with respect to the main diagonal) and $\mathcal{U}^{-1}$ denoting the inverse of $\mathcal{U}$ ($\mathcal{U}\mathcal{U}^{-1} = \mathbf{1}$, where $\mathbf{1}$ is the 2×2 identity matrix).

Example:

$$\mathcal{U} = \begin{pmatrix} a+ib & c+id \\ e+if & g+ih \end{pmatrix} \quad \mathcal{U}^\dagger = \begin{pmatrix} a-ib & e-if \\ c-id & g-ih \end{pmatrix}$$

The usual notation for the group of unitary 2×2 matrices is $\mathcal{U}(2)$, and is $\mathcal{SU}(2)$ for its subgroup with elements $\mathcal{U}$ of $\mathcal{U}(2)$ satisfying det $(\mathcal{U}) = 1$. As for the $\mathcal{O}(2)$ case, an exponential expression can be obtained for any $\mathcal{U}$ in $\mathcal{SU}(2)$:

$$\mathcal{U} = e^{i(a\sigma_1 + b\sigma_2 + c\sigma_3)} \qquad a, b, c \in \mathbf{R} \tag{31}$$

where the three 2×2 σ matrices are the so called Pauli matrices defined as follows:

$$\sigma_1 = \begin{pmatrix} 0 & 1 \\ 1 & 0 \end{pmatrix} \qquad \sigma_2 = \begin{pmatrix} 0 & -i \\ i & 0 \end{pmatrix} \qquad \sigma_3 = \begin{pmatrix} 1 & 0 \\ 0 & -1 \end{pmatrix}$$

and satisfying the commutation relations:

$$[\sigma_1, \sigma_2] = \sigma_1\sigma_2 - \sigma_2\sigma_1 = 2i\sigma_3 \tag{32}$$

$$[\sigma_2, \sigma_3] = 2i\sigma_1 \quad [\sigma_3, \sigma_1] = 2i\sigma_2 \tag{33}$$

In $\mathcal{SU}(2)$ there are three real continuous parameters, namely a, b and c and then three infinitesimal generators: $\sigma_1, \sigma_2, \sigma_3$.

In any Lie group, the infinitesimal generators form a basis of the Lie algebra of the corresponding Lie group.

Let us give the general definition of a Lie algebra A:

Definition 5: A Lie algebra is an algebra, that is first a linear vector space on $\mathbf{R}$ (or on $\mathbf{C}\ldots$) with a second internal law @ satisfying (if we denote $+$ the first law of the vector space):

$$\forall x, y, z \in \mathrm{A}\,,: (x + y)@z = x@z + y@z \qquad x@(y + z) = x@z + y@z \tag{34}$$

Moreover, this second law satisfies:

$$\forall x, y, z \in \mathrm{A} : x@y = -y@x \qquad x@(y@z) + y@(z@x) + z@(x@y) = 0 \tag{35}$$

One can easily note that, in our case, the @ internal law is just the commutator, denoted by [., .]:

$$[x,\, y] = x.y \;-\; y.x \tag{36}$$

The property of a Lie group (or continuous group of transformations satisfying some more analytical properties that we will not consider here in order not to overload this first introductory lesson)element to be written as an exponential of an element of its Lie algebra is particularly useful. Indeed, it will now be possible to work most of the time (at least when topological questions are not on purpose) with the Lie algebra, that is replacing tedious and enormous computations on the group by computations involving mainly linear algebras! That will be particularly precious for constructing and studying representations of Lie groups.

Actually, the $\mathcal{SU}(2)$ group is also known as the "spin" group in elementary particle physics.

Now, let us consider a little more to notion of representations, as mathematically defined by Definition 3. Indeed, the same group can act non trivially on spaces of different dimensions. The $\mathcal{SU}(2)$ group is defined by the set of 2×2 unitary matrices: then its natural space of representation is the 2-dim. complex plane. We can say that the "fundamental" representation of $\mathcal{SU}(2)$ is given by the 2×2 unitary matrices of determinant $= 1$, acting on the 2 dimensional complex plane $\mathbf{C}_2$, which we call the "representation" space.

We will now construct other $\mathcal{SU}(2)$ representations. But, before, let us remark that any element $\mathcal{U} = exp(a\sigma_1 + b\sigma_2 + c\sigma_3)$ in $\mathcal{SU}(2)$ can be transformed into another element of $\mathcal{SU}(2)$: $V = U\prime\ U\ (U\prime)^{-1}$ which is diagonal. Indeed, the matrix $H = a\sigma_1 + b\sigma_2 + c\sigma_3$ is Hermitian, that is $H = H^\dagger$ (see definition above) and so can be diagonalized by an unitary matrix $U\prime$, whence V diagonal. Note that U and V are mathematically equivalent in the sense that there is a change of basis in $\mathbf{C}_2$ — actually given by $U\prime$ — which will allow to see U as a diagonal matrix. And we will have:

$$V = exp(d\sigma_3) \qquad d\ real$$

At the Lie algebra level, the diagonal generator $(1/2\sigma_3)$ has two eigenvalues: $+1/2$ and $-1/2$ associated to the two eigenvectors (1, 0) and (0,−1): for a physicist, these are the two spin states $\uparrow$ and $\downarrow$. Moreover, one can see that the matrices: $\sigma_\pm = \sigma_1 \pm i\sigma_2$ transforms the vector (0,1) into (1,0) and (1,0) into the null vector (0,0) (resp. (1,0) into (0,1) and (0,1) into (0,0)) (they are called "raising and lowering operators"). We also note the commutation relations:

$$[\sigma_3, \sigma_+] = +2\sigma_+ \qquad [\sigma_3, \sigma_-] = +2\sigma_- \qquad [\sigma_+, \sigma_-] = 4\sigma_3 \tag{37}$$

But one knows that there are not only particles of spin 1/2, there are also particles of spin 0, 1, 3/2, 2,..They will lie in other representations of the group $\mathcal{SU}(2)$. Let us see how to construct them. For such a purpose we need to define the tensorial product of two vector spaces.

Definition 6: Let $\mathbf{V}$ and $\mathbf{V}\prime$ two vector spaces of respective dimensions n and $n\prime$, and respective basis $(\vec{e_1}, \ldots, \vec{e_n})$ and $(\vec{e_1}, \ldots, \vec{e_{n}\prime})$. We define the tensor product $\mathbf{V} \otimes \mathbf{V}\prime$ as the vector space of dimension $n \times n\prime$ with elements: $\sum_{(i,j)} \alpha(i,j)[\vec{e_i} \otimes \vec{e_j\prime}]$ with $i = 1, \ldots, n$ and $j = 1, \ldots, n\prime$ and $\alpha(i,j) \in \mathbf{R}$. If the group $\mathcal{G}$ acts on $\mathbf{V}$ via the representation D and on $\mathbf{V}\prime$ via the representation $D\prime$ such that $\forall g \in \mathcal{G} \to D(g)$ acts on $\mathbf{V}$ and

$\forall g \in \mathcal{G} \rightarrow D\prime(g)$ acts on $\mathbf{V}\prime$, on $\mathbf{V} \otimes \mathbf{V}\prime$ one can define the $\mathcal{G}$-epresentation $D \otimes D\prime$ such that: $\forall g \in \mathcal{G} \rightarrow (D \otimes D\prime)(g) = D(g) \otimes D\prime(g)$.

Now, let us take as $\mathbf{V}$ the 2-dimensional complex vector space $\mathbf{C}_2$ and consider the action of $\mathcal{G} = \mathcal{SU}(2)$ on $\mathbf{V} \otimes \mathbf{V}$. Then any $U \in \mathcal{G}$ acts on any vector $v \otimes v\prime$ of $\mathbf{V} \otimes \mathbf{V}$ as:

$$\forall U \in \mathcal{G} \rightarrow U(v) \otimes U(v\prime)$$

But we know that we can rewrite U as:

$$U = \exp(\mathbf{M}) = \mathbf{1} + \mathbf{M} + \ldots$$

Then, infinitesimally, we will have:

$$(\mathbf{1}+\mathbf{M}+\ldots)(\mathbf{v}) \otimes (\mathbf{1}+\mathbf{M}+\ldots)(\mathbf{v}\prime) = (\mathbf{1}\otimes\mathbf{1}+\mathbf{M}\otimes\mathbf{1}+\mathbf{1}\otimes\mathbf{M}+\ldots)(\mathbf{v}\otimes\mathbf{v}\prime) \quad (38)$$

therefore, infinitesimally — or in other words: at the Lie algebra level, we will have:

$$v \otimes v\prime \longrightarrow M(v) \otimes v\prime + v \otimes M(v\prime) \quad (39)$$

So, let us start with the vector: $\uparrow \otimes \uparrow$. The σ_--action will give:

$$(\uparrow \otimes \uparrow) \longrightarrow (\uparrow \otimes \downarrow + \downarrow \otimes \uparrow) \longrightarrow (\downarrow \otimes \downarrow) \longrightarrow 0 \quad (40)$$

But we know that we have four vectors in the tensor space under consideration. After some simple computation, one can see that the vector: $(\uparrow \otimes \downarrow - \downarrow \otimes \uparrow)$ is such that the action of σ_- as well as the action of σ_+ on it gives 0. Thus, the four dimensional space $\mathbf{C}_2 \otimes \mathbf{C}_2$ is indeed a good representation space for $\mathcal{G} = \mathcal{SU}(2)$, but it splits into two subspaces, one of dimension 3 and one of dimension 1, each of them being a good representation of $\mathcal{G}$. Actually, each of these two $\mathbf{C}_2 \otimes \mathbf{C}_2$ subspaces are "invariant subspaces under $\mathcal{G}$", and we have obtained what are called "irreducible representations of $\mathcal{G}$ ". Let us define correctly these objects:

Definition 7: Let D be a representation of the group $\mathcal{G}$ in $\mathbf{V}$. The subspace $\mathbf{E} \subset \mathbf{V}$ is an invariant subspace of $\mathbf{V}$ under D if:

$$\forall g \in \mathcal{G} : (D(g))\mathbf{E} \subset \mathbf{E}$$

Definition 8: The representation D of $\mathcal{G}$ in V is irreducible if there is no invariant subspace, except the trivial one (i.e.: 0). If not D is said reducible.

In the just considered case, the representation $\mathbf{C}_2 \otimes \mathbf{C}_2$ of $\mathcal{SU}(2)$, is reducible and decomposes in two separate representations, we have a "partition" of the space)) irreducible $\mathcal{SU}(2)$ representations, of dimension 3 and 1 respectively. Remark: The basis σ_+, σ_- and σ_3 is more likely associated to the Lie algebra of the group $\mathcal{S}l(2, R)$, often — and incorrectly — written

$\mathcal{S}l(2)$, and defined as the group of 2×2 real matrices with determinant = 1. It is a non-compact form of $\mathcal{SU}(2)$; both $\mathcal{S}l(2)$, and $\mathcal{SU}(2)$ possess the same set of irreducible finite dimensional representations.

Conclusions from the above discussion:

- from the 2-dimensional representation of $\mathcal{SU}(2)$ we have constructed the (irreducible) 3-dimensional representation and the 1-dimensional (or trivial)one. The first one corresponds to the "spin" one representation, with the three states with eigenvalue 1, 0, and -1, while the second is the "spin 0" representation with only one state with 0 eigenvalue;
- we have also "appreciated" the powerfulness of a Lie group to have a Lie algebra which allows easier computations (a Lie algebra satisfying the property of a linear algebra).

More on (finite dimensional) representations of the $\mathcal{SU}(2)$ group: Any irreducible finite dimensional representation of $\mathcal{SU}(2)$ is usually denoted $D(j)$ with j being a positive (or null) integer or half integer. The $D(j)$ representation contains $(2j+1)$ states, each state being an eigenstate of the generator corresponding to σ_3 with eigenvalue: $j, j-1, \ldots, -j+1, -j$ respectively. The "spin 1" representation above discussed is then D(1) with three states associated to the eigenvalue $+1, 0, -1$ respecively, while the "spin 1/2" representation $D(1/2)$ contains two states with σ_3 eigenvalue $+1/2$ and $-1/2$. Let us add that the product $D(j) \otimes D(j\prime)$ of the two representations $D(j)$ and $D(j\prime)$ of $\mathcal{SU}(2)$ decomposes as the sum of the irreducible representations:

$$D(j) \otimes D(j\prime) = D(j + j\prime) \oplus D(j + j\prime - 1) \oplus \ldots \oplus D(|j - j\prime|) \tag{41}$$

Finally, let us mention that the main Lie groups, at least with the most simple properties, are the following (they are called "simple groups" but we will not overload our text with the definition of a simple group)

- $\mathcal{SO}(n)$: orthogonal groups in n-dimensional real space (i.e. group of real $n \times n$ orthogonal matrices);
- $\mathcal{SU}(n)$: unitary groups in n-dimensional complex space (i.e. group of $n \times n$ complex unitary matrices);
- $\mathcal{S}p(n)$: group of $2n \times 2n$ symplectic matrices.
- there are also 5 "exceptional" groups: the word exceptional is used because they do not enter in infinite series as the above ones.

More details and informations — and complete definitions — on this section can be found in[12].

5.2. *A II: Quantum groups and crystal basis*

It is of course not our purpose to develop in detail the theory of quantum groups as it appeared in the works of V. Drinfeld on one hand and of M. Jimbo on the other hand in the middle of the eighties, but to provide the minimum of definitions and properties of a quantum group $\mathcal{U}_q(g)$, g being the Lie algebra of a Lie group $\mathcal{G}$ and q denoting the corresponding deformation parameter. We have noticed in the first part of the Appendix that a Lie algebra G of a Lie group $\mathcal{G}$ has a structure of linear vector space. Let us consider now the universal enveloping algebra of this Lie algebra, i.e. the space of polynomials and formal power series in $g \in G$ on which we apply the commutation relations appropriate for that Lie algebra. Then the quantum group $\mathcal{U}_q(g)$ will be a deformation relative to the parameter q of the universal enveloping algebra g. More explicitly, let us consider the example of the $\mathcal{U}_q((\mathcal{S}l(2))$ quantum group and let us denote J_+, J_- and J_3 the generators corresponding σ_+, σ_- and σ_3 in the 2-dimensional space representation, then we have:

$$[J_3,\, J_\pm] = \pm J_\pm \qquad [J_+,\, J_-] = \frac{q^{J_3} - q^{-J_3}}{q^{1/2} - q^{-1/2}} \tag{42}$$

We remark that when the parameter $q \to 1$, one recovers the $\mathcal{S}l(2)$commutation relations[f]

$$[J_3,\, J_\pm] = \pm J_\pm \qquad [J_+,\, J_-] = 2J_3 \tag{43}$$

Another important limit is the one corresponding to $q \to 0$. A detailed study of this case has first been done by M. Kashiwara[8] who found particularly well behaved base called "Crystal Base". Particularly interesting for our purpose is the rule providing the product of two irreducible representations of $\mathcal{U}_q(G)$ when $q \to 0$. A remarkable property of such a rule is that the elements in the obtained representation spaces arising from the product of two representations are not linear combinations of states, as in the case of an usual group as $\mathcal{S}l(2)$ (see the example considered in Appendix A I), but only made of a single product of the form $u\prime \otimes v\prime$ with $u\prime \in B_1$ and $v\prime \in B_2$: see Theorem 2.1 in Subsection 2.1.

[f] The commutation relations of Eq. (43) follow from Eq. (37) defining $J_i = \sigma_i/2$, $i = 1, 2, 3$.

We can make more explicit the way of computing such states by considering the example of the product $D(3/2) \otimes D(1)$ which decomposes as — see above in Appendix AI:

$$D(3/2) \otimes D(1) = D(5/2) \oplus D(3/2) \oplus D(1/2)$$

In Fig. 1, we have representedby black points on an horizontal line the four J_3 eigenstates $3/2, 1/2, -1/2, -3/2$ of $D(3/2)$ and on a vertical line the three J_3 eigenstates $1, 0, -1$ of $D(1)$. Using the above theorem, the six states in the obtained representation $D(5/2)$ appear as the black points in the upper elbow constituted by the two — one horizontal and one vertical — segments, the same for $D(3/2)$ in the lower elbow, while finally the two states of the $D(1/2)$ representation states show up in the small horizontal segment below:

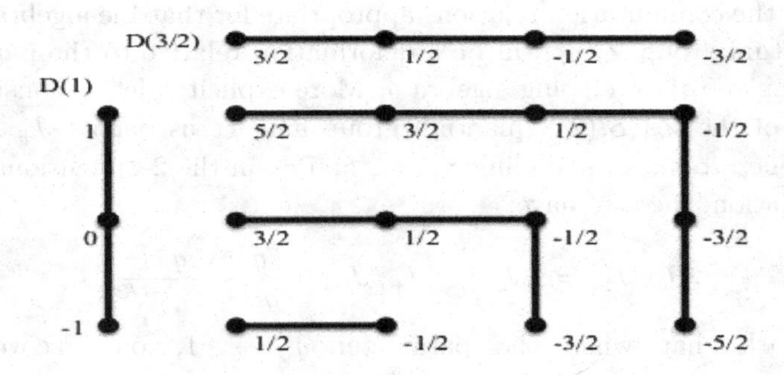

Figure 1. Diagram of the tensor product of irreps. $D(3/2) \otimes D(1)$.

It is this property which is intensively used in our symmetry approach of the genetic code. In our model, the used quantum group is $\mathcal{U}_q(Sl(2) \oplus Sl(2))$, $q \to 0$. Let us make more explicit the construction of dinuclotides from the product:

$$(\frac{1}{2}, \frac{1}{2}) \otimes (\frac{1}{2}, \frac{1}{2}) = (1,1) \oplus (0,1) \oplus (0,1) \oplus (0,0)$$

and following the prescription given in the scheme(1)in Subsection 2.1. Starting from the state CC, the action of J_- in $Sl(2)_H$ will provide UC and UU successively, while the action of J_- in $Sl(2)_V$ gives GC and GG from CC, AC and AG from UC, and finally AU and AA from UU. Using once more the diagrammatic rule above — that is the Kashiwara theorem — one gets CU as a singlet of $Sl(2)_H$ but member of a triplet of $Sl(2)_V$, in the same way CG as a singlet of $Sl(2)_V$ but in a triplet of $Sl(2)_H$ and finally CA as a singlet of both $Sl(2)$.

Acknowledgments

P.S. would like to express his gratitude to Professor Mondaini, President of the BIOMAT Consortium, for his kind invitation to present our results at the BIOMAT 2016 International Symposium held at the Chern Institute of Mathematics, Nankai University, Tianjin, China. He is indebted to Professors Jishou Ruan and Richard Kerner of the Organizing Committee for contributing so efficiently to the success of this very interesting conference. His thanks are also to Professor Chengming Bai, Director of the Chern Institute of Mathematics for his warm welcome.

References

1. E. Schrödinger, *What is life?*, Cambridge University Press (1944).
2. L. Frappat, A. Sciarrino and P. Sorba, *Phys. Lett. A* **250**, 214 (1998).
3. L. Frappat, A. Sciarrino and P. Sorba, *Phys. Lett. A* **311**, 264 (2003).
4. L. Frappat, A. Sciarrino and P. Sorba, *J. Biol. Phys.* **28**, 17 (2002).
5. A. Sciarrino and P. Sorba, *BioSystems* **107**, 113 (2012).
6. A. Sciarrino and P. Sorba, *BioSystems* **111**, 175 (2013).
7. A. Sciarrino and P. Sorba, *BioSystems* **141**, 20 (2016).
8. M. Kashiwara, *Commun. Math. Phys.* **133**, 249 (1990).
9. L. Frappat, A. Sciarrino and P. Sorba, *J. Biol. Phys.* **28**, 27 (2002).
10. L. Frappat, A. Sciarrino and P. Sorba, *Physica A* **351**, 461 (2005).
11. D. Cocurullo and A. Sciarrino, *arXiv* **1609.02141v1** (2016).
12. L. Frappat, A. Sciarrino and P. Sorba, *Dictionary on Lie algebras and Superalgebras*, Academic Press (2000).
13. A. Sciarrino, *arXiv* **1703.00445v1** (2017).
14. A. Sciarrino, *BioSystems* **69**, 1 (2003).

INDEX

Printed in the United States
By Bookmasters